Managing soil health for sustainable agriculture

Volume 1: Fundamentals

It is widely recognised that agriculture is a significant contributor to global warming and climate change. Agriculture needs to reduce its environmental impact and adapt to current climate change whilst still feeding a growing population, i.e. become more 'climate-smart'. Burleigh Dodds Science Publishing is playing its part in achieving this by bringing together key research on making the production of the world's most important crops and livestock products more sustainable.

Based on extensive research, our publications specifically target the challenge of climate-smart agriculture. In this way we are using 'smart publishing' to help achieve climate-smart agriculture.

Burleigh Dodds Science Publishing is an independent and innovative publisher delivering high quality customer-focused agricultural science content in both print and online formats for the academic and research communities. Our aim is to build a foundation of knowledge on which researchers can build to meet the challenge of climate-smart agriculture.

For more information about Burleigh Dodds Science Publishing simply call us on +44 (0) 1223 839365, email info@bdspublishing.com or alternatively please visit our website at www.bdspublishing.com.

Related titles:

Managing soil health for sustainable agriculture Volume 2: Monitoring and management
Print (ISBN 978-1-78676-192-7) Online (ISBN 978-1-78676-195-8; 978-1-78676-194-1)

Improving organic crop cultivation
Print (ISBN 978-1-78676-184-2); Online (ISBN 978-1-78676-186-6, 978-1-78676-187-3)

Integrated weed management for sustainable agriculture
Print (ISBN 978-1-78676-164-4); Online (ISBN 978-1-78676-166-8, 978-1-78676-167-5)

Water management for sustainable agriculture
Print (ISBN 978-1-78676-176-7); Online (ISBN 978-1-78676-178-1, 978-1-78676-179-8)

Chapters are available individually from our online bookshop: https://shop.bdspublishing.com

climate•SMART•publishing
IN AGRICULTURAL SCIENCE

BURLEIGH DODDS SERIES IN AGRICULTURAL SCIENCE

NUMBER 48

Managing soil health for sustainable agriculture

Volume 1: Fundamentals

Edited by Dr Don Reicosky, Soil Scientist Emeritus
USDA-ARS and University of Minnesota, USA

burleigh dodds
SCIENCE PUBLISHING

Published by Burleigh Dodds Science Publishing Limited
82 High Street, Sawston, Cambridge CB22 3HJ, UK
www.bdspublishing.com

Burleigh Dodds Science Publishing, 1518 Walnut Street, Suite 900, Philadelphia, PA 19102-3406, USA

First published 2018 by Burleigh Dodds Science Publishing Limited

Library of Congress Control Number: 2018947509

British Library Cataloguing in Publication Data
A catalogue record for this book is available from the British Library

ISBN 978-1-78676-188-0 (Print)
ISBN 978-1-78676-191-0 (PDF)
ISBN 978-1-78676-190-3 (ePub)
ISSN 2059-6936 (print)
ISSN 2059-6944 (online)

DOI 10.19103/AS.2017.0033.1

Typeset by Deanta Global Publishing Services, Chennai, India
Printed by CPI Group (UK) Ltd, Croydon CR0 4YY

Contents

Part 3 Soil dynamics

Series list

Title	Series number
Achieving sustainable cultivation of maize - Vol 1 From improved varieties to local applications Edited by: Dr Dave Watson, CGIAR Maize Research Program Manager, CIMMYT, Mexico	001
Achieving sustainable cultivation of maize - Vol 2 Cultivation techniques, pest and disease control Edited by: Dr Dave Watson, CGIAR Maize Research Program Manager, CIMMYT, Mexico	002
Achieving sustainable cultivation of rice - Vol 1 Breeding for higher yield and quality Edited by: Prof. Takuji Sasaki, Tokyo University of Agriculture, Japan	003
Achieving sustainable cultivation of rice - Vol 2 Cultivation, pest and disease management Edited by: Prof. Takuji Sasaki, Tokyo University of Agriculture, Japan	004
Achieving sustainable cultivation of wheat - Vol 1 Breeding, quality traits, pests and diseases Edited by: Prof. Peter Langridge, The University of Adelaide, Australia	005
Achieving sustainable cultivation of wheat - Vol 2 Cultivation techniques Edited by: Prof. Peter Langridge, The University of Adelaide, Australia	006
Achieving sustainable cultivation of tomatoes Edited by: Dr Autar Mattoo, USDA-ARS, USA & Prof. Avtar Handa, Purdue University, USA	007
Achieving sustainable production of milk - Vol 1 Milk composition, genetics and breeding Edited by: Dr Nico van Belzen, International Dairy Federation (IDF), Belgium	008
Achieving sustainable production of milk - Vol 2 Safety, quality and sustainability Edited by: Dr Nico van Belzen, International Dairy Federation (IDF), Belgium	009
Achieving sustainable production of milk - Vol 3 Dairy herd management and welfare Edited by: Prof. John Webster, University of Bristol, UK	010
Ensuring safety and quality in the production of beef - Vol 1 Safety Edited by: Prof. Gary Acuff, Texas A&M University, USA & Prof. James Dickson, Iowa State University, USA	011
Ensuring safety and quality in the production of beef - Vol 2 Quality Edited by: Prof. Michael Dikeman, Kansas State University, USA	012
Achieving sustainable production of poultry meat - Vol 1 Safety, quality and sustainability Edited by: Prof. Steven C. Ricke, University of Arkansas, USA	013
Achieving sustainable production of poultry meat - Vol 2 Breeding and nutrition Edited by: Prof. Todd Applegate, University of Georgia, USA	014
Achieving sustainable production of poultry meat - Vol 3 Health and welfare Edited by: Prof. Todd Applegate, University of Georgia, USA	015

Acknowledgement

We wish to acknowledge the following for their help in reviewing particular chapter:

- Chapter 6: Dr Michael S. Strickland, Virginia Tech University, USA

Introduction

There has been growing concern that both intensive agriculture in the developed world and rapid expansion of crop cultivation in developing countries is damaging the health of the soils which are the foundation of farming. Healthy soils are also critical for addressing finite resources for food production, an expanding global population, climate change, ecosystem services, environmental quality, and the overall quality of life which depends on food security. At the same time, we are discovering much more about how complex soils are as living biological systems and the need for biological solutions to effective soil management. These issues are addressed in the two volumes of *Managing soil health for sustainable agriculture*:

- Volume 1: Fundamentals
- Volume 2: Monitoring and management

Volume 1 reviews the latest research on soil science. After an overview of the role of soil as a provider of ecosystem services, the book reviews soil structure and chemistry as well organic matter, soil microorganisms and fauna. The third part of the book discusses soil dynamics, from water and nutrient cycles to soil erosion mechanisms.

Part 1 Overview

The first part of the volume offers an overview of different aspects of soil health. Chapter 1 offers a general overview of soil health, charting the shift from a utilitarian focus on soil quality for crop production to a more holistic concept of soil health which sees soil as an ecosystem in which soil biology, moderated by physical and chemical properties, is central to its functionality. This concept highlights the potential difference between performance and inherent potential. The chapter looks at the implications of this broader concept of soil health for different ways of measuring soil health, whether in terms of its physical and chemical properties, functional performance, soil organic carbon or biological status.

Building on the foundation provided by Chapter 1, Chapter 2 provides an overview of the range of ecosystem services provided by soil. Ecosystem services provided by soil can be supporting (e.g. provisioning primary production and biodiversity) or regulatory (e.g. erosion control, water infiltration, nutrient retention, atmospheric gas regulation, and pest control). The chapter explains how ecosystem services benefit human welfare through these functions, addressing the soil's role in such areas as the production of food, fiber and energy, erosion control, carbon storage, pest control and biodiversity.

The subject of Chapter 3 is the critical nexus of soil health and climate change. A healthy soil has the capacity to sustain biological activity, maintain environmental quality and promote plant and animal health. The chapter reviews the impact of changing climate on the key components of soil health, including soil physical, chemical, and biological properties and such risks as increasing acidity and salinity. The chapter also reviews some of the key management practices that have been demonstrated to improve soil health and deliver climate benefits, including conservation agricultural techniques such as zero/minimum tillage, retention of crop residues, rotations and cover crops which both protect soil from the effects of climate change and improve its capacity to store carbon.

As highlighted earlier, soil function is an interplay of physical, chemical, and biological processes, with soil microbes playing a direct role in driving soil chemical and physical processes that are important for overall ecosystem function. Building on the foundations of the previous chapters, Chapter 4 provides an overview of how best to achieve soil conservation and regeneration through integrated soil health management. After highlighting some of the known impacts of management strategies on soil health, the chapter introduces an Integrated Soil Health Management (ISHM) framework that may be used to design, implement, and evaluate management decisions for soil conservation and regeneration.

The final chapter in Part 1, Chapter 5, provides an overview of the economics of soil health. Soil heath management practices and systems can be important tools for farmers to realize on-farm benefits associated with decreased erosion, improving levels of soil organic matter, and improved soil structure and function. The chapter describes existing research on the economics of soil health, with a particular focus on the adoption of soil health management practices by farmers. The chapter covers such issues as the costs and benefits of soil health practices, public benefits from soil health and soil health management practices, and barriers to the adoption of soil health practices by farmers. The chapter provides some detailed case studies and evaluates the role of government regulations, policies and incentive programs.

Part 2 Soil structure and composition

The theme of the second part of the volume is soil structure and composition. The focus of Chapter 6 is on the role of soil texture and structure in soil health. The chapter examines the relationship between particle size and soil health, affecting such areas as cation exchange capacity, hydraulic conductivity, gas transport function and mechanical strength. The chapter then considers what determines optimum particle size and the relationship between soil health and soil aggregate formation.

Moving from soil texture to its chemical composition, Chapter 7 begins by reviewing the use of key chemical indicators for soil health such as pH, salinity and sodicity. It then focuses on research on major nutrients such as nitrogen, phosphorus and potassium. In each case, the chapter discusses what we know about its function in plant growth, fixation in soil, loss mechanisms and factors affecting availability to plants. The chapter also discusses other nutrients such as calcium, magnesium and sulfur. Building on this foundation, the chapter discusses management practices to improve nutrient availability and reduce the reliance on inorganic fertilizers. The chapter includes a case study from India that show how integrated nutrient management can be used to optimize soil health.

The theme of Chapter 8 is the role of soil microorganisms in soil health. Soil is a complex environment that supports the largest, most diverse and resilient microbial community on the planet, essential for nutrient cycling and plant growth. The details of many soil microorganisms remain unknown, but the recent improvements in molecular methods for microbial ecology have made it possible to view the soil microbiome as a whole, providing information on its responses to changes in land-use and microbially-mediated functions relevant to agricultural sustainability and major geochemical cycles. The chapter discusses the contribution of soil microorganisms to nutrient cycling in soil, and the impact of human activities on the soil microbiome. The chapter considers the influence of the microbiome on crop health and yields and discusses the potential for future exploitation of the soil microbiome.

The final chapter of the section, Chapter 9, addresses the role of soil fauna in soil health and the delivery of ecosystem services. Soil fauna includes microscopic organisms such as nematodes (microfauna), mites and springtails (mesofauna), up to larger invertebrates such as worms, spiders, ants, termites and beetles (macrofauna) and vertebrates (megafauna) such as amphibians, reptiles and mammals. The chapter explains how their activity is important for soil health and can result in various ecosystem services, including provision of food and primary production, nutrient cycling and organic matter dynamics, water availability, gas exchanges, soil formation, pollination and seed dispersal. The chapter classifies soil fauna and reviews currently available information on soil fauna biodiversity and functional roles worldwide, their role in soil health and the importance of management in determining soil fauna communities.

Part 3 Soil dynamics

The third part of the volume addresses the theme of soil dynamics. Chapter 10 considers the role of soil hydrology in soil health. Soil hydrology stands at the forefront of soil health due to its critical importance in regulating physical, chemical, and biological processes in soils. Not only do the interactions between soil and water affect a number of soil processes and properties, soil hydrology also links closely to soil types and landscape features as well as management practices that are closely tied to soil health. The chapter discusses various soil hydrologic processes and properties that influence soil health along with crop growth, ecosystem services, and environmental impacts. The chapter describes how soil physical, chemical, and biological processes are often mediated by soil hydrologic conditions. The chapter relates these processes and properties to soil hydrologic functions in the context of soil health and management.

Complementing the preceding chapter's focus on soil hydrology, the focus of Chapter 11 is nutrient cycling in soils. Sustainable agricultural systems must enable profitable production of sufficient high-quality crops to meet human demand, while simultaneously minimising off-site impacts. Although simple chemical equilibria play a part, nutrient cycling processes in soil are largely the outcome of biochemical processes resulting from the activity and interactions of a huge diversity of soil organisms. The chapter outlines the underlying concepts of nutrient cycling, before giving reviewing current research on carbon, nitrogen, phosphorus other nutrient cycles in soil.

Building on the themes of the two previous chapters, Chapter 12 offers an overview of plant-soil interactions. Plant-soil interactions are the prime mechanism for integrating the impacts of the aerial environment, animals (including plant shoot symbionts and pathogens), plant shoot physiology, plant root physiology and structure, the soil microbiome and soil chemistry and structure into what is commonly termed soil health. The chapter addresses three aspects of this chain: interactions of the shoot with the roots, interactions between the roots and the microbiome, and interactions between the root system and physical and structural characteristics of the soil.

The subject of the volume's concluding chapter, Chapter 13, is mechanisms of soil erosion and degradation. The chapter offers an overview of soil erosion processes, focusing on the most common kind (erosion by water). However, the chapter also deals with other mechanisms of soil erosion, before going on to consider the consequences of erosion. The chapter relates soil erosion to crop productivity and sustainable agriculture, and examines the role of soil conservation in mitigating soil erosion.

Part 1

Overview

Chapter 1

Soil and soil health: an overview

Mark G. Kibblewhite, Cranfield University, UK and Landcare Research, New Zealand

1 Introduction

Soil security is critical to human security (Minami, 2009) and underpins food and water security (McBratney et al., 2014). Humans have observed variations in soil properties and especially its fertility since the advent of settled agricultural societies. Until the emergence of modern science in the seventeenth century, this assessment was subjective and experiential, and this remains an important approach for many farmers (Romig et al., 1995). However, science has revealed insights into the working of the soil system and its role in supporting agriculture and the wider land system, leading to more objective and quantitative assessments of soil quality and health. Interest and studies in this area have been intense over recent decades, but a final understanding of how best to assess soil status remains elusive. This chapter outlines some theories of the soil system and applies this to clarify the meaning of 'quality' and 'health' as they apply to soil. It explores how to assess soil health and provides a commentary on which indicators of soil health may be useful in practical agriculture. It underlines the importance of soil governance and how this may be optimized.

http://dx.doi.org/10.19103/AS.2017.0033.01

2 Constructs of soil quality and health: utilitarian and holistic

The consequence of variation in soil-forming factors over space and time is that soil properties vary widely between and across landscapes and give rise to distinct soil types. Soil taxonomic systems are used as legends for maps and spatial inventories of soils. The practical reason for collecting this soil information has been mainly to inform land capability assessment for agriculture. Different soil types have different potentials for agricultural production, for example depending on their depth, texture, wetness and acidity. Thus soils can be assigned to different categories of 'quality' for agriculture. From the early nineteenth century to the late twentieth century, land capability assessment was widely applied as part of the introduction of modern farming techniques to existing agricultural land and its widespread extension to natural and semi-natural lands. Therefore, the focus was on the inherent properties and potential of soils to support modern agriculture. Subsequently, a new concern arose about the loss of soil condition from degradation caused by inappropriate agricultural exploitation, often described as a loss of soil quality. Thus, alongside its older meaning, the concept of soil quality developed as a descriptor of soil condition in terms of its fitness for agricultural production relative to its potential.

The utilitarian concept of soil quality as a measure of fitness for agricultural production developed when the dominant construct of soil was as a medium for plant growth in which the driving processes were physical and chemical. However, a different and richer concept of the soil system has emerged. This views soil as an ecosystem in which soil biology, moderated by physical and chemical properties, is central to its functionality. Moreover, especially following the Millennium Ecosystem Assessment (2005) and the introduction of the concept of natural capital, new emphasis is being given to the full range of ecosystem services that are supported by soil, including but not confined to food and fibre production. At its heart, this perspective is informed by a cosmology that is holistic and that at some level encompasses the idea that humans have a stewardship role for life on Earth, rather than that nature exists solely for humans to exploit. In this context, a holistic construct of soil health has emerged, as an integrative measure of the condition of a multifunctional living system (Doran and Zeiss, 2000). At the same time, causing some confusion, the utilitarian meaning of soil quality has been extended to include its fitness to support the full range of ecosystem services. A critical message in this chapter is that soil quality and health are distinct concepts and should not be confused (Lal, 2016): soil quality refers to the inherent potential capacity of a particular soil to support specified services (Karlen et al., 1997) while soil health describes its actual condition as a system for supporting those services (McBratney et al., 2014). Therefore, at least from a utilitarian perspective, soil health indicates how close the condition of a soil is to its optimal one for supporting specified services, that is, those that define its inherent quality. It focuses assessment towards the quantity and quality of specified services that the soil is able to support; for example, if the specified service is to support agricultural production then the assessment of soil health will be directed towards measurement of agricultural yields and the soil properties that control these outputs. By contrast, the holistic approach to soil health (Doran and Zeiss, 2000; Van Bruggen and Semenov, 2000) considers overall multifunctional capacity and focuses on the status of the soil system to deliver all its functions. These two constructs of soil health are equally valid but present different challenges when considering options for assessing soil health. The utilitarian approach focuses particularly on factors that impact on the level of specified services that a soil is currently capable of supporting, relative to those that can be supported by a completely healthy soil of the same quality. The holistic approach focuses more on the current status of the biotic community in response to land use and management and its impact on soil processes and in turn their level of support for multifunctionality. The utilitarian approach is the one mainly adopted in this chapter, since here the focus is towards soil as a resource for agriculture.

3 The soil system and its performance

3.1 The soil system

The soil system is a sub-system of the overall land system. Its main functions are organic matter decomposition, nutrient cycling, soil structure maintenance and community moderation, and associated pest and disease control (Van Bruggen and Semenov, 2000; Kibblewhite et al., 2008). These functions support ecosystem services (see Fig. 1) within the wider land system, some of which deliver final goods and services to the human economy, including agricultural products.

The soil system has both abiotic and biotic components and processes. Distinctive habitats characterised by their physical and chemical properties evolve within different soils depending on soil-forming factors (Jenny, 1941) including time, parent materials, climate, natural vegetation and human land use and land management. Although soil systems operate across a wide range of scales, from micrometres to kilometres, arguably, the scale that is most relevant to soil health extends from a few microns to millimetres, as this is the one at which the soil architecture moderates biotic activity (Young and Crawford, 2004). Of particular importance in determining this architecture is firstly the soil pore structure, as characterised by overall pore volume, dimensions, continuity and tortuosity, and secondly, the chemistry of clay and other surfaces within the soil and their interaction with solution chemistry. The resulting soil habitats define the living space of the ecological community that is a phenotypic expression of the genetic information within the many different species present in soil. A key function of soil is the decomposition of plant and animal residues and the release of nutrients for uptake by plants. Importantly, this decomposition harvests energy for the living soil system and so underpins the delivery of

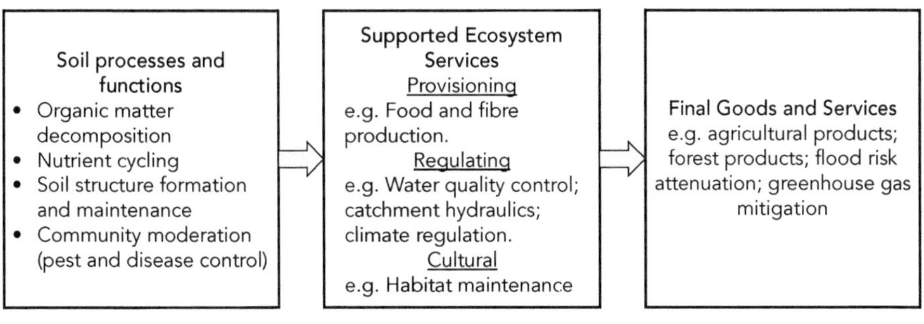

Figure 1 Cascading of functions, services and final goods from the soil system.

all of its functions. The uptake of carbon from plant residues and its progressive oxidation, as it moves through a complex food web and is respired, drives the different processes required for the maintenance of soil structure, nutrient cycling and community moderation, as well as organic matter decomposition. Thus the levels and forms of soil organic carbon are critical to the functioning of soil and, by implication, to its health.

3.2 System performance

The performance of systems can be viewed as a measure of the efficiency of conversion of inputs to outputs. In general, a performance curve (Fig. 2) can be described with a 'working range' extending to a point at which there is no further increase in output for an increment in inputs (and beyond which further inputs likely lead to a deterioration of performance). Applying this approach, inherent quality is represented by the working range of a fully functional system, that is a completely healthy system, while an indicator of system health is provided by the current working range, expressed as a percentage of the optimal one. The working range of a system can also be related to its resilience, or how well the performance recovers from a disturbance (short-term pressure) or stress (long-term or chronic pressure) after this is removed (Fig. 3). A completely resilient system is one whose performance returns to that immediately before the introduction of the disturbance or stress (Szabolcs, 1994). If the system state remains within its working range, resilience is expected. However, if the working range is exceeded this may damage system components, altering its state and leading to a new and degraded performance curve and a loss of health. Thus resilience to loading is reduced if the working range is exceeded and degree of resilience is related to system health.

3.3 Applying the general systems approach to soil

If the general systems approach described above is applied to soil, the health of a soil depends on its current working range relative to its potential one, as defined by its quality.

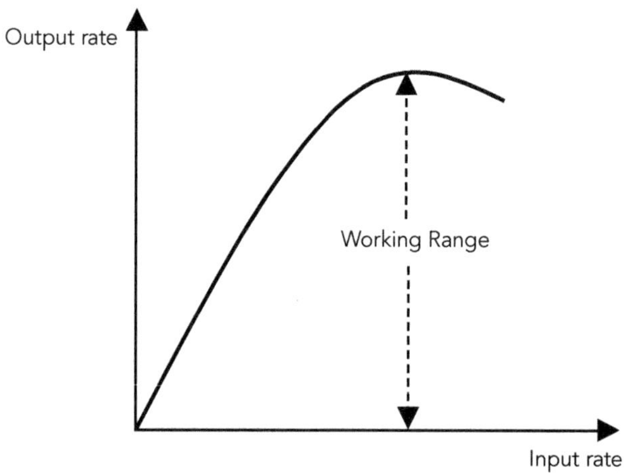

Figure 2 System performance curve and working range.

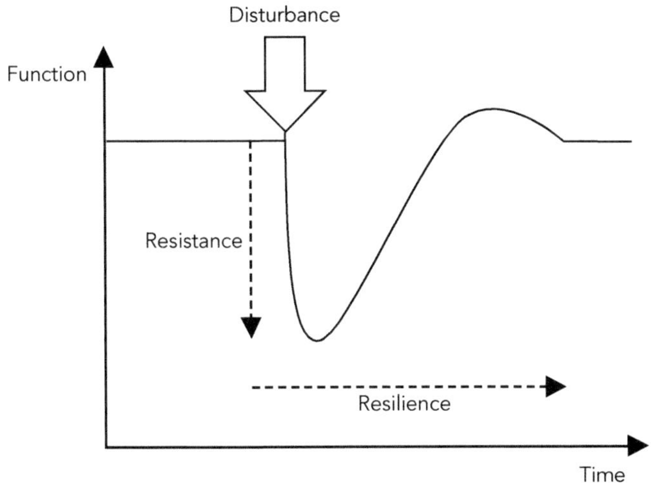

Figure 3 System response and resilience (after Griffiths and Philippot, 2012).

And the healthier the soil, the more likely it is that its performance will be resilient to management challenges such as tillage, topsoil compaction, or a wetting and drying cycle under irrigation. However, a loss of performance may be reversible or not, depending on its cause and form. Where the working range is not recoverable, a permanent reduction in the potential of the soil to deliver its functions means there will be a loss of quality rather than health. For example, soil quality is reduced when a soil system is stressed by inappropriate land use and there is a permanent loss of functionality from soil erosion and reduced soil depth, or from salinisation.

4 Soil health and its assessment

4.1 Approaches

Considering soil health as an integrative property, its assessment could be approached directly, in principle, by estimation of the working ranges for delivery of the different functions. One method would be to estimate the limiting functional output (e.g. respiration and nutrient release) when the supply of carbon substrate is not limited; another could be estimation of the range over which functionality is resilient. Alternatively, soil health may be inferred from information about the system that supports it, by assessment of the condition of the abiotic habitat (e.g. aggregate stability, bulk density and acidity) and/or observation of critical biotic components and processes. Considering the latter, overall microbial biomass provides information about the biotic state of the soil system, but more specific information is accessible from observation of the 'soil genome' and its phenotypic expression, as represented by, for example, gene diversity, community structure, the populations of individual species and/or enzyme activities. Further, the quantities of different forms of soil organic carbon are likely indicative of soil health because these provide 'buffer stocks of energy' in the food chain that connect the system's components and their levels will affect resilience. However, no single approach is adequate by itself. A very large number of indicators have been proposed for assessing soil health and the complexity of the soil system means that several are needed to properly describe its condition. A selection of indicators is needed to provide information about different critical features. This selection is itself complicated and requires a systematic framework (Ritz et al., 2009). Only a brief overview of some indicators is provided here with an emphasis on those that might be applied more routinely in practical agriculture. More comprehensive reviews (Van Bruggen and Semenov, 2000; Bastida et al., 2008; Ritz et al., 2009) cover indicators and measurements relevant to researching the soil system and its environmental responses. To be of practical use, indicators should provide meaningful information to support soil management decisions. This requires indicator thresholds corresponding to different levels of soil health. In this regard, generic guidance on what levels of indicators indicate good and poor soil health is often of limited value because the variety and variability of soils and their properties and qualities means there is also variation in threshold values representing acceptable and unacceptable health. Ideally, thresholds need to be established for different soils depending on their land use and land management, but these data are not generally available, especially for biological indicators (Pulleman et al., 2012). However, even where meaningful thresholds are not available, trends in indicator values are an indication of the trajectory of soil health.

4.2 Physical and chemical properties

Some physical and chemical properties, including texture, are essentially fixed and strongly influence the potential functionality of soil (i.e. its quality), and while they are important for soil management they are not strictly relevant as indicators of soil health. Other physical and chemical properties vary with soil health and provide useful information about changes in the condition of the habitat within which the biotic soil system operates. These include bulk density and water-holding capacity (which both reflect soil pore characteristics), aggregate stability, pH, electrical conductivity (as a measure of the salt content of the soil solution) and estimates of the available inorganic pool of nutrients (e.g. available phosphate and potassium).

4.3 Functional performance

The respiration rates of isotopically labelled organic materials added to incubated soil samples can provide information about the key function of soil to decompose organic residues (Jenkinson, 1971). However, this is somewhat complicated and costly for general application. Simple measurement of soil respiration is valuable but includes respiration of carbon in soil organic matter as well as plant residues and therefore is only a partial measure of the rate of plant residue decomposition. The quantity of nitrogen mineralised on incubation of soil (Keeney, 1982) provides information on the nutrient cycling function and this is relatively straightforward to do and gives valuable information about a soil's ability to provide nitrogen to crops. Direct assessment of rates of soil structure formation and maintenance is problematic, but the current aggregate stability of the soil and its resistance and resilience to pressures may provide an indirect estimation of this functionality.

4.4 Soil organic carbon

Total soil organic carbon provides a measure of the chemical energy derived from photosynthesis that is stored in the soil system. Its level reflects the relative rates of inputs of organic carbon to respired carbon. Increasing total soil organic carbon indicates that the supply of carbon substrate exceeds demand, while a decreasing trend indicates the reverse and that the energy requirements of the system are not being met fully by external inputs. A rising trajectory of total soil organic carbon is indicative of less stress on the soil system than a falling one, meaning that an upward trend is likely associated with improving soil health and downward one with reducing soil health. Therefore, the trend in total soil organic carbon and its rate of change are useful indicators of whether soil health is improving or deteriorating. Thus there is a rationale for the measurement and monitoring of trends in soil organic carbon in agricultural soils as a basic indicator of soil health.

Knowledge remains incomplete about the forms of soil organic carbon in terms of molecular structures, their association with soil surfaces and inorganic components and the effects of pore structure on their physical accessibility to microbes. The half-life of organic compounds in soil depends only partially on their molecular structure, but also on their combination with inorganic materials and their potential occlusion from microbes within the soil matrix (Dungait et al., 2012; Lal, 2016). A possible order of energy accessibility and yield for different soil organic matter fractions is highest for sugars and other low-molecular-weight compounds in the soil solution, less for polysaccharides including polyuronic acids that are weakly bound to soil surfaces and not occluded in soil pores, and even less for polyphenolic and other complex polymers that are strongly bound to soil surfaces or occluded in the soil matrix. However, simple sugars that are strongly bound to soil surfaces or occluded within the soil matrix may be inaccessible and not available as substrate. The distribution of organic carbon in terms of its molecular structure and its locus within the physical soil architecture may be indicative of soil health since as a soil system is stressed, by lowered inputs of organic carbon or high functional demands, forms of existing soil organic carbon that are more accessible and that have higher energy yields will be progressively drawn down with greater use being made of less-favourable energy sources. Therefore, the proportion of total organic carbon within accessible and higher energy-yielding forms may be an indicator of soil health. The rate of respiration when soil is incubated provides a basic indicator of this carbon. Another more valuable indicator of accessible carbon is the proportion of microbial biomass carbon to total soil organic carbon (Bastida et al., 2008), which allows an estimation of the carbon being utilised per unit of microbial biomass carbon, either from within the active microbial food chain or from 'accessible' soil organic carbon.

Selective extraction of soil organic carbon has been proposed as an indicator of soil health. Cold water extraction (Chantigny, 2003) removes organic matter in the soil solution, which is readily accessible to microbes; hot water extraction removes more organic matter, including polysaccharides which are an energetically preferred source of substrate relative to more recalcitrant compounds. Other extractions proposed for estimating 'labile' soil organic carbon rely on partial oxidation (e.g. with $KMnO_4$) or target particular organo-mineral combinations (e.g. by reducing Fe oxides). These methods continue a long tradition of using solvents and chemical solutions to extract apparently meaningful fractions of soil organic matter. While some of these methods for estimating labile carbon may be shown to be useful empirically, the mechanistic rationale for them is less certain because the fraction of soil organic matter they extract is likely heterogeneous and includes materials within a wide range of microbial accessibility.

4.5 Biological status

The level of active soil biomass provides only a crude indirect indicator of the gross capacity of the soil system to deliver its functions. Recognition that the soil system is an ecosystem with a community operating over several trophic levels with large and varied populations of species, suggests that characterising the ecological community in terms of its diversity may be a useful tool for assessing soil health. Some methods for characterisation of community composition are mature (e.g. phospholipid fatty acids (PLFA) profiling or DNA-based techniques such as terminal restriction fragment length polymorphism) while other emergent techniques (metagenomics and metaproteomics) offer potential (Ritz et al., 2009). The results from application of these techniques demonstrate that community composition is dynamic and that observable trajectories can be related to changes in land use and land management (e.g. Jangid et al., 2011). However, an answer to the question: 'What is a healthy community composition?' is not easily provided. One consideration is the extent to which diversity confers functional resilience relative to other factors (Orwin and Wardle, 2004) such as the evenness of the microbial community composition (Griffiths and Philippot, 2012). This unmet need to provide interpretative guidelines for community composition limits the practical application of its measurement in agriculture. Enzymatic activity (e.g. dehydrogenase and phenol oxidase) has been correlated with indicators of soil condition (Veum et al., 2014). Estimation of populations of particular species (Römbkea et al., 2006) as sentinels for soil health is an alternative approach. Species in higher trophic levels, for example, worms, mites and nematodes are likely useful indicators as their populations depend in good part on the health of lower trophic levels. Worm populations are quite easily estimated in the field and therefore may provide a useful indicator of trends in soil health that can be applied in practical agriculture.

5 Practical assessment and governance of soil health

5.1 Practical assessment of soil health in agricultural operations

In any scheme for assessing soil health, a fundamental indicator is soil organic carbon level. Unfortunately, the spatial and temporal variability of its rate of change means that changes are usually only confidently observable over several years. To partly overcome this lack of temporal sensitivity, its measurement can be supplemented by estimation of an active or accessible carbon pool, for which microbial biomass carbon and its ratio to total soil organic carbon appear of most value.

Effective soil management requires observation focused on indicators of soil health and their trends. The indicators chosen for field use by farmers and other land managers should be meaningful for field management of agricultural soils, straightforward to apply in the field or available from routine testing laboratories at reasonable cost. Table 1 lists an example of a set of indicators recommended as an aid to assessing soil health (Friedman et al., 2001), some of which are applicable in the field, although most require laboratory-based testing. The current state-of-the-art, which reflects previous indicator selections but is based on a systematic trial of indicators as applied to long-term field

Table 1 Example of a minimum data set of indicators for soil health (Friedman et al., 2001)

Indicator	Relationship to soil health
Soil organic matter (SOM)	Soil fertility, structure, stability, nutrient retention, soil erosion and available water capacity
Physical	
Soil structure	Retention and transport of water and nutrients, habitat for microbes and soil erosion
Depth of soil and rooting	Estimate of crop productivity potential, compaction and plough pan
Infiltration and bulk density	Water movement, porosity and workability
Water-holding capacity	Water storage and availability
Chemical	
pH	Biological and nutrient availability
Electrical conductivity	Plant growth, microbial activity and salt tolerance
Extractable nitrogen (N), phosphorus (P) and potassium (K)	Plant-available nutrients and potential for nitrogen (N) and phosphorus (P) loss
Biological	
Microbial biomass carbon (C) and N	Microbial catalytic potential and repository for C and N
Potentially mineralisable N	Soil productivity and N-supplying potential
Soil respiration	Microbial activity measure

experiments, is the Cornell Soil Health Assessment (Omololu et al., 2008; Moebius-Clune et al., 2016) which focuses on the soil processes relevant to the crop production function of soil. A set of soil tests (see Table 2) were selected from 39 candidates and a scoring algorithm developed for each to provide a set of scores related to soil properties together with the total score. The same set of tests were used to develop an Ontario Soil Health Assessment (Congreves et al., 2015) but with weighting of different test scores to reflect the sensitivity of these to soils in Ontario. In addition to laboratory testing of soil, there is advantage from encouraging integration of farmers' experiential observation and knowledge of their soils with scientific understanding by identifying field tests that are applicable without specialist training and that have a supporting scientific rationale. Candidate tests for this include in-field soil structure assessment (Ball et al., 2007) and earthworm counts. However, the use of proximate field sensors such as near-infrared reflectance spectroscopy (Stenberg, 2010) is already generating real-time information about soil conditions and this technology, combined with advanced data analysis and fusion methods, promises to provide farmers with a rich and sophisticated source of interpreted information on the health of their soils.

5.2 Governance of soil health

Soil is multifunctional and supports the interests of many different actors in society. While individual farmers and land managers should take practical action to optimise the health of their soils, institutions at all levels from local to regional to national to transnational have

Table 2 Soil quality indicators included in the Cornell soil health test (Omolulu et al., 2008)

Soil indicator	Related soil process
Physical	
Soil texture and stone content	All
Aggregate stability	Aeration, infiltration, shallow rooting and crusting
Available water capacity	Plant-available water retention
Soil strength (penetrometer)	Rooting
Biological	
Organic matter content	Energy/carbon storage, water and nutrient retention
Active carbon content	Organic materials to support biological functions
Potentially mineralisable nitrogen	Ability to supply nitrogen
Root health rating	Soil-borne pest pressure
Chemical	
pH	Toxicity, nutrient availability
Extractable phosphorus	Phosphorus availability, environmental loss potential
Extractable potassium	Potassium availability
Minor element contents	Micronutrient availability

a critical role in developing and implementing agricultural and environmental policies that support soil health. Some relevant policy measures are fiscal incentives, regulation, training, monitoring and research. As well as the public interest, private commercial actors in upstream agricultural supply and downstream food industry chains have a direct interest in ensuring that primary agricultural production has continuing access to healthy soil resources, so that the agriculture sector is sustainable and continues to support their activities. Additionally, commercial operators have a wider social responsibility to support sustainability including via appropriate management of soil resources, which is emphasised by evidence that by far the majority of the costs of soil degradation are incurred off-farm and borne by the wider community (Graves et al., 2015). Therefore, effective governance of soil health requires the connected and congruent participation of many different public and private actors to achieve the common good as realisable within prevailing social norms and political contexts. Specifically, effective legal frameworks for optimising soil health require integration of bottom-up with top-down measures (Kibblewhite et al., 2013).

Some encouraging progress has been made towards achieving the common good of healthy soils. The Global Soil Partnership (Montanarella, 2014) and its constituent regional soil partnerships offer technical support for global soil governance. Continental-scale initiatives include those of the European Union (European Commission, 2006) and the US Department of Agriculture (Natural Resources Conservation Service, 2017). An increasing number of regions, as represented by national or state authorities and agencies, have soil security programmes that support soil health. There are already many local initiatives on soil health often led by farmer organisations and their number is increasing. Some key lessons can be drawn from the experience gained to date, as follows.

Well-found soil monitoring is necessary to provide evidence both to justify public investment in interventions that encourage good soil management practices and to demonstrate their effectiveness. Unfortunately, the rate of change of soil properties at policy-relevant landscape scales is slow and this means that results are not always timely in relation to policy review cycles of typically only a few years. However, effective soil monitoring schemes are essential to provide data to direct efficient investment in soil health.

The variety and variation in soil types, conditions, use and management is immense and policy measures that are top-down and adopt a 'one-size-fits-all' approach prove inefficient. They also have the potential to discourage action and innovation by farmers and land managers because of obvious mismatches between compliance requirements and local knowledge and understanding. Therefore, a more successful approach is to set broad objectives and provide facilitating resources for a 'bottom-up' approach, informed by farmers and land managers' experiential understanding.

6 Conclusions

1. Soil health is a measure of the current performance of a soil relative to its inherent capability, which defines its quality.
2. Soil health is an integrative property of a living system that comprises biotic and abiotic components; it describes how well the soil system can deliver functions (organic residue decomposition, nutrient cycling, soil structure maintenance and ecological community moderation) to support outputs from the wider land system.
3. Experimental assessment of soil health is complicated by the great complexity and multifunctionality of the soil system. At present, understanding of this system is not sufficiently complete to confirm which components and processes are most critical to its functions and inform an unequivocal choice of biological parameters that are indicative of soil health status. However, knowledge of the system is increasing quickly and with it the prospect of such biological indicators.
4. Soil organic carbon levels and trajectories are indicative of soil health. A rising level of total soil organic carbon indicates that the soil system is adequately supplied with substrate carbon whereas a falling one is indicative of stress due to an insufficient supply of energy substrate. Trends in the ratio of active microbial carbon to total soil organic carbon also provide useful indications.
5. Monitoring to support management of soil health in the field should make use of the experiential knowledge of farmers and land managers. This can be supported by in-field soil structural assessment and observation of sentinel organisms, for example, earthworms and routine laboratory-based testing.
6. Soil health is critical to soil security, which is a shared responsibility from local to global levels and for public institutions and private actors. Effective governance of soil health requires investment in monitoring and needs to be participatory with an emphasis on bottom-up as well as top-down policy initiatives.

7 Where to look for further information

The website of the Natural Resources Conservation Service of the US Department of Agriculture (see Natural Resources Conservation Service, 2017) includes many informative pages devoted to soil health. These provide commentaries on different aspects of soil health and include a set of literature reviews.

8 Acknowledgements

This chapter has benefited greatly from many discussions with colleagues, especially at Cranfield University, over the past decade. Its development was helped considerably by the discussions at a workshop on soil health held at Landcare Research in Lincoln, New Zealand, in January 2017 and the author thanks all the participants in that workshop and especially its organiser, Professor David Whitehead.

9 References

Ball, B. C., Batey, T. and Munkholm, L. J. (2007) 'Field assessment of soil structural quality – A development of the Peerlkamp test', *Soil Use and Management*, 23, 329–37.

Bastida, F., Zsolnay, A., Hernández, T. and García, C. (2008) 'Past, present and future of soil quality indices: A biological perspective', *Geoderma*, 147, 159–71.

Chantigny, M. H. (2003) 'Dissolved and water-extractable organic matter in soils: a review on the influence of land use and management practices', *Geoderma*, 113, 357–80.

Congreves, K. A., Hayes, A., Verhallen, E. A. and Van Eerd, L. L. (2015) 'Long-term impact of tillage and crop rotation on soil health at four temperate agroecosystems', *Soil and Tillage Research*, 152, 17–28.

Doran, J. W. and Zeiss, M. R. (2000) 'Soil health and sustainability: managing the biotic component of soil quality', *Applied Soil Ecology* 15, 3–11.

Dungait, J. A. J., Hopkins, D. W., Gregory A. S. and Whitmore A. P. (2012) 'Soil organic matter turnover is governed by accessibility not recalcitrance', *Global Change Biology*, 18, 1781–96.

European Commission. (2006) *Thematic Strategy for Soil Protection*, COM (2006) 0231 Final. Brussels: Commission of the European Communities.

Friedman, D., Hubbs, M., Tugel, A., Seybold, C. and Sucik, M. (2001) Guidelines for Soil Quality Assessment in Conservation Planning, Soil Quality Institute, Natural Resources Conservation Service, United States Department of Agriculture: Washington DC.

Graves, A. R., Morris, J., Deeks, L. K., Rickson, R. J., Kibblewhite, M. G., Harris, J. A., Farewell, T. and Truckle, I. (2015) 'The total costs of soil degradation in England and Wales', *Ecological Economics*, 119, 399–413.

Griffiths, B. S. and Philippot, L. (2012) 'Insights into the resistance and resilience of the soil microbial community', *FEMS Microbiology Reviews*, 37, 112–29.

Jangid, K., Williams, M. A., Franzluebbers, A. J., Schmidt, T. M., Coleman, D. C. and Whitman, W. B. (2011) 'Land-use history has a stronger impact on soil microbial community composition than above ground vegetation and soil properties', *Soil Biology and Biochemistry*, 43, 2184–93.

Jenkinson, D. S. (1971) 'Studies on the decomposition of C14 labelled organic matter in soil', *Soil Science*, 111, 64–70.

Jenny, H. (1941) Factors of soil formation. A System of Quantitative Pedology, MCraw-Hill: New York.

Karlen, D. L., Mausbach, M. J., Doran, J. W., Cline, R. G., Harris, R. F. and Schuman, G. E. (1997) Soil quality: A concept, definition, and framework for evaluation, *Soil Science Society of America Journal*, 61, 4–10.

Keeney, D. R. (1982) 'Nitrogen—availability indices', In Page, A. L., Miller, R. H. and Keeney, D. R. (Eds), Methods of Soil Analysis. *Part 2. Chemical and Microbiological Properties*, American Society of Agronomy and Soil Science Society of America: Madison, pp. 711–33.

Kibblewhite, M. G., Ritz, K. and Swift, M. J. (2008) 'Soil health in agricultural systems', *Philosophical Transactions of the Royal Society. B*, 363, 685–701.

Kibblewhite. M. G., Miko, L. and Montanarella, L. (2012) 'Legal frameworks for soil protection: current development and technical information requirements', *Current Opinion in Environmental Sustainability*, 4, 573–7.

Lal, R. (2016) 'Soil health and carbon management', *Food and Energy Security*, 5, 212–22.

McBratney, A., Fielda, D. J. and Koch, A. (2014) 'The dimensions of soil security', *Geoderma*, 213, 203–13.

Millennium Ecosystem Assessment (2005). Ecosystems and Human Well-being: Synthesis. Island Press: Washington, DC.

Minami, K. (2009) 'Soil and humanity: Culture, civilization, livelihood and health', *Soil Science and Plant Nutrition*, 55, 603–15

Moebius-Clune, B. N., Moebius-Clune, D. J., Gugino, B. K., Idowu, O. J., Schindelbeck, R. R., Ristow, A. J., van Es, H. M., Thies, J. E., Shayler, H. A., McBride, M. B., Wolfe, D. W. and Abawi, G. S. (2016). Comprehensive Assessment of Soil Health – The Cornell Framework Manual, Edition 3.1, Cornell University: Geneva, NY.

Montanarella, L. (2014) 'The Global Soil Partnership', In *Soil Change Matters 2014, IOP Conf. Series: Earth and Environmental Science*, 25, 012001. doi:10.1088/1755-1315/25/1/012001.

Natural Resources Conservation Service (2017) Soil Health, https://www.nrcs.usda.gov/wps/portal/nrcs/main/national/soils/health/Accessed 6 April 2017.

Omololu, J., Idowu, O. J., van Es, H. M., Abawi, G. S., Wolfe, D. W., Ball, J. I., Gugino, B. K., Moebius, B. N., Schindelbeck, R. R. and Bilgili, A. V. (2008) 'Farmer-oriented assessment of soil quality using field, laboratory, and VNIR spectroscopy methods', *Plant and Soil*, 307, 243–53.

Orwin, K. H. and Wardle, D. A. (2004) 'New indices for quantifying the resistance and resilience of soil biota to exogenous disturbances', *Soil Biology and Biochemistry*, 36, 1907–12.

Pulleman, M., Creamer, R., Hamer, U., Helder, J., Pelosi, C., Peres, G. and Rutgers, M. (2012) 'Soil biodiversity, biological indicators and soil ecosystem services—an overview of European approaches', *Current Opinion in Environmental Sustainability*, 4, 529–38.

Ritz, K., Black, H. I. J., Campbell, C. D., Harris, J. A. and Wood, C. (2009) 'Selecting biological indicators for monitoring soils: A framework for balancing scientific and technical opinion to assist policy development', *Ecological Indicators*, 9, 1212–21.

Römbkea, J., Sousab, J.-P., Schoutenc, T. and Riepertd, F. (2006) 'Monitoring of soil organisms: a set of standardized field methods proposed by ISO', *European Journal of Soil Biology*, 42, S61–4.

Romig, D. E., Garlynd, M. J., Harris, R. F. and McSweeney, K. (1995) 'How farmers assess soil health and quality', *Journal of Soil and Water Conservation*, 50, 229–36.

Stenberg, B., Rossel, R. A., Mouazen, A. M. and Wetterlind, J. (2010) 'Visible and near infrared spectroscopy in soil science', In Sparks, D. L. (Ed.), *Advances in Agronomy*, 107, 163–215.

Szabolcs, I. (1994) 'The concept of soil resilience', In Greenland, D. J. and Szabolcs, I. (Eds), *Soil Resilience and Sustainable Land Use*. CAB International: Wallingford, pp. 33–9.

Van Bruggen, A. H. C. and Semenov, A. M. (2000) 'In search of biological indicators for soil health and disease suppression', *Applied Soil Ecology*, 15, 13–24.

Veum, K. S., Goyne, K. W., Kremer, R. J., Miles, R. J. and Sudduth, K. A. (2014) 'Biological indicators of soil quality and soil organic matter characteristics in an agricultural management continuum', *Biogeochemistry*, 117, 81–99.

Young, I. M. and Crawford J. W. (2004) 'Interactions and Self-Organization in the Soil-Microbe Complex', *Science*, 304, 1634–7.

Chapter 2

Soil ecosystem services: an overview

Sara G. Baer, Southern Illinois University, USA; and Hannah E. Birgé[1]*, University of Nebraska, USA*

1 Introduction

Soil is the foundation of terrestrial ecosystems and the majority of ecosystem services needed for human survival arise from soil (Kibblewhite et al., 2008). Ecosystem services provided by soil can be supporting (e.g. primary production and biodiversity) or regulatory (e.g. erosion control, water infiltration, nutrient retention, atmospheric gas regulation and pest control). By definition, ecosystem services benefit human welfare and represent nature's capital (Costanza et al., 1997; Robinson et al., 2012). For example, the economic value of soil microbial metabolic pathways in removing greenhouse gases from the atmosphere, abating nutrients, eradicating pathogens and degrading organic pollutants has been estimated to be double that of the gross annual product (Guimaraes et al., 2010). Many ecosystem services inherently depend on soil health and biodiversity of the soil biota (Barrios, 2007; Brussaard, 2013). Soil health refers to the capacity of soil to function, meaning to sustain or improve productivity and health of plants and higher trophic levels, as well as air and water quality in natural and managed ecosystems (Kibblewhite et al., 2008).

Exploitation of soil has resulted in the collapse of civilizations and regime shifts that have irreversibly reduced the productive capacity of some ecosystems (Holling and Meffe,

[1] The Nature Conservancy, North America Region, Nebraska, USA.

http://dx.doi.org/10.19103/AS.2017.0033.02

1996; Scheffer et al., 2001; Walker and Salt, 2006; Birgé et al., 2016). These events have been instrumental in revealing the most obvious ecosystem service provided by soil: production of food, fibre and fuel for a burgeoning human population. Soil degradation from extractive farming practices eventually reduces crop yields as a result of erosion, reduced nutrient supply from soil with low organic matter stocks and degraded soil structure, all of which can aggravate drought stress (Lal, 2009). The second most obvious ecosystem service provided by soil is infiltration of water. Conversion of wetlands to agricultural and urban environments exacerbates flooding and damage to people's lives, property and local economies. Disturbance regulation (e.g. flood control) is an ecosystem service provided by undegraded soil in natural systems with an estimated value of US $1.8 trillion a year (Costanza et al., 1997).

Other ecosystem services provided by soil are less obvious to the general public because more interdisciplinary knowledge (i.e. chemistry, physics, atmospheric science, soil science, biology, natural history and ecology) is needed to recognize the role of soil in regulating greenhouse gases, reducing nutrient export from agricultural land, controlling pests and supporting biodiversity that provides many of these services to society. Verstraete and Mertens (2004) summarized the general public's lack of awareness of the importance of soil vitality and critical functions that benefit plant and animal health as the largest threat to soil. Thus, the plight soil scientists and ecologists face is the challenge of educating non-scientists about how soil influences quality of human life, future human lives and global sustainability of agricultural and natural ecosystems (Doran, 2002), such that conserving and managing this resource become paramount.

2 Production of food, fibre and energy

Soil is a heterogeneous mixture of minerals, decomposing organic matter and living biomass that supplies nutrients for plant growth and higher trophic levels. Plant production can be used directly for food and indirectly as food for livestock and converted to fuel (lignocellulosic energy) and other consumer products (e.g. lumber, paper and textiles). Thus, many basic requirements for life and human comfort depend on soil and its fertility, health and management to sustain its productive capacity (MEA, 2005).

Land usable for agriculture has increased through advances in crop and livestock genetics, better knowledge of nutrient management, and irrigation (Ellis and Ramankutty, 2008). This has enabled the agricultural industry to meet demands of an increasing human population without a proportional increase in land area used for production (Cassman, 1999). The productive capacity of soil, however, can be compromised if agricultural practices cross degradation thresholds and tipping points, to result in state transitions that are difficult if not impossible to reverse (Peters et al., 2015).

Causes of state transitions that reduce land potential include soil salinization and desertification. For example, removal of deep-rooted perennial vegetation to cultivate shallow-rooted annual wheat in Australia mobilized deep salt deposits to the soil surface and caused persistent declines in production (McFarlane et al., 1992; Pankhurst et al., 2001; Gordon et al., 2003). Irrigating land with water containing trace solutes can also cause soil salinization in arid and semi-arid agroecosystems (Hart et al., 1991; Sumner, 1995; Rietz and Haynes, 2003), and it is estimated that nearly 50% of agricultural soils are susceptible to or experience some degree of salinization. Management practices to counteract soil salinization cost US $12 billion annually (Smedema and Shiati, 2002; Pitman and Läuchli, 2002). Desertification can result from overgrazing (sometimes proximally triggered by drought) and is characterized by a shift from perennial grass cover to more bare ground and shrub cover, soil destabilization and reduced potential for land to support livestock (Peters et al., 2015). Desertification can result in a state change that is difficult to reverse by changing management if it does not revert plant communities (e.g. shrubs to grasses). This phenomenon is exacerbated when restoration is limited by propagule supply, stable soil for plant establishment and water for plant growth. It is estimated that 12 million hectares of land are newly desertified each year (UNCCD, 2013) at a cost of US $3 trillion per year (Berry et al., 2003).

The sustainability of food, fuel and fibre production will ultimately require simultaneous improvement in soil functions of organic matter accrual, nutrient conservation, infiltration and pest control, also referred to as, 'multifunctionality' (Wall et al., 2004; Gamfeldt et al., 2008; Maestre et al., 2012; Wagg et al., 2014). Practices recommended to sustain the productive capacity of soil include shifting from conventional to no-till cropping systems, managing residue, adopting cover crops and using more precise nutrient application regimes (Lal, 2009). There is copious evidence supporting the fact that soil conservation practices increase soil fertility and soil health, reduce erosion and abate available nutrients with high potential for export to aquatic systems, but effects on other ecosystem services are not always consistent (Palm et al., 2014). Thus, providing multiple ecosystem services from agricultural soil is largely accomplished by cover crops, fertilizer management and no-till cropping systems (Syswerda and Robertson, 2014).

3 Erosion control

Erosion is the movement of soil by wind or water. Although a natural process, land-use change and disturbance to soil have accelerated the rate of erosion worldwide (Walling, 1999). The global population consumes food produced from just 11% of the terrestrial surface (FAO, 2011) and 75 billion tons of topsoil is lost annually to erosion from arable land (Eswaran et al., 2002). Soil disturbance from agriculture and urbanization contributes 46 billion metric tons of sediment to the oceans per year (Milliman and Farnsworth, 2011). This sediment contains minerals, nutrients, heavy metals and dissolved organic carbon that change the functioning of aquatic ecosystems. Thus, land-use practices that promote development of a stable soil structure are prerequisite for reducing the export of particle-bound nutrients from terrestrial watersheds to aquatic ecosystems.

Removal of vegetation and physical disturbance to soil facilitate erosion. Eradicating perennial plants reduces surface heterogeneity created by plant stems and litter that stabilize soil, as well as a boundary layer that protects soil from wind. Physical disturbance to soil like conventional tillage disrupts aggregates that hold soil particles together, degrading its structure. The activity, turnover, necromass and exudates of the soil biota contribute to binding particles of sand, silt and clay around organic matter into aggregates that stabilize soil (Oades and Waters, 1991). This is why no-till cropping systems contain a more stable aggregate structure relative to conventionally tilled soil (Six et al., 2000; Stockmann et al., 2012).

The Dust Bowl demonstrates how incomplete knowledge of the effects of agricultural practices on soil resulted in catastrophic movement of soil in the Great Plains, with dire consequences for crop production and rural economies. Replacing deeply rooted plants in the prairie ecosystem with shallow-rooted annual crops displaced soil from a vast area of the Great Plains during prolonged drought in the 1930s. This led to bankruptcy, farm abandonment and mass emigration from the region (Warrick, 1980; Hurt, 1981; Gregory, 1989; McLeman et al., 2014; Zhang et al., 2016). It is now common knowledge that practices such as cover cropping, contour ploughing, conservation tillage, no-till and crop rotation reduce erosion from cultivated land. Non-arable soils in arid ecosystems are also more prone to erosion if physically disturbed. In these ecosystems, biological crusts comprised of lichens, mosses and cyanobacteria reduce soil susceptibility to wind erosion (Belnap and Gillett, 1997). Thus, undisturbed soils perform the ecosystem service of erosion prevention.

Erosion disproportionately removes the light and soluble fraction of organic carbon (C) from the terrestrial ecosystem (Lal, 2003; Quine and van Oost, 2007). As dissolved organic C is transported through soil and to aquatic systems, it becomes prone to decomposition (Dungait et al., 2012). Respiration of labile organic C compounds to carbon dioxide (CO_2) alters C inputs and C:nutrient ratios in terrestrial and aquatic ecosystems. Erosion-driven export of C from terrestrial ecosystems to oceans ranges from 4 to 6 Pg per year (Lal, 2003), of which 20–100% of this C is respired to CO_2 in inland waters prior to reaching the ocean (Smith et al., 2001; Schlesinger, 1995). Carbon exported from eroded systems, however, may be recovered. Recently eroded soils can be efficient C sinks, mitigating erosion-driven C loss through 'dynamic replacement' (Stallard, 1998; Harden et al., 1999, 2002; Quine and van Oost, 2007). Further, erosion of some soils may expose silicate-rich parent material, which when chemically weathered consumes CO_2 (Gabet and Mudd, 2009), although soils with calcareous parent material can be a source of CO_2 (Lobb et al., 2002).

Stream phosphorus (P) concentrations have doubled in the last century (Filippelli, 2008) and erosion-driven fluxes of P from terrestrial to inland waters and the ocean are estimated at 22.5 and 4-6 Tg per year (Quinton et al., 2010), respectively. This flux represents a major global translocation of a critical nutrient with clear implications for terrestrial and aquatic ecological functioning. Phosphorus is known to cause eutrophication of freshwater ecosystems and contribute to hypoxia where major rivers enter the ocean (Rabalais et al., 1996; Diaz and Rosenberg, 2008). Agricultural runoff contains phosphate ($PO_4^{=}$) derived from fertilizers that is readily taken up by aquatic primary producers. Phosphorus is also exported from terrestrial to aquatic systems with sediment, bound to minerals and contained in particulate organic matter. Mineral-bound or occluded P can be desorbed and organic P mineralized to inorganic P in aquatic ecosystems. There are no biological means to remove P from aquatic ecosystems, only uptake by plants and microorganisms, and sedimentation of P-containing organic matter (accretion), which reduce the availability and transport of P downstream. For this reason, protecting soil with high erosion potential from disturbance, preserving riparian zones, creating filter strips of vegetation along waterways and managing agricultural soil to make it less susceptible to loss via wind and water can reduce P export from terrestrial to aquatic ecosystems (Schindler, 1978; Correll, 1998; Schindler et al., 2008; Baer et al., 2016). Because crop P demands (much of which are produced for livestock) exceed P weathering rates and P is not recycled, mining of apatite for fertilizers is depleting global reserves of P needed to sustain agricultural production over the long term (Elser and Bennett, 2011).

In addition to C and nutrient mobilization, erosion transports fine sediment to river systems. Suspended sediments eventually fall out and settle behind dams that create reservoirs. The build-up of sediments behind dams reduces their capacity for water storage, flood control and drought mitigation. Trapped sediments also increase water turbidity and change the energy flow pathways in river systems. For example, sediment containing allochthonous organic matter deposited on river and lake bottoms shifts stream production towards heterotrophy (Stall, 1972), which can stymie fish production and increase water treatment costs (Bilotta and Brazer, 2008). Additionally, erosion selectively transports finer grained soil mineral particles (<63 μm; Koiter et al., 2017). Thus, managing soil to reduce erosion using a landscape or watershed approach is needed to sustain ecosystem services that aquatic habitats provide.

It should be noted that current rates of erosion associated with industrial farming far exceed pedogenesis, and sustainable agricultural practices have done little to ameliorate the mismatch between soil development and loss (Montgomery, 2007). Additionally, mechanisms driving larger watershed-scale sedimentation appear to differ from more poorly understood basin-scale mechanisms (Sharpley et al., 2002; de Vente and Poesen, 2005; Pietron et al., 2017), which when combined with widespread lack of monitoring data (Milliman and Farnsworth, 2011), create uncertainty surrounding management effects on local sedimentation and runoff rates.

4 Nutrient abatement

Fertilizer inputs, livestock production and urban development are major sources of nitrogen (N) and P to surface waters. These nutrients are responsible for eutrophication (Carpenter et al., 1998) and are the leading cause of hypoxia at the mouth of major rivers (Turner and Rabalais, 2003; Diaz and Rosenberg, 2008). The global availability of reactive N has increased as a result of industrial–fixation for fertilizer production, increased biological N fixation through mass production of legumes, land clearing and combustion of fossil fuels (Gruber and Galloway, 2008). Globally, N transported by erosion is estimated at 23–42 Tg per year, and the FAO (2015) predicts an increase in annual demand for N fertilizer to result in an additional 4–6 Tg of reactive N inputs to the biosphere before 2020. The primary sources of P in surface waters are from fertilizer in overland flow, erosion and livestock containments. Hence, reducing N and P loading in aquatic ecosystems will require better land management practices to reduce the availability of these nutrients for loss from terrestrial ecosystems.

Soil biota can remove N from the ecosystem. Denitrifying microorganisms can reduce nitrate (NO_3^-) to molecular N (N_2) using NO_3^- as an alternate electron acceptor in cellular respiration (Knowles, 1982). This metabolic pathway (dissimilatory nitrate reduction or denitrification) is conducted by facultative anaerobic bacteria when O_2 becomes limiting, NO_3^- is available and labile C is present for oxidation to synthesize ATP (Tiedje et al., 1984). Strategically locating and managing wetlands for periodic inundation has the potential to remove significant amounts of reactive N from agricultural and urban runoff through denitrification (Mitsch et al., 2001; Hey et al., 2012). Riparian systems may need management to maximize available C in soil to sustain denitrification (Orr et al., 2007). Restoring cultivated land to perennial vegetation improves soil structure, increases labile C pools and reduces available inorganic N susceptible to leaching (Baer et al., 2002;

Rosenzweig et al., 2016; Baer et al., 2016; Scott et al., 2017). Management that improves soil structure will increase the residence time of nitrate in soil for plant uptake or removal via denitrification.

Reducing nutrient export from whole watersheds is needed to improve water quality in agricultural regions and will require cooperation between government agencies that can offer incentives for better land management and the private sector in willingness to achieve a common goal. For example, reducing high nutrient loads into Lake Erie required targeting half of the cropland area with highest P export and a combination of more strategic P application (subsurface) to reduce P loss, planting cover crops to reduce erosion and take up on-farm available P, and installing buffer strips in riparian areas to take up P exported from agricultural land (Scavia et al., 2017). Once in aquatic ecosystems, the availability of P can be reduced through plant uptake, sedimentation, chemical precipitation and adsorption, but the deposition of organic debris (accretion) containing assimilated P is the primary mechanism by which P is made less available, and thus, P export to rivers can be reduced by strategically locating wetlands (Hey et al., 2012).

5 Infiltration and disturbance regulation

Undisturbed soil with a complex hierarchical aggregate structure has higher infiltration and water-holding capacity than physically disturbed soil (Kemper et al., 2011), a fact which reduces runoff, sedimentation and nutrient pollution. Long-term tillage, operation of heavy equipment, topsoil removal and high livestock densities compact soil. Soil compaction impedes the infiltration of water, reduces water residence time for nutrient abatement and increases the quantity of water in surface runoff. Thus, protecting undisturbed soils containing natural vegetation, especially in floodplains, can reduce flood severity through greater infiltration into the soil and water loss via transpiration through plants (Christen and Dalgaard, 2013). Further, managing soil using practices targeted to improve soil health will increase soil organic matter, and concomitantly, water-holding capacity (Hudson, 1994; Rawls et al., 2003). In drier regions, managing soil to promote water-holding capacity and infiltration can lessen drought effects (Palm et al., 2014). In these ways, infiltration moderates severity of disturbance (drought or flooding), increases water supply, improves water quality and promotes primary production.

Soil macroinvertebrates influence infiltration through their effect on soil porosity. Numerous studies demonstrate earthworms increase yield in agroecosystems by increasing infiltration through the creation of casts, incorporating organic matter into the soil, stimulating microbial growth and exuding organic binding substrates that promote aggregate stability and soil water-holding capacity (Lavelle et al., 2006; Johnson-Maynard, 2016; van Groenigen et al., 2014). Termites increase infiltration by cultivating fungi, mounding and sheeting activities, particularly in barren soil (Lavelle et al., 1997; Kaiser et al., 2017), although some termite mounds can reduce infiltration. Soil macroinvertebrate density in Amazonia has been positively correlated with soil macroporosity and water availability (Marichal et al., 2014); and in over-grazed systems with compacted soil, dung beetles reduce soil bulk density and improve hydrologic properties (Brown et al., 2010).

Land managers can enhance infiltration and water-holding capacity to protect terrestrial and aquatic ecosystems from numerous agricultural disservices. Adopting reduced tillage, cover crops, planting buffer strips, managing residue and land restoration are practices that promote infiltration (Sargody, 1993; Walling and Fang, 2003; Hoffman et al., 2010; Grudens-Schuck et al., 2017). Farmscaping, the creation of multiple on-farm habitats to increase biodiversity and landscape heterogeneity, is used to promote infiltration and agroecosystem multifunctionality (Smukler et al., 2010).

6 Greenhouse gas and climate regulation

Global warming projections are attributed mostly to increasing levels of CO_2, nitrous oxide (N_2O) and methane (CH_4) in the atmosphere (IPCC, 2014). Although agriculture has contributed a relatively small amount of CO_2 to the atmosphere compared to combustion of fossil fuels, it is estimated that agricultural practices are responsible for more than 50% and 80% of anthropogenic CH_4 and N_2O emissions, respectively (Smith et al., 2008). Despite this disservice, changing agricultural practices and restoring cropland to perennial vegetation cover represent means to remove CO_2 from the atmosphere and reduce greenhouse gas emissions from soil to the atmosphere (Gelfand et al., 2016). The effect of warming on decomposition (soil respiration), however, imparts uncertainty to climate change prediction models, but there is evidence that warming will increase CO_2 emissions through microbial degradation of soil C pools (Knorr et al., 2005). Whether or not microbial processes in soil result in positive or negative feedback on CO_2 flux from soil that accelerates or mitigates climate change, respectively, will depend on how much C in soil is available for oxidation (Bardgett et al., 2008).

Agricultural soils generally have a higher capacity to sequester C from the atmosphere into long-term stable pools relative to never-cultivated soils because C stocks in long-term cultivated soils are below their capacity to store C (Paustian et al., 1997; Lal et al., 2011). Cultivation reduces soil C stocks in the tilled depth by 20–80% (Mann, 1986). Carbon saturation theory, developed from an analysis of soil C gain relative to inputs across a wide range of agroecosystems, predicts that soils with greater C saturation deficit have more capacity to accumulate C relative to soils with C stocks closer to saturation (Stewart et al., 2007). For C to become sequestered in soil, not only do inputs have to exceed outputs, but organic matter must become stabilized and less susceptible to decomposition. Organic matter stabilization mechanisms include sorption to clay minerals (physico-chemical protection), protection within aggregates and biochemical recalcitrance (Six et al., 2002). For example, long-term cultivated sandy soil restored to grassland did not develop an improved aggregate structure or accrue C on a decadal timescale, whereas grassland restored on soil containing more clay in the same region showed an increase in mean aggregate diameter corresponding with C accrual, demonstrating the key role of soil texture in C accumulation (Baer et al., 2010). Further, similarly textured restored grassland soil in different continents showed more rapid accumulation of C where soil C saturation deficit was greater (Baer et al., 2015). Restoring perennial vegetation on degraded soil, implementing on-farm soil conservation practices (e.g. cover crops and reduced tillage) and biochar application to soil have the most potential to sequester C and help offset increasing atmospheric CO_2 (Paustian et al., 1998; Lal, 2004; Stavi and Lal, 2013). Biochar can also increase crop yields, soil water-holding capacity and soil fertility (Spokas et al., 2012).

Although the concentration of N_2O in the atmosphere is orders of magnitude less than CO_2, this greenhouse gas has approximately three-hundred times the radiative forcing as CO_2 (IPCC, 2014). Nitrous oxide presents an additional concern because it is inert in the troposphere. As a result, N_2O has a very long residence time and mixes with the stratosphere where it acts as a catalyst in the destruction of protective ozone (Warneck, 2000). Nitrous oxide is a biogenic trace gas produced mainly from two microbial metabolic pathways: nitrification and denitrification (Davidson, 1992). Although soil is a natural source of N_2O (the only sink for N_2O is mixing with the stratosphere), land-use change has increased nitrification. For example, recently cleared forests exhibit increased nitrification rates in soil and export of NO_3^- to stream ecosystems (Likens et al., 1970). Statistical models estimate that the global emissions of N_2O from fertilized cropland are more than four times that of natural systems (Stehfest and Bouwman, 2006). Compared to a wide variety of cropping systems, N_2O flux was 3–10 times lower in agricultural soil converted to forest cover or allowed to recover through successional processes (Robertson et al., 2000). Overall, conversion of conventionally cultivated soil to no-till reduces global warming potential mostly through C sequestration, but maximum greenhouse gas mitigation can be achieved by removing marginal lands from crop production due to the reduction in CO_2-producing agricultural activities, lower emissions of N_2O and greater potential for CH_4 oxidation in soil (Robertson et al., 2000).

Methane is a greenhouse gas with about 20 times the global warming potential as CO_2 (IPCC, 2014). The only biological sink for CH_4 is microbial oxidation by methanotrophs, which can consume as much as 90% of the CH_4 produced by methanogens in aerobic soils (Chowdhury and Dick, 2013). Thus, understanding factors that influence methanotrophy is needed to protect this ecosystem service provided by soil. Few reviews on this topic are available, but it is generally understood that agriculture increases CH_4 emissions from soil, high levels of N inhibit methanotrophy, and there is uncertainty about the role of available P and N:P ratios in soil on CH_4 oxidation (Veraart et al., 2015).

7 Pest control

Agricultural intensification reduces natural enemies of crop pests (Zhao et al., 2015), which are responsible for the destruction of roughly one-fifth of global crop production each year (FAO, 1999). To combat these losses, synthetic pesticides are used nearly ubiquitously in most non-organic cropping systems despite the cost associated with recurrent pest resistance, secondary pest emergences, adverse human health effects, environmental impacts and frequent obsolescence (Power, 1982). 'Conservation biological control' entails adopting farming practices and manipulating the environment in ways to promote populations of natural enemies of agricultural pests (Barbosa, 1998). Managing agroecosystems and surrounding areas to harbour natural enemies of agricultural pests is a low-cost, safe method to mitigate crop damage from pests while simultaneously supporting other soil ecosystem services that arise from a diverse belowground community.

Bacteria, fungi, nematodes, microarthropods and larger arthropods in soil are known to control agricultural pests and pathogens. Bacteria regulate plant disease through pathogen antagonism in the rhizosphere (Weller, 1988). Entomopathogenic fungi (Ascomycota) are a natural enemy of many agricultural insect pests in temperate regions. Entomopathogenic fungal populations and crops benefit under no-till and other conservation tillage practices

(Meyling and Eilenberg, 2007). Soil nematodes have been used to biocontrol thrips, small-winged insects that transmit fungal infections and viruses to crops worldwide (Pappu et al., 2009). Even a small initial population of the nematode *Thripinema* sp. can proliferate and control thrips populations following thrips invasion (Arthurs and Heinz, 2006), reducing the need for pesticides or other intensive management interventions. Fungivorous microarthropods (collembolans and mites) and nematophagous mites can also perform pest-control services by feeding on plant pathogenic fungi and plant-feeding nematodes, respectively (Coleman et al., 2004). Earthworms benefit plant growth by dispersing microorganisms that are antagonistic to root pathogens (Scheu, 2003). In the absence of dung beetles, livestock pastures accumulate faeces that foul forage and result in more enteric parasites. In fact, the estimated net value of dung beetles to the US beef cattle industry is US $380 million a year (Nichols et al., 2008).

Integrated pest management incorporates ecological knowledge into management strategies and farming practices to maintain yield and minimize the use of pesticides. This approach emphasizes ecosystem management for biodiversity to promote natural enemies of agricultural pests and minimize application of pesticides that are dangerous to humans and other animals. The FAO (2017) recommends crop rotation or inter-cropping, proper sanitation techniques, pest-resistant cultivars, managing soil to promote C storage and minimal use of pesticides.

8 Biodiversity

Although the abundance of bacteria, fungi, protozoans and animals in soil is astounding, most taxa and their natural history have hardly been described (Coleman et al., 2004). The most abundant group of organisms in soil are bacteria. One gram of soil can contain up to 10^9 bacterial cells (Gans et al., 2005), constituting up to 10^4 species (Curtis et al., 2002). Bacteria can benefit plant growth directly by stimulating root branching and root hair development (Hayat et al., 2010; Vacheron et al., 2013) and indirectly by dispersing pathogens (Weller, 1988). Bacteria also regulate biogeochemical cycling (Falkowski et al., 2008), influence atmospheric chemistry and climate through the production and consumption of trace gases (Conrad, 1996) and degrade organic contaminants in soil (Guimaraes et al., 2010). The abundance of viruses actually exceeds prokaryotes (Weinbauer, 2004), and as such, viral-induced bacteria mortality has consequence for biogeochemical cycling and other ecosystem services provided by soil (Weinbauer, 2004).

Eukaryotes are also abundant in soil. One gram of soil can also contain >100 m of fungal hyphae (Leake et al., 2004). Although some fungi are pathogenic to plants, their hyphae entangle soil and play an important role in soil aggregation (Oades et al., 1991; Wilson et al., 2009) and plant nutrient use (Brussaard et al., 2007). Ten to hundreds of thousands of protozoa (single-celled eukaryotes) can also be found in a gram of soil, and their feeding on bacteria and fungi stimulates nutrient turnover (Bouwman and Zwart, 1994). Nematodes can be the most numerous of all animals in the soil and occupy all consumer trophic levels as bacteria feeders, fungi feeders, root herbivores, predators and omnivores – contributing to the great complexity of soil foodwebs (Coleman et al., 2004). Microarthropod (collembolans and mites) abundances range from 10^3 to 10^5 individuals per square metre of soil, and despite the key role they play in decomposition (Heneghan et al., 1999) only 10% of the species have been explored or described (Andre et al., 2002).

Enchytraeids are miniature, unpigmented earthworms that ingest organic matter, soil particles and associated bacteria and fungi (Coleman et al., 2004). Enchytraeid abundances in soil range from 10^3 to 10^5 individuals per square metre and these species can influence C and N dynamics during decomposition, depending on whether they stimulate or inhibit microbial activity (van Vliet, 2000; van Vliet et al., 2004).

The soil macrofauna include insects, isopods, earthworms, spiders, centipedes, millipedes and other invertebrates >10 mm in length (Coleman et al., 2004). Soil macroarthropods influence primary productivity, pedogenesis, belowground food webs, litter decomposition, nutrient cycling, soil structure, water infiltration, soil fertility and greenhouse gas emissions (Lavelle et al., 2006; Osler and Somnerkom, 2007; Folgarait, 1998; Eisenhauer et al., 2009). For example, bioturbation by dung beetles increases water infiltration, N mineralization and plant productivity (Nichols et al., 2008). Soil-dwelling ants benefit plant growth by modifying biotic and abiotic aspects of soil (e.g. structure, moisture content, and nutrient availability) and increasing mycorrhizal colonization (Folgarait, 1998; Cammeraat et al., 2002; Dauber et al., 2008).

Earthworms have been studied extensively because they are ecosystem engineers (Lavelle et al., 1997; Blouin et al., 2013) and can increase yield in agroecosystems (van Groenigen et al., 2014). Earthworms have a positive effect on plant growth by improving soil structure to retain more water, stimulating microbial turnover, increasing N mineralization and dispersing microorganisms antagonistic to root pathogens (Bohlen and Edwards, 1995; Scheu, 2003). Earthworms can also modify the soil seed bank and composition of germinating seeds (Eisenhauer et al., 2009). Their exudates and physical incorporation of organic matter in soil promotes aggregate formation and stability (Johnson-Maynard and Strawn, 2016). Earthworm-mediated formation of large macroaggregates that contain stable microaggregates is an important mechanism in organic matter protection (Bossuyt et al., 2004). While soil bacteria and fungi ultimately mediate many soil processes that underpin ecosystem services, their functioning is influenced by macrofauna that reinforce a system state favourable to those organisms and the ecosystem services they provide (Lavelle, 2006).

Agricultural intensification compromises ecosystem services through negative impacts to soil health and soil biodiversity (Giller et al., 1997; Brussaard, 1997; de Fries et al., 2004; Bisset et al., 2011; Al-Kaisi and Lowery, 2017). Cultivation favours the development of bacteria-dominated communities. Bacteria contribute less to C sequestration in soil relative to fungi due to the key role of mycorrhizal fungi in soil aggregation (Wilson et al., 2009) and more recalcitrant necromass of fungi relative to bacteria (Six et al., 2006). A comparison of cultivated soil with a bacteria-dominated food web to a managed grassland with a fungi-dominated food web demonstrated that tillage lowered the resistance and resilience of the soil food web to drought (de Vries et al., 2012). Whether or not microbial diversity *per se* increases soil resistance and resilience cannot be generalized (Griffiths and Philippot, 2013), but biological stability in soil is determined by soil structure and its effect on the composition and physiology of microbes (Griffiths et al., 2008). If C flow, however, is not maintained through the system, available C substrates for microbial metabolism will become depleted and the organic matter required to maintain soil structure will be reduced (Janzen, 2015). Some negative impacts of agricultural intensification on biodiversity, however, can be alleviated. Higher microbial and macrofauna diversity and biomass was reported in a corn–legume intercropping system relative to traditional monoculture corn (Blanchart et al., 2006). Soils under intercropping also had higher organic matter, lower runoff and higher corn yield, emphasizing the complexity of distinguishing causal, correlative or synergistic relationships between belowground biodiversity and functioning.

It is important to remember that although this chapter is focused on ecosystem services provided by soil, the influence of plants on the diversity of soil biota and services provided by soil cannot be ignored. Plant–soil feedback can occur through variation in the quality and quantity of litter inputs, production of allelopathic compounds, culturing a soil microbial community that promotes or inhibits the growth of other plant species, or altering nutrient availability in ways that promote or inhibit other species (Ehrenfeld et al., 2005). Plants influence the diversity of microorganisms in soil (Thiele-Bruhn et al., 2012) because most soil microbiota are concentrated around plant roots, in the rhizosphere. There is also increasing awareness that plant hosts culture the microbiome around the root (rhizobiome) to increase plant function (Berendsen et al., 2012; Wubs et al., 2016), for example, drought tolerance (Marasco et al., 2012). In fact, the soil microbiome differs between the rhizosphere and bulk soil, among genotypes within a plant species and among fields (Peiffer et al., 2013). Thus, illuminating the black box of soil biodiversity, functional repertoire of the soil microbiome and interactions with multiple dimensions of plant diversity (genetic, functional and species) may be key to improving soil health and understanding the relationship between plant biodiversity, microbial diversity and ecosystem functioning in terrestrial ecosystems (de Deyn and van der Putten, 2005; Chaparro et al., 2012; Philoppot et al., 2013).

9 Conclusion and future trends

Agriculture has been a source of ecosystem disservices by causing increased erosion, greenhouse gas emissions from soil, nutrient pollution and loss of biodiversity. Thus, realizing the potential for soil to provide ecosystem services involves reducing disservices caused by agriculture (Power, 2010). This will require conserving soil for sustainable production of raw materials that human lives depend on; adopting conservation-based cropping methods to reduce erosion; precision nutrient application in cropping systems to reduce nutrient loss from agricultural systems; land-use changes that promote nutrient conservation and removal in the terrestrial system to reduce export to aquatic ecosystems; and conversion of marginal agricultural lands to more natural systems that conserve soil, water, nutrients and biodiversity. The abundance, diversity, activity and composition of soil biota regulate many ecosystem functions (and multifunctionality) that underpin ecosystem services provided by soil (Wagg et al., 2014). Knowledge of the diversity and redundancy within the soil biota is needed to predict their role in promoting resilience of ecosystem services provided by soil in the face of many global environmental changes.

To meet the global food requirement by 2050, there must be at least a 70% increase in agricultural production (FAO, 2009), which can only be achieved by agricultural intensification of existing cropland and marginal lands (Foley et al., 2011). Only 3% of cropland is classified as 'highly fertile', meaning resilient to disturbance and capable of maintaining high yields (FAO, 2003). Large increases in global yield in past decades have resulted mostly from agricultural intensification on the most fertile croplands (Foley, 2011). Yield increases have also resulted from agricultural expansion into natural ecosystems, particularly tropical forests, grasslands and savannas (Ellis and Ramankutty et al., 2008). While affecting a relatively small fraction of Earth's surface, agriculture is a major driver of soil degradation, globally, and the direct effects and feedbacks of agricultural intensification and expansion combined with other global change drivers will determine the fate of soil health and human civilization. Thus, sustainable intensification, defined as increase in agricultural yields without negative impact to the environment or conversion of non-agricultural land to agricultural production (Pretty and Bharucha, 2014), is needed for global sustainability.

Long-term data are needed to predict how directional change or variability in climate, coupled with increasing atmospheric CO_2, will influence feedbacks between plants and soil (properties, processes and biota) to better understand, improve and foster multiple potential ecosystem services provided by agroecosystems. Our rapidly expanding knowledge of soil microbial diversity and the influence of plant chemical signals on the composition and function of the rhizobiome suggests that future management of agroecosystems for sustainable yields will require particular attention to interactions between plant genotypes and rhizobiomes that promote tolerance to environmental stress, nutrient acquisition and disease resistance. Gaining a better understanding of belowground thresholds and tipping points in response to changing environmental conditions will also be needed to predict and prevent agroecosystem state transitions that may be difficult to manage and, in the worst-case scenario, impossible to reverse.

10 Where to look for further information

This chapter contains an exhaustive review of the literature on ecosystem services provided by soil and many articles cited review different aspects of this subject. A comprehensive collection of reviews can be found in a synthetic edited book entitled *Soil Ecology and Ecosystem Services* (Wall et al., 2012). A basic understanding of soil ecology [the soil biota and how they interact with other organisms and the environment] is prerequisite to understanding the role of state factors and soil organisms in providing ecosystem services. The seminal book on this subject was authored by David C. Coleman, D. A. Crossley and Paul F. Hendrix (Coleman et al., 2004). In 2011, Diana Wall initiated the Global Soil Biodiversity Initiative with the aim of providing a 'coherent platform for promoting the translation of expert knowledge on soil biodiversity into environmental policy and sustainable land management for the protection and enhancement of ecosystem services' (https://globalsoilbiodiversity.org/). This led to the production of *Global Soil Biodiversity Atlas* (Orgiazzi et al., 2016) in honour of the 2015 United Nations International Year of Soils. The atlas is a visually stimulating and information-rich resource on soil that covers a wide range of topics including soil formation processes, diversity and natural history of organisms in soil, the role of soil biodiversity in ecosystem processes, human impacts on soil and its biodiversity, and how we can intervene to protect and improve ecosystem services provided by soil.

11 References

Al-Kaisi, M. and Lowery, B. (2017), *Soil Health and Intensification of Agroecosystems*, Academic Press, London.

Andre, H. M., Ducarme, X. and Lebrun, P. (2002), 'Soil biodiversity: Myth, reality or conning?', *Oikos*, 96(1), pp. 3–24.
Arthurs, S. and Heinz, K. M. (2006), 'Evaluation of the nematodes *Steinernema feltiae* and *Thripinema nicklewoodi* as biological control agents of western flower thrips *Frankliniella occidentalis* infesting chrysanthemum', *Biocontrol Science and Technology*, 16(2), pp. 141–55.
Baer, S. G. (2016), 'Nutrients as determinants and endpoints in ecological restoration', pp. 333–64. In: Palmer, M. A., Zedler, J. and Falk, D. (Eds), *Foundations of Restoration Ecology*, 2nd edition. Island Press, Washington DC.
Baer, S. G., Bach, E. M., Meyer, C. K., du Preez, C. C. and Six, J. (2015), 'Belowground ecosystem recovery during grassland restoration: South African highveld compared to US tallgrass prairie', *Ecosystems*, 18(3), pp. 390–403.
Baer, S. G., Kitchen, D. J., Blair, J. M. and Rice, C. W. (2002), 'Changes in ecosystem structure and function along a chronosequence of restored grasslands', *Ecological Applications*, 12(6), pp. 1688–701.
Baer, S. G., Meyer, C. K., Bach, E. M., Klopf, R. P. and Six, J. (2010), 'Contrasting ecosystem recovery on two soil textures: implications for carbon mitigation and grassland conservation', *Ecosphere*, 1(1), pp. 1–22.
Barbosa, P. (1998), *Conservation Biological Control*, Academic Press, London.
Bardgett, R. D., Freeman, C. and Ostle, N. J. (2008), 'Microbial contributions to climate change through carbon cycle feedbacks', *Isme Journal*, 2(8), pp. 805–14.
Barrios, E. (2007), 'Soil biota, ecosystem services and land productivity', *Ecological Economics*, 64(2), pp. 269–85.
Belnap, J. and Gillette, D. A. (1997), 'Disturbance of biological soil crusts: Impacts on potential wind erodibility of sandy desert soils in southeastern Utah', *Land Degradation & Development*, 8(4), pp. 355–62.
Berendsen, R. L., Pieterse, C. M. J. and Bakker, P. (2012), 'The rhizosphere microbiome and plant health', *Trends in Plant Science*, 17(8), pp. 478–86.
Berry, L., Olson, J. M. and Campbell, D. J. (2003), *Assessing the Extent, Cost and Impact of Land Degradation at the National Level: Findings and Lessons Learned from Seven Pilot Case Studies*, Washington, DC: World Bank.
Bilotta, G. S. and Brazier, R. E. (2008), 'Understanding the influence of suspended solids on water quality and aquatic biota', *Water Research*, 42(12), pp. 2849–61.
Birgé, H. E., Allen, C. R., Garmestani, A. S. and Pope, K. L. (2016), 'Adaptive management for ecosystem services', *Journal of Environmental Management*, 183, pp. 343–52.
Bissett, A., Richardson, A. E., Baker, G. and Thrall, P. H. (2011), 'Long-term land use effects on soil microbial community structure and function', *Applied Soil Ecology*, 51, pp. 66–78.
Blanchart, E., Villenave, C., Viallatoux, A., Barthes, B., Girardin, C., Azontonde, A. and Feller, C. (2006), 'Long-term effect of a legume cover crop (*Mucuna pruriens* Var. *utilis*) on the communities of soil macrofauna and nematofauna, under maize cultivation, in Southern Benin', *European Journal of Soil Biology*, 42, pp. 136–44.
Blouin, M., Hodson, M. E., Delgado, E. A., Baker, G., Brussaard, L., Butt, K. R., Dai, J., Dendooven, L., Peres, G., Tondoh, J. E., Cluzeau, D. and Brun, J. J. (2013), 'A review of earthworm impact on soil function and ecosystem services', *European Journal of Soil Science*, 64(2), pp. 161–82.
Bohlen, P. J. and Edwards, C. A. (1995), 'Earthworm effects on N dynamics and soil respiration in microcosms receiving organic and inorganic nutrients', *Soil Biology & Biochemistry*, 27(3), pp. 341–8.
Bossuyt, H., Six, J. and Hendrix, P. F. (2004), 'Rapid incorporation of carbon from fresh residues into newly formed stable microaggregates within earthworm casts', *European Journal of Soil Science*, 55(2), pp. 393–9.
Bouwman, L. A. and Zwart, K. B. (1994), 'The ecology of bacterivorous protozoans and nematodes in arable soil', *Agriculture Ecosystems & Environment*, 51(1–2), pp. 145–60.

Brown, J., Scholtz, C. H., Janeau, J. L., Grellier, S. and Podwojewski, P. (2010), 'Dung beetles (Coleoptera: Scarabaeidae) can improve soil hydrological properties', *Applied Soil Ecology*, 46(1), pp. 9–16.

Brussaard, L. (2013), 'Ecosystem Services Provided by the Soil Biota', *Soil Ecology and Ecosystem Services*, pp. 45–58.

Brussaard, L., de Ruiter, P. C. and Brown, G. G. (2007), 'Soil biodiversity for agricultural sustainability', *Agriculture Ecosystems & Environment*, 121(3), pp. 233–44.

Brussaard, L., Brussaard, L., Behan-Pelletier, V. M., Bignell, D. E., Brown, V. K., Didden, W., Folgarait, P., Fragoso, C., Freckman, D. W., Gupta, v., Hattori, T., Hawksworth, D. L., Klopatek, C., Lavelle, P., Malloch, D. W., Rusek, J., Soderstrom, B., Tiedje, J. M. and Virginia, R. A. (1997), 'Biodiversity and ecosystem functioning in soil', *Ambio*, 26(8), pp. 563–70.

Cammeraat, L. H., Willott, S. J., Compton, S. G. and Incoll, L. D. (2002), 'The effects of ants' nests on the physical, chemical and hydrological properties of a rangeland soil in semi-arid Spain', *Geoderma*, 105(1–2), pp. 1–20.

Carpenter, S. R., Caraco, N. F., Correll, D. L., Howarth, R. W., Sharpley, A. N. and Smith, V. H. (1998), 'Nonpoint pollution of surface waters with phosphorus and nitrogen', *Ecological Applications*, 8(3), pp. 559–68.

Cassman, K. G. (1999), 'Ecological intensification of cereal production systems: Yield potential, soil quality, and precision agriculture', *Proceedings of the National Academy of Sciences of the United States of America*, 96(11), pp. 5952–9.

Chaparro, J. M., Sheflin, A. M., Manter, D. K. and Vivanco, J. M. (2012), 'Manipulating the soil microbiome to increase soil health and plant fertility', *Biology and Fertility of Soils*, 48(5), pp. 489–99.

Chowdhury, T. R. and Dick, R. P. (2013), 'Ecology of aerobic methanotrophs in controlling methane fluxes from wetlands', *Applied Soil Ecology*, 65, pp. 8–22.

Christen, B. and Dalgaard, T. (2013), 'Buffers for biomass production in temperate European agriculture: A review and synthesis on function, ecosystem services and implementation', *Biomass & Bioenergy*, 55, pp. 53–67.

Coleman, D. C., Crossley, D. A. Jr. and Hendrix, P. F. (2004), *Fundamentals of Soil Ecology*, Elsevier, Burlington MA.

Conrad, R. (1996), 'Soil microorganisms as controllers of atmospheric trace gases (H_2, CO, CH_4, OCS, N_2O, and NO)', *Microbiological Reviews*, 60(4), pp. 609–40.

Correll, D. L. (1998), 'The role of phosphorus in the eutrophication of receiving waters: A review', *Journal of Environmental Quality*, 27(2), pp. 261–6.

Costanza, R., dArge, R., deGroot, R., Farber, S., Grasso, M., Hannon, B., Limburg, K., Naeem, S., O'Neill, R. V., Paruelo, J., Raskin, R. G., Sutton, P. and vandenBelt, M. (1997), 'The value of the world's ecosystem services and natural capital', *Nature*, 387(6630), pp. 253–60.

Curtis, T. P., Sloan, W. T. and Scannell, J. W. (2002), 'Estimating prokaryotic diversity and its limits', *Proceedings of the National Academy of Sciences of the United States of America*, 99(16), pp. 10494–9.

Dauber, J., Niechoj, R., Baltruschat, H. and Wolters, V. (2008), 'Soil engineering ants increase grass root arbuscular mycorrhizal colonization', *Biology and Fertility of Soils*, 44(5), pp. 791–6.

Davidson, E. A. (1992), 'Sources of nitric-oxide and nitrous oxide following wetting of dry soil', *Soil Science Society of America Journal*, 56(1), pp. 95–102.

de Deyn, G. B. and van der Putten, W. H. (2005), 'Linking aboveground and belowground diversity', *Trends in Ecology & Evolution*, 20(11), pp. 625–33.

de Fries, R. S., Foley, J. A. and Asner, G. P. (2004), 'Land-use choices: Balancing human needs and ecosystem function', *Frontiers in Ecology and the Environment*, 2(5), pp. 249–57.

de Vente, J. and Poesen, J. (2005), 'Predicting soil erosion and sediment yield at the basin scale: Scale issues and semi-quantitative models', *Earth-Science Reviews*, 71(1–2), pp. 95–125.

de Vries, F. T., Liiri, M. E., Bjornlund, L., Bowker, M. A., Christensen, S., Setala, H. M. and Bardgett, R. D. (2012), 'Land use alters the resistance and resilience of soil food webs to drought', *Nature Climate Change*, 2(4), pp. 276–80.

Diaz, R. J. and Rosenberg, R. (2008), 'Spreading dead zones and consequences for marine ecosystems', *Science*, 321(5891), pp. 926–9.

Doran, J. W. (2002), 'Soil health and global sustainability: Translating science into practice', *Agriculture Ecosystems & Environment*, 88(2), pp. 119–27.

Dungait, J. A. J., Hopkins, D. W., Gregory, A. S. and Whitmore, A. P. (2012), 'Soil organic matter turnover is governed by accessibility not recalcitrance', *Global Change Biology*, 18(6), pp. 1781–96.

Ehrenfeld, J. G., Ravit, B. and Elgersma, K. (2005), 'Feedback in the plant-soil system', *Annual Review of Environment and Resources*, 30, pp. 75–115.

Eisenhauer, N., Schuy, M., Butenschoen, O. and Scheu, S. (2009), "Direct and indirect effects of endogeic earthworms on plant seeds', *Pedobiologia*, 52(3), pp. 151–62.

Ellis, E. C. and Ramankutty, N. (2008), 'Putting people in the map: Anthropogenic biomes of the world', *Frontiers in Ecology and the Environment*, 6(8), pp. 439–47.

Elser, J. and Bennett, E. (2011), 'A broken biogeochemical cycle', Nature, 478(7367), pp. 29–31.

Eswaran, H., Lal, R., Reich, P. F. (2002), 'Land degradation: An overview', In: Bridges, E. M., Hannam, I. D., Oldeman, L. R., Pening de Vries, F. W. T., Scherr, S. J., Sompatpanit, S. (Eds), *Responses to Land Degradation: Proceedings of the 2nd International Conference on Land Degradation and Desertification*. Khon Kaen, Thailand, 2001, New Delhi, India: Oxford Press.

Falkowski, P. G., Fenchel, T. and Delong, E. F. (2008), 'The microbial engines that drive Earth's biogeochemical cycles', *Science*, 320(5879), pp. 1034–9.

Filippelli, G. M. (2008), 'The global phosphorus cycle: Past, present, and future', *Elements*, 4(2), pp. 89–95.

Foley, J. A., Ramankutty N., Brauman, K. A., Cassidy, E. S., Gerber, J. S., Johnston, M., Mueller, N. D., O'Connell, C., Ray, D. K., West, P. C., Balzer, C., Bennett, E. M., Carpenter, S. R., Hill, J., Monfreda, C., Polasky, S., Rockström, J., Sheehan, J., Siebert, S., Tilman, D. and Zaks, D. P. M. (2011), 'Solutions for a cultivated planet', *Nature* 478(7369), pp. 337–42.

Folgarait, P. J. (1998), 'Ant biodiversity and its relationship to ecosystem functioning: A review', *Biodiversity and Conservation*, 7(9), pp. 1221–44.

Food and Agricultural Organization (FAO) (1999), 'Insect damage: post-harvest operations, INPho – Post-harvest Compendium', available: http://www.fao.org/3/a-av013e.pdf.

Food and Agricultural Organization (FAO) (2003), '*Compendium of Agricultural-environmental Indicators 1989–91 to 2000*', Rome: Statistics Analysis Service.

Food and Agriculture Organization (FAO) (2009), 'How to feed the world in 2050. World Summit on Food Security, Executive Report', available http://www.fao.org/fileadmin/templates/wsfs/docs/expert_paper/How_to_Feed_the_World_in_2050.pdf.

Food and Agricultural Organization (FAO) (2011), 'The State of the World's Land and Water Resources for Food and Agriculture (SOLAW) – Managing Systems at Risk', Rome and London: The United Nations and Earthscan.

Food and Agricultural Organization (FAO) (2015), 'World fertilizer trends and outlook to 2018', available: http://www.fao.org/3/a-i4324e.pdf.

Food and Agricultural Organization (FAO) (2017), 'More about Integrated Pest Management (IPM)', available: http://www.fao.org/agriculture/crops/thematic-sitemap/theme/pests/ipm/more-ipm/en/.

Gabet, E. J. and Mudd, S. M. (2009), 'A theoretical model coupling chemical weathering rates with denudation rates', *Geology*, 37(2), pp. 151–4.

Gamfeldt, L., Hillebrand, H. and Jonsson, P. R. (2008), 'Multiple functions increase the importance of biodiversity for overall ecosystem functioning', *Ecology*, 89(5), pp. 1223–31.

Gans, J., Wolinsky, M. and Dunbar, J. (2005), 'Computational improvements reveal great bacterial diversity and high metal toxicity in soil', *Science*, 309(5739), pp. 1387–90.

Gelfand, I., Shcherbak, II, Millar, N., Kravchenko, A. N. and Robertson, G. P. (2016), 'Long-term nitrous oxide fluxes in annual and perennial agricultural and unmanaged ecosystems in the upper Midwest USA', *Global Change Biology*, 22(11), pp. 3594–607.

Giller, K. E., Beare, M. H., Lavelle, P., Izac, A. M. N. and Swift, M. J. (1997), 'Agricultural intensification, soil biodiversity and agroecosystem function', *Applied Soil Ecology*, 6(1), pp. 3–16.

Gordon, L., Dunlop, M. and Foran, B. (2003), 'Land cover change and water vapour flows: Learning from Australia', *Philosophical Transactions of the Royal Society of London Series B-Biological Sciences*, 358(1440), pp. 1973–84.

Gregory, J. N. (1989), *American Exodus: The Dust Bowl Migration and Okie Culture in California*, Oxford University Press.

Griffiths, B. S. and Philippot, L. (2013), 'Insights into the resistance and resilience of the soil microbial community', *Fems Microbiology Reviews*, 37(2), pp. 112–29.

Griffiths, B. S., Hallett, P. D., Kuan, H. L., Gregory, A. S., Watts, C. W. and Whitmore, A. P. (2008), 'Functional resilience of soil microbial communities depends on both soil structure and microbial community composition', *Biology and Fertility of Soils*, 44(5), pp. 745–54.

Gruber, N. and Galloway, J. N. (2008), 'An Earth-system perspective of the global nitrogen cycle', *Nature*, 451(7176), pp. 293–6.

Grudens-Schuck, N., Helmers, M. J., Youngquist, T. D. and Johnson, M. S. (2017), 'Prairie strips for sediment and nutrient control and biodiversity', *Journal of Extension* 55 (1), pp. 1–6.

Guimaraes, B. C. M., Arends, J. B. A., van der Ha, D., Van de Wiele, T., Boon, N. and Verstraete, W. (2010), 'Microbial services and their management: Recent progresses in soil bioremediation technology', *Applied Soil Ecology*, 46(2), pp. 157–67.

Harden, J. W., Fries, T. L. and Pavich, M. J. (2002) Cycling of beryllium and carbon through hillslope soils in Iowa', *Biogeochemistry*, 60(3), pp. 317–35.

Harden, J. W., Sharpe, J. M., Parton, W. J., Ojima, D. S., Fries, T. L., Huntington, T. G. and Dabney, S. M. (1999), 'Dynamic replacement and loss of soil carbon on eroding cropland', *Global Biogeochemical Cycles*, 13(4), pp. 885–901.

Hart, B. T., Bailey, P., Edwards, R., Hortle, K., James, K., McMahon, A., Meredith, C. and Swadling, K. (1991), 'A review of the salt sensitivity of the Australian fresh-water biota', *Hydrobiologia*, 210(1–2), pp. 105–44.

Hayat, R., Ali, S., Amara, U., Khalid, R. and Ahmed, I. (2010), 'Soil beneficial bacteria and their role in plant growth promotion: A review', *Annals of Microbiology*, 60(4), pp. 579–98.

Heneghan, L., Coleman, D. C., Zou, X., Crossley, D. A. and Haines, B. L. (1999), 'Soil microarthropod contributions to decomposition dynamics: Tropical-temperate comparisons of a single substrate', *Ecology*, 80(6), pp. 1873–82.

Hey, D. L., Kostel, J. A., Crumpton, W. G., Mitsch, W. J. and Scott, B. (2012), 'The roles and benefits of wetlands in managing reactive nitrogen', *Journal of Soil and Water Conservation*, 67(2), pp. 47A–53A.

Hoffmann, T., Thorndycraft, V. R., Brown, A. G., Coulthard, T. J., Damnati, B., Kale, V. S., Middelkoop, H., Notebaert, B. and Walling, D. E. (2010), 'Human impact on fluvial regimes and sediment flux during the Holocene: Review and future research agenda', *Global and Planetary Change*, 72(3), pp. 87–98.

Holling, C. S. and Meffe, G. K. (1996), 'Command and control and the pathology of natural resource management', *Conservation Biology*, 10(2), pp. 328–37.

Hudson, B. D. (1994), 'Soil organic-matter and available water capacity', *Journal of Soil and Water Conservation*, 49(2), pp. 189–94.

Hurt, R. D. (1981), '*The Dust Bowl: An Agricultural and Social History*', Nelson-Hall.

Intergovernmental Panel on Climate Change, IPCC (2014), *Climate Change 2014: Mitigation of Climate Change: Working Group III Contribution to the IPCC Fifth Assessment Report*, Cambridge University Press, New York, NY.

Janzen, H. H. (2015), 'Beyond carbon sequestration: Soil as conduit of solar energy', *European Journal of Soil Science*, 66(1), pp. 19–32.

Johnson-Maynard, J. L. and Strawn, D. G. (2016), 'Linking physical and biogeochemical properties and processes in the drilosphere', *Soil Science*, 181(3–4), pp. 126–32.

Kaiser, D., Lepage, M., Konate, S. and Linsenmair, K. E. (2017), 'Ecosystem services of termites (Blattoidea: Termitoidae) in the traditional soil restoration and cropping system Zai in northern Burkina Faso (West Africa)', *Agriculture Ecosystems & Environment*, 236, pp. 198–211.

Kemper, W. D., Schneider, N. N. and Sinclair, T. R. (2011), 'No-till can increase earthworm populations and rooting depths', *Journal of Soil and Water Conservation*, 66(1), pp. 13A–17A.

Kibblewhite, M. G., Ritz, K. and Swift, M. J. (2008), 'Soil health in agricultural systems', *Philosophical Transactions of the Royal Society B-Biological Sciences*, 363(1492), pp. 685–701.

Knorr, W., Prentice, I. C., House, J. I. and Holland, E. A. (2005), 'Long-term sensitivity of soil carbon turnover to warming', *Nature*, 433(7023), pp. 298–301.

Knowles, R. (1982), 'Denitrification', *Microbiological Reviews*, 46(1), pp. 43–70.

Koiter, A. J., Owens, P. N., Petticrew, E. L. and Lobb, D. A. (2017), 'The role of soil surface properties on the particle size and carbon selectivity of interrill erosion in agricultural landscapes', *Catena*, 153, pp. 194–206.

Lal, R. (2003), 'Offsetting global CO_2 emissions by restoration of degraded soils and intensification of world agriculture and forestry', *Land Degradation & Development*, 14(3), pp. 309–22.

Lal, R. (2004), 'Soil carbon sequestration to mitigate climate change', *Geoderma*, 123(1–2).

Lal, R. (2009), 'Soils and food sufficiency. A review', *Agronomy for Sustainable Development*, 29(1), pp. 113–33.

Lal, R., Delgado, J. A., Groffman, P. M., Millar, N., Dell, C. and Rotz, A. (2011), 'Management to mitigate and adapt to climate change', *Journal of Soil and Water Conservation*, 66(4), pp. 276–85.

Lavelle, P., Bignell, D., Lepage, M., Wolters, V., Roger, P., Ineson, P., Heal, O. W. and Dhillion, S. (1997), 'Soil function in a changing world: The role of invertebrate ecosystem engineers', European *Journal of Soil Biology*, 33(4), pp. 159–93.

Lavelle, P., Decaens, T., Aubert, M., Barot, S., Blouin, M., Bureau, F., Margerie, P., Mora, P. and Rossi, J. P. (2006), 'Soil invertebrates and ecosystem services', *European Journal of Soil Biology*, 42, pp. S3–S15.

Leake, J. R., Johnson, D., Donnelly, D. P., Muckle, G. E., Boddy, L. and Read, D. J. (2004), 'Networks of power and influence: The role of mycorrhizal mycelium in controlling plant communities and agroecosystem functioning', *Canadian Journal of Botany-Revue Canadienne De Botanique*, 82(8), pp. 1016–45.

Likens, G. E., Bormann, F. H., Johnson, N. M., Fisher, D. W. and Pierce, R. S. (1970), 'Effects of forest cutting and herbicide treatment on nutrient budgets in the Hubbard Brook watershed–ecosystem', *Ecological Monographs* 40(1), 23–47.

Lobb, D. A., Burton, D. L., Lindstrom, M. J., Reicosky, D. C. (2002), 'Impacts of soil erosion on the production and emission of greenhouse gases and carbon sequestration in the Canadian prairies', In: Smith, C. A. S. (Ed.), *Soil Organic Carbon and Agriculture: Developing Indicators for Policy Analyses: Proceedings of OECD expert meeting on Soil Organic Indicators for Agricultural Lands.* Ottawa, Canada, October 2002, Ottawa: Organization for Economic Co-operation and Development, pp. 235–44.

Maestre, F. T., Castillo-Monroy, A. P., Bowker, M. A. and Ochoa-Hueso, R. (2012), 'Species richness effects on ecosystem multifunctionality depend on evenness, composition and spatial pattern', *Journal of Ecology*, 100(2), pp. 317–30.

Mann, L. K. (1986), 'Changes in soil carbon storage after cultivation', Soil Science, 142(5), pp. 279–88.

Marasco, R., Rolli, E., Ettoumi, B., Vigani, G., Mapelli, F., Borin, S., Abou-Hadid, A. F., El-Behairy, U. A., Sorlini, C., Cherif, A., Zocchi, G. and Daffonchio, D. (2012), 'A drought resistance-promoting microbiome is selected by root system under desert farming', *Plos One*, 7(10), p. e48479.

Marichal, R., Grimaldi, M., Feijoo, A. M., Oszwaldd, J., Praxedes, C., Cobo, D. H. R., Hurtado, M. D., Desjardins, T., da Silva, M. L., Costag, L. G. D., Miranda, I. S., Oliveira, M. N. D., Brown, G. G., Tselouiko, S., Martins, M. B., Decaens, T., Velasquez, E. and Lavelle, P. (2014), 'Soil macroinvertebrate communities and ecosystem services in deforested landscapes of Amazonia', *Applied Soil Ecology*, 83, pp. 177–85.

McFarlane, D. J., Loughran, R. J. and Campbell, B. L. (1992), 'Soil-erosion of agricultural land in western Australia estimated by cesium-137', *Australian Journal of Soil Research*, 30(4), pp. 533–46.

McLeman, R. A., Dupre, J., Ford, L. B., Ford, J., Gajewski, K. and Marchildon, G. (2014), 'What we learned from the Dust Bowl: Lessons in science, policy, and adaptation', *Population and Environment*, 35(4), pp. 417–40.

Meyling, N. V. and Eilenberg, J. (2007), 'Ecology of the entomopathogenic fungi *Beauveria bassiana* and *Metarhizium anisopliae* in temperate agroecosystems: Potential for conservation biological control', *Biological Control*, 43(2), pp. 145–55.

Millennium Ecosystem Assessment, MEA. (2005), '*Ecosystems and Human Well-being: Biodiversity Synthesis*', Washington, DC, USA: World Resources Institute.

Milliman, J. D. and Farnsworth, K. L. (2011), *River Discharge to the Coastal Ocean: A Global Synthesis*, Cambridge University Press, New York, NY.

Mitsch, W. J., Day, J. W., Gilliam, J. W., Groffman, P. M., Hey, D. L., Randall, G. W. and Wang, N. M. (2001), 'Reducing nitrogen loading to the Gulf of Mexico from the Mississippi River Basin: Strategies to counter a persistent ecological problem', *Bioscience*, 51(5), pp. 373–88.

Montgomery, D. R. (2007), 'Soil erosion and agricultural sustainability', *Proceedings of the National Academy of Sciences of the United States of America*, 104(33), pp. 13268–72.

Nichols, E., Spector, S., Louzada, J., Larsen, T., Amequita, S., Favila, M. E. and Scarabaeinae Res, N. (2008), 'Ecological functions and ecosystem services provided by Scarabaeinae dung beetles', *Biological Conservation*, 141(6), pp. 1461–74.

Oades, J. M. and Waters, A. G. (1991), 'Aggregate hierarchy in soils', *Australian Journal of Soil Research*, 29(6), pp. 815–28.

Orgiazzi, A., Bardgett, R. D., Barrios, E., Behan-Pelletier, V., Briones, M. J. I., Chotte, J., De Deyn, G. B., Eggleton, P., Fierer, N., Fraser, T., Hedlund, K., Jeffery, S., Johnson, N. C., Jones, A., Kandeler, E., Kaneko, N., Lavelle, P., Lemanceau, P., Miko, L., Montanarella, L., Moreira, F. M. S., Ramirez, K. S., Scheu, S., Singh, B. K., Six, J., van der Putten, W. H., Wall, D. H. (2016), *Global Soil Biodiversity Atlas*, 1st edition, Luxembourg: European Union, Office des publications.

Orr, C. H., Stanley, E. H., Wilson, K. A. and Finlay, J. C. (2007), 'Effects of restoration and reflooding on soil denitrification in a leveed midwestern floodplain', *Ecological Applications*, 17(8), pp. 2365–76.

Osler, G. H. R. and Sommerkorn, M. (2007), 'Toward a complete soil C and N cycle: Incorporating the soil fauna', *Ecology*, 88(7), pp. 1611–21.

Palm, C., Blanco-Canqui, H., DeClerck, F., Gatere, L. and Grace, P. (2014), 'Conservation agriculture and ecosystem services: An overview', *Agriculture Ecosystems & Environment*, 187, pp. 87–105.

Pankhurst, C. E., Yu, S., Hawke, B. G. and Harch, B. D. (2001), 'Capacity of fatty acid profiles and substrate utilization patterns to describe differences in soil microbial communities associated with increased salinity or alkalinity at three locations in South Australia', *Biology and Fertility of Soils*, 33(3), pp. 204–17.

Pappu, H. R., Jones, R. A. C. and Jain, R. K. (2009), 'Global status of tospovirus epidemics in diverse cropping systems: Successes achieved and challenges ahead', Virus Research, 141(2), pp. 219–36.

Paustian, K., Andren, O., Janzen, H. H., Lal, R., Smith, P., Tian, G., Tiessen, H., van Noordwijk, M. and Woomer, P. L. (1997), 'Agricultural soils as a sink to mitigate CO_2 emissions', *Soil Use and Management*, 13(4), pp. 230–44.

Paustian, K., Cole, C. V., Sauerbeck, D. and Sampson, N. (1998), 'CO_2 mitigation by agriculture: An overview', *Climatic Change*, 40(1), pp. 135–62.

Peiffer, J. A., Spor, A., Koren, O., Jin, Z., Tringe, S. G., Dangl, J. L., Buckler, E. S. and Ley, R. E. (2013), 'Diversity and heritability of the maize rhizosphere microbiome under field conditions', *Proceedings of the National Academy of Sciences of the United States of America*, 110(16), pp. 6548–53.

Peters, D. P. C., Havstad, K. M., Archer, S. R. and Sala, O. E. (2015), 'Beyond desertification: New paradigms for dryland landscapes', *Frontiers in Ecology and the Environment*, 13(1), pp. 4–12.

Philippot, L., Raaijmakers, J. M., Lemanceau, P. and van der Putten, W. H. (2013) 'Going back to the roots: The microbial ecology of the rhizosphere', *Nature Reviews Microbiology*, 11(11), pp. 789–99.

Pietron, J., Chalov, S. R., Chalova, A. S., Alekseenko, A. V. and Jarsjo, J. (2017), 'Extreme spatial variability in riverine sediment load inputs due to soil loss in surface mining areas of the Lake Baikal basin', *Catena*, 152, pp. 82–93.

Pitman, M. G. and Läuchli, A. (2002), ''Global impact of salinity and agricultural ecosystems', In: Läuchli, A. and Luttge, U. (Eds), *Salinity: Environment – Plants – Molecules*. Springer, Dordrecht.

Power, A. G. (1982), 'Cropping systems, insect movement, and the spread of insect-transmitted diseases in crops', In: Gliessman, S. (Ed.), *Agroecology: Researching the Ecological Basis for Sustainable Agriculture*. New York: Springer-Verlag, pp. 47–69.

Power, A. G. (2010), 'Ecosystem services and agriculture: tradeoffs and synergies', *Philosophical Transactions of the Royal Society B-Biological Sciences*, 365(1554), pp. 2959–71.

Pretty, J. and Bharucha, Z. P. (2014), 'Sustainable intensification in agricultural systems', *Annals of Botany*, 114(8), pp. 1571–96.

Quine, T. A. and van Oost, K. (2007), 'Quantifying carbon sequestration as a result of soil erosion and deposition: retrospective assessment using caesium-137 and carbon inventories', *Global Change Biology*, 13(12), pp. 2610–25.

Quinton, J. N., Govers, G., Van Oost, K. and Bardgett, R. D. (2010), 'The impact of agricultural soil erosion on biogeochemical cycling', *Nature Geoscience*, 3(5), pp. 311–14.

Rabalais, N. N., Wiseman, W. J., Turner, R. E., SenGupta, B. K. and Dortch, Q. (1996), 'Nutrient changes in the Mississippi River and system responses on the adjacent continental shelf', *Estuaries*, 19(2B), pp. 386–407.

Rawls, W. J., Pachepsky, Y. A., Ritchie, J. C., Sobecki, T. M. and Bloodworth, H. (2003), 'Effect of soil organic carbon on soil water retention', *Geoderma*, 116(1–2), pp. 61–76.

Rietz, D. N. and Haynes, R. J. (2003), 'Effects of irrigation-induced salinity and sodicity on soil microbial activity', *Soil Biology & Biochemistry*, 35(6), pp. 845–54.

Robertson, G. P., Paul, E. A. and Harwood, R. R. (2000), 'Greenhouse gases in intensive agriculture: Contributions of individual gases to the radiative forcing of the atmosphere', *Science*, 289(5486), pp. 1922–5.

Robinson, D. A., Hockley, N., Dominati, E., Lebron, I., Scow, K. M., Reynolds, B., Emmett, B. A., Keith, A. M., de Jonge, L. W., Schjonning, P., Moldrup, P., Jones, S. B. and Tuller, M. (2012), 'Natural capital, ecosystem services, and soil change: Why soil science must embrace an ecosystems approach', *Vadose Zone Journal*, 11(1), p. 6.

Rosenzweig, S. T., Carson, M. A., Baer, S. G. and Blair, J. M. (2016), 'Changes in soil properties, microbial biomass, and fluxes of C and N in soil following post-agricultural grassland restoration', *Applied Soil Ecology*, 100, pp. 186–94.

Sargody, J. A. (1993), 'Pollution by sediments', In: Prevention of Water Pollution by Agriculture and Related Activities: Proceedings of the FAO Expert Consultation. Santiago, Chile, 1992, Rome: FAO, pp. 19–26.

Scavia, D., Kalcic, M., Muenich, R. L., Read, J., Aloysius, N., Bertani, I., Boles, C., Confesor, R., DePinto, J., Gildow, M., Martin, J., Redder, T., Robertson, D., Sowa, S., Wang, Y. C. and Yen, H. (2017), 'Multiple models guide strategies for agricultural nutrient reductions', *Frontiers in Ecology and the Environment*, 15(3), pp. 126–32.

Scheffer, M., Carpenter, S., Foley, J. A., Folke, C. and Walker, B. (2001), 'Catastrophic shifts in ecosystems', *Nature*, 413(6856), pp. 591–6.

Scheu, S. (2003), 'Effects of earthworms on plant growth: patterns and perspectives', *Pedobiologia*, 47(5–6), pp. 846–56.

Schindler, D. W. (1978), 'Factors regulating phytoplankton production and standing crop in the world's freshwaters', *Limnology and Oceanography* 23(3), 478–86.

Schindler, D. W., Hecky, R. E., Findlay, D. L., Stainton, M. P., Parker, B. R., Paterson, M. J., Beaty, K. G., Lyng, M. and Kasian, S. E. M. (2008), 'Eutrophication of lakes cannot be controlled by reducing nitrogen input: Results of a 37-year whole-ecosystem experiment', *Proceedings of the National Academy of Sciences of the United States of America*, 105(32), pp. 11254–8.

Schlesinger, W. H. (1995), 'Soil respiration and changes in soil carbon stocks', In: Woodwell, G. M., Mackenzie, G. M. (Eds), *Biotic Feedbacks in the Global Climatic System: Will the Warming Feed the Warming?* Oxford University Press, Oxford.

Scott, D. A., Baer, S. G. and Blair, J. M. (2017), 'Recovery and relative influence of root, microbial, and structural properties of soil on physically sequestered carbon stocks in restored grassland', *Soil Science Society of America Journal*, 81(1), pp. 50–60.

Sharpley, A. N., Kleinman, P. J. A., McDowell, R. W., Gitau, M. and Bryant, R. B. (2002), 'Modeling phosphorus transport in agricultural watersheds: Processes and possibilities', *Journal of Soil and Water Conservation*, 57(6), pp. 425–39.

Six, J., Conant, R. T., Paul, E. A. and Paustian, K. (2002), 'Stabilization mechanisms of soil organic matter: Implications for C-saturation of soils', *Plant and Soil* 241(2), pp. 155–76.

Six, J., Elliott, E. T. and Paustian, K. (2000), 'Soil macroaggregate turnover and microaggregate formation: A mechanism for C sequestration under no-tillage agriculture', *Soil Biology & Biochemistry* 32(14), pp. 2099–103.

Six, J., Frey, S. D., Thiet, R. K. and Batten, K. M. (2006), 'Bacterial and fungal contributions to carbon sequestration in agroecosystems', *Soil Science Society of America Journal*, 70(2), pp. 555–69.

Smedema, L. M. and Shiati, K. (2002), 'Irrigation and salinity: A perspective review of the salinity hazards of irrigation development in the arid zone', *Irrigation and Drainage Systems* 16, pp. 161–74.

Smith, P., Martino, D., Cai, Z., Gwary, D., Janzen, H., Kumar, P., McCarl, B., Ogle, S., O'Mara, F., Rice, C., Scholes, B., Sirotenko, O., Howden, M., McAllister, T., Pan, G., Romanenkov, V., Schneider, U., Towprayoon, S., Wattenbach, M. and Smith, J. (2008), 'Greenhouse gas mitigation in agriculture', *Philosophical Transactions of the Royal Society B-Biological Sciences*, 363(1492), pp. 789–813.

Smith, S. V., Renwick, W. H., Buddemeier, R. W. and Crossland, C. J. (2001), 'Budgets of soil erosion and deposition for sediments and sedimentary organic carbon across the conterminous United States', *Global Biogeochemical Cycles*, 15(3), pp. 697–707.

Smukler, S. M., Sanchez-Moreno, S., Fonte, S. J., Ferris, H., Klonsky, K., O'Geen, A. T., Scow, K. M., Steenwerth, K. L. and Jackson, L. E. (2010), 'Biodiversity and multiple ecosystem functions in an organic farmscape', *Agriculture Ecosystems & Environment*, 139(1–2), pp. 80–97.

Spokas, K. A., Cantrell, K. B., Novak, J. M., Archer, D. W., Ippolito, J. A., Collins, H. P., Boateng, A. A., Lima, I. M., Lamb, M. C., McAloon, A. J., Lentz, R. D. and Nichols, K. A. (2012), 'Biochar: A synthesis of its Agronomic impact beyond carbon sequestration', *Journal of Environmental Quality*, 41(4), pp. 973–89.

Stall, J. B. (1972), 'Effects of sediment on water quality', *Journal of Environmental Quality*, 1, pp. 353–60.

Stallard, R. F. (1998), 'Terrestrial sedimentation and the carbon cycle: Coupling weathering and erosion to carbon burial', *Global Biogeochemical Cycles*, 12(2), pp. 231–57.

Stavi, I. and Lal, R. (2013), 'Agroforestry and biochar to offset climate change: A review', *Agronomy for Sustainable Development*, 33(1), pp. 81–96.

Stehfest, E. and Bouwman, L. (2006), 'N_2O and NO emission from agricultural fields and soils under natural vegetation: Summarizing available measurement data and modeling of global annual emissions', *Nutrient Cycling in Agroecosystems*, 74(3), pp. 207–28.

Stewart, C. E., Paustian, K., Conant, R. T., Plante, A. F. and Six, J. (2007), 'Soil carbon saturation: Concept, evidence and evaluation', *Biogeochemistry*, 86(1), pp. 19–31.

Stockmann, U., Adams, M. A., Crawford, J. W., Field, D. J., Henakaarchchi, N., Jenkins, M., Minasny, B., McBratney, A. B., de Courcelles, V. D., Singh, K., Wheeler, I., Abbott, L., Angers, D. A., Baldock, J., Bird, M., Brookes, P. C., Chenu, C., Jastrow, J. D., Lal, R., Lehmann, J., O'Donnell, A. G., Parton, W. J., Whitehead, D. and Zimmermann, M. (2013), 'The knowns, known unknowns and unknowns of sequestration of soil organic carbon', *Agriculture Ecosystems & Environment*, 164, pp. 80–99.

Sumner, M. E. (1995), 'Sodic soils: New perspectives', In: Naidu, R., Sumner, M. E. and Rengasamy, P. (Eds), *Australian Sodic Soils: Distribution, Properties and Management*. Melbourne: CSIRO, pp. 1–34.

Syswerda, S. P. and Robertson, G. P. (2014), 'Ecosystem services along a management gradient in Michigan (USA) cropping systems', *Agriculture Ecosystems & Environment*, 189, pp. 28–35.

Thiele-Bruhn, S., Bloem, J., de Vries, F. T., Kalbitz, K. and Wagg, C. (2012), 'Linking soil biodiversity and agricultural soil management', *Current Opinion in Environmental Sustainability*, 4(5), pp. 523–8.

Tiedje, J. M., Sexstone, A. J., Parkin, T. B., Revsbech, N. P. and Shelton, D. R. (1984), 'Anaerobic processes in soil', *Plant and Soil*, 76(1–3), pp. 197–212.

Turner, R. E. and Rabalais, N. N. (2003), 'Linking landscape and water quality in the Mississippi river basin for 200 years', *Bioscience*, 53(6), pp. 563–72.

United Nations Convention to Combat Desertification (UNCCD) (2013), Economic assessment of desertification, sustainable land management and resilience of arid, semi-arid and dry sub-humid areas, UNCCD 2nd Scientific Conference, Conference Outcomes, available: http://2sc.unccd.int/conference-documents/.

Vacheron, J., Desbrosses, G., Bouffaud, M. L., Touraine, B., Moenne-Loccoz, Y., Muller, D., Legendre, L., Wisniewski-Dye, F. and Prigent-Combaret, C. (2013), 'Plant growth-promoting rhizobacteria and root system functioning', *Frontiers in Plant Science*, 4, p. 19.

van Groenigen, J. W., Lubbers, I. M., Vos, H. M. J., Brown, G. G., De Deyn, G. B. and van Groenigen, K. J. (2014), 'Earthworms increase plant production: A meta-analysis', *Scientific Reports*, 4, Article number 6365.

van Vliet, P. C. J. (2000), 'Enchytraieds', In: Sumner, M. E. (Ed.), '*Handbook of Soil Science*', Boca Raton: CRC Press, pp. 70–7.

van Vliet, P. C. J., Beare, M. H., Coleman, D. C. and Hendrix, P. F. (2004), 'Effects of enchytraeids (Annelida: Oligochaeta) on soil carbon and nitrogen dynamics in laboratory incubations', *Applied Soil Ecology*, 25(2), pp. 147–60.

Veraart, A. J., Steenbergh, A. K., Ho, A., Kim, S. Y. and Bodelier, P. L. E. (2015), 'Beyond nitrogen: The importance of phosphorus for CH_4 oxidation in soils and sediments', *Geoderma*, 259, pp. 337–46.

Verstraete, W. and Mertens, B. (2004), 'The key role of soil microbes', In: Doelman, P. and Eisjasckers, H. J. (Eds), *Developments in Soil Science: Vital Soil Functions, Values and Properties*. Amsterdam: Elsevier, pp. 127–58

Wagg, C., Bender, S. F., Widmer, F. and van der Heijden, M. G. A. (2014), 'Soil biodiversity and soil community composition determine ecosystem multifunctionality', *Proceedings of the National Academy of Sciences of the United States of America*, 111(14), pp. 5266–70.

Walker, B. H. and Salt, D. (2006), *Resilience Thinking: Sustaining Ecosystems and People in a Changing World*, Island Press.

Wall, D. H., Bardgett, R. D., Covich, A. R. and Snelgrove, P. V. R. (2004), 'The need for understanding how biodiversity and ecosystem functioning affect ecosystem services in soils and sediments', *Sustaining Biodiversity and Ecosystem Services in Soils and Sediments*, 64, pp. 1–12.

Wall, D. H., Bargett, R. D., Behan-Pelletier, V., Herrick, J. E., Jones, H., Ritz, K., Six, J., Strong, D. R., van der Putten, W. H. (2012), '*Soil Ecology and Ecosystem Services*', Oxford University Press.
Walling, D. E. (1999), 'Linking land use, erosion and sediment yields in river basins', *Hydrobiologia*, 410, pp. 223–40.
Walling, D. E. and Fang, D. (2003), 'Recent trends in the suspended sediment loads of the world's rivers', *Global and Planetary Change*, 39(1–2), pp. 111–26.
Warneck, P. (2000), '*Chemistry of the Natural Atmosphere*', Elsevier.
Warrick, R. A. (1980), 'Drought in the Great Plains: A case study of research on climate and society in the USA', In: Ausubel, J. H. and Biswas, A. K. (Eds), *Climate Constraints and Human Activities*. Laxenburg: Pergamon Press Ltd, pp. 93–125.
Weinbauer, M. G. (2004), 'Ecology of prokaryotic viruses', *Fems Microbiology Reviews*, 28(2), pp. 127–81.
Weller, D. M. (1988), 'Biological-control of soilborne plant-pathogens in the rhizosphere with bacteria', *Annual Review of Phytopathology*, 26, pp. 379–407.
Wilson, G. W. T., Rice, C. W., Rillig, M. C., Springer, A. and Hartnett, D. C. (2009), 'Soil aggregation and carbon sequestration are tightly correlated with the abundance of arbuscular mycorrhizal fungi: Results from long-term field experiments', *Ecology Letters*, 12(5), pp. 452–61.
Wubs, E. R. J., van der Putten, W. H., Bosch, M. and Bezemer, T. M. (2016), 'Soil inoculation steers restoration of terrestrial ecosystems', *Nature Plants*, 2(8).
Zhang, X. L., Zhao, L. J., Tong, D. Q., Wu, G. J., Dan, M. and Teng, B. (2016), 'A systematic review of global desert dust and associated human health effects', *Atmosphere*, 7(12), 158.
Zhao, Z. H., Hui, C., He, D. H. and Li, B. L. (2015), 'Effects of agricultural intensification on ability of natural enemies to control aphids', *Scientific Reports*, 5.

Chapter 3

Soil health and climate change: a critical nexus

Promil Mehra, New South Wales Department of Primary Industries, Australia; Bhupinder Pal Singh[1], New South Wales Department of Primary Industries, University of Newcastle and University of New England, Australia; Anitha Kunhikrishnan, New South Wales Department of Primary Industries and University of Newcastle, Australia; Annette L. Cowie, New South Wales Department of Primary Industries and University of New England, Australia; and Nanthi Bolan, University of Newcastle, Australia

1 Introduction

Soil health refers to the capacity of the soil to perform a range of agronomic and ecosystem functions, thus to sustain biological productivity, maintain environmental quality and promote plant and animal health (Kibblewhite et al., 2008; Lal, 2011). Healthy soils are thus the foundation of sustainable and productive agroecosystems. The maintenance of healthy soils under sustainable agricultural systems can be accomplished by employing the principles of soil health: protecting the soil surface with growing plants or mulch; disturbing the soil profile as little as possible; increasing diversity of plant species through rotations, cover crops and polycultures; enhancing soil biota; building soil organic matter (SOM); and integrating livestock in farming systems, where practically possible (Brussaard et al., 2007; Al-Kaisi and Lal, 2017). Applying these soil health principles improves the key functions of soils in agroecosystems, namely, nutrient cycling and availability, water availability and storage, soil structural stability and root growth, carbon (C) storage, biodiversity conservation, and filtering/buffering of potential water pollutants (Keesstra et al., 2012; Kibblewhite et al., 2008; Nannipieri et al., 2003).

1. Corresponding Author (bp.singh@dpi.nsw.gov.au).

http://dx.doi.org/10.19103/AS.2017.0033.04

Under current pressures from land degradation and climate change, soils have become one of the most endangered natural resources in the world (Titeux et al., 2016). Each year, an estimated 25–40 billion tons of fertile soil are lost globally (FAO and ITPS, 2015). Hence, improving soil health through sustainable land management should be a common goal for farmers and land managers, to protect, maintain and build their most vital resource – soils. Soils are the major reservoir of C in terrestrial ecosystems, and soil C plays a dynamic role in influencing the global C cycle and climate change (Fig. 1) while regulating soil health and productivity (Mehra et al., 2018; Singh et al., 2011, 2018). Soil contains C in two forms: soil organic C (SOC) and soil inorganic C (SIC), with most soils (except calcareous soils) having more SOC than SIC (Fig. 1). Thus far, enormous scientific progress has been attained in understanding soil functional characteristics relating to SOC stocks and C dynamics in agroecosystems (Stockmann et al., 2013). The stock of SOC is determined by the net balance between C inputs (including plant and animal residues, decaying roots) and C outputs (including C emissions as a result of soil respiration, leaching of dissolved organic C and loss of C through soil erosion) (Mu et al., 2014; Izaurralde et al., 2007). These C transfer processes occur vertically (through leaching or emissions) and laterally (through erosion and deposition) as illustrated in Fig. 2. Lack of adequate resources for monitoring SOC stocks at national and global levels is hindering the design and the implementation of policies to protect and enhance soil health.

According to the Intergovernmental Panel on Climate Change (IPCC), globally farmers are facing two major risks, that is, climate change and climate variability. Climate change refers to long-term (decadal or longer) change in the mean and/or variability of climate properties (IPCC, 2012). In contrast, climate variability refers to short-term fluctuations in temperature, rainfall distribution and other climatic variables over a day, month, season or

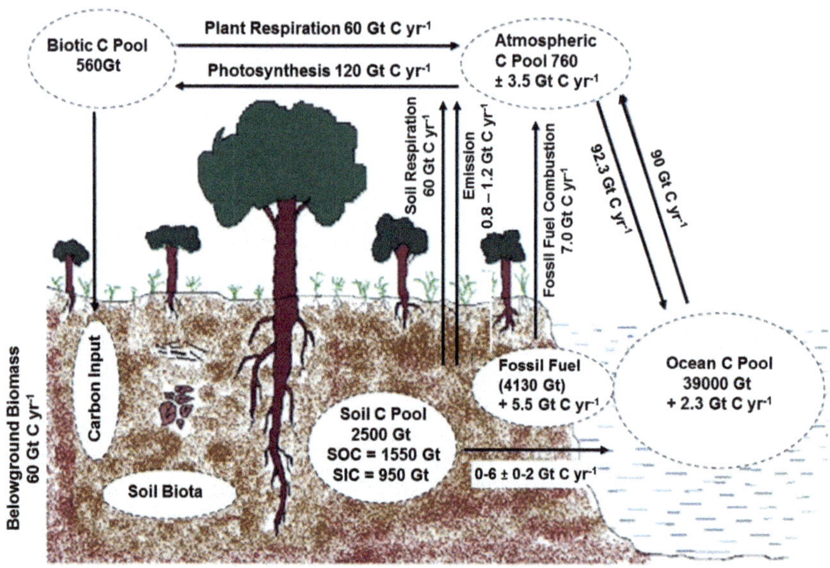

Figure 1 Sources and sinks of carbon from different pools under terrestrial and aquatic ecosystems. Adapted from Mehra et al. (2018).

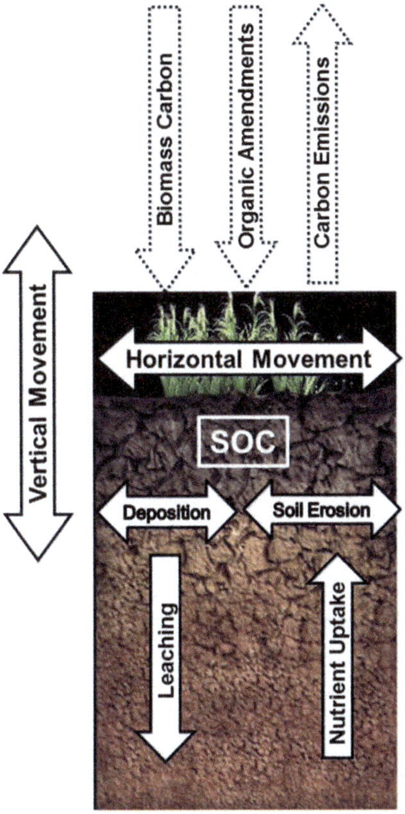

Figure 2 Pictorial representation of carbon transfer processes influencing the soil organic carbon (SOC) pool. Adapted from Mehra et al. (2018).

year(s) (Thornton et al., 2014; IPCC, 2012). Moreover, there is unanimity among the scientific community that the global environmental changes (i.e. climate change, land degradation and biodiversity loss) present serious threats to soil health and agricultural productivity. Considering the rising population trends, it is estimated that by 2050, world population will reach to 9.8 billion (FAO, 2016) and will impede the food self-sufficiency and security of many nations. According to Food and Agriculture Organization Statistics (FAOSTAT) 2014, of the total ~13.4 billion ha of world land area, around 10.9% (1.4 billion ha) is used for crop production (Table 1) (FAO, 2016). Climate records show increasing evidence of global warming, with the year 2016 ranked as the warmest year on record (Potter et al., 2017), indicating a 125% increase in the global temperature anomaly since 2000 (from 0.44°C to 0.99°C in 2016) (GISTEMP Team, 2018).

Soil can act both as a source and a sink for greenhouse gas (GHG) emissions. As a source, the microbial decomposition of SOM leads to the release of carbon dioxide (CO_2). Rising soil temperatures *via* climate change can increase the decomposition of SOM (Singh et al., 2017; Fang et al., 2017; Ghee et al., 2013), and influence the priming of SOM and its temperature sensitivity (Thiessen et al., 2013; Fang et al., 2017). Rising temperatures

Table 1 Arable land use (M ha) globally in 2017

Region	Arable land	Permanent crops	Permanent meadows and pastures
Africa	219.6	31.3	885.1
Americas	366.4	28.2	817.6
Asia	486.8	75.1	1090.5
Europe	278.5	15.8	179.0
Oceania	47.9	1.5	390.7
Total	1399.2	151.9	3362.9

Adapted from FAO (2016).

resulting from climate change could thus cause soils to release more CO_2, creating a feedback mechanism that can lead to even more global warming (Davidson and Janssens, 2006). There have been widespread recent climate extreme events across the globe, such as the European heat wave in 2003 and the forest fires in Russia in 2010 (Bastos et al., 2013), the Pakistan floods in 2010 (Trenberth and Fasullo, 2012) and the 2010–11 La Niña rains over Australia (Poulter et al., 2014). Such climate extreme events impede soil health not only through immediate direct impacts but possibly also *via* impacting plant growth and soil biota and creating biophysical stresses in soil. This chapter outlines the knowns, known-unknowns and unknowns of soil health–climate interactions. The conceptual flowchart (Fig. 3) highlights the influence of climate change on plant and soil properties/processes that are aligned with soil health.

2 Impacts of climate change on the key components of soil health

Climate change impacts our daily lives across the world in many ways, some of which are subtle, and some severe. One of the important impacts of climate change is that it hinders various processes of natural ecosystems, leading to reduced supply and quality of fresh water, loss of species, decline in biodiversity and degradation of landscapes (Christensen et al., 1996; Rogers and McCarty, 2000). Some key soil properties are highly sensitive to climate change and climate variability (Karmakar et al., 2016; Mitchell and Soga, 2005; Thornton et al., 2014), as discussed in the following section. As climate change places soils under stress, it is important to understand the climate- and management-driven changes in soil physical, chemical and biological health and functions (Figs. 3 and 4).

2.1 Soil physical health

Soil physical health is largely determined by the impact of management and climate factors on SOM, soil structure, bulk density and air and water movement through the soil profile (Luo et al., 2017). One of the most important attributes of a healthy soil is 'soil structural stability', which affects the movement and management of water, oxygen and nutrients (Ussiri et al., 2009; Kladivko, 2001; Datiri and Lowery, 1991). Thus, changes in soil structure that result from anthropogenic activities and indirectly from climate

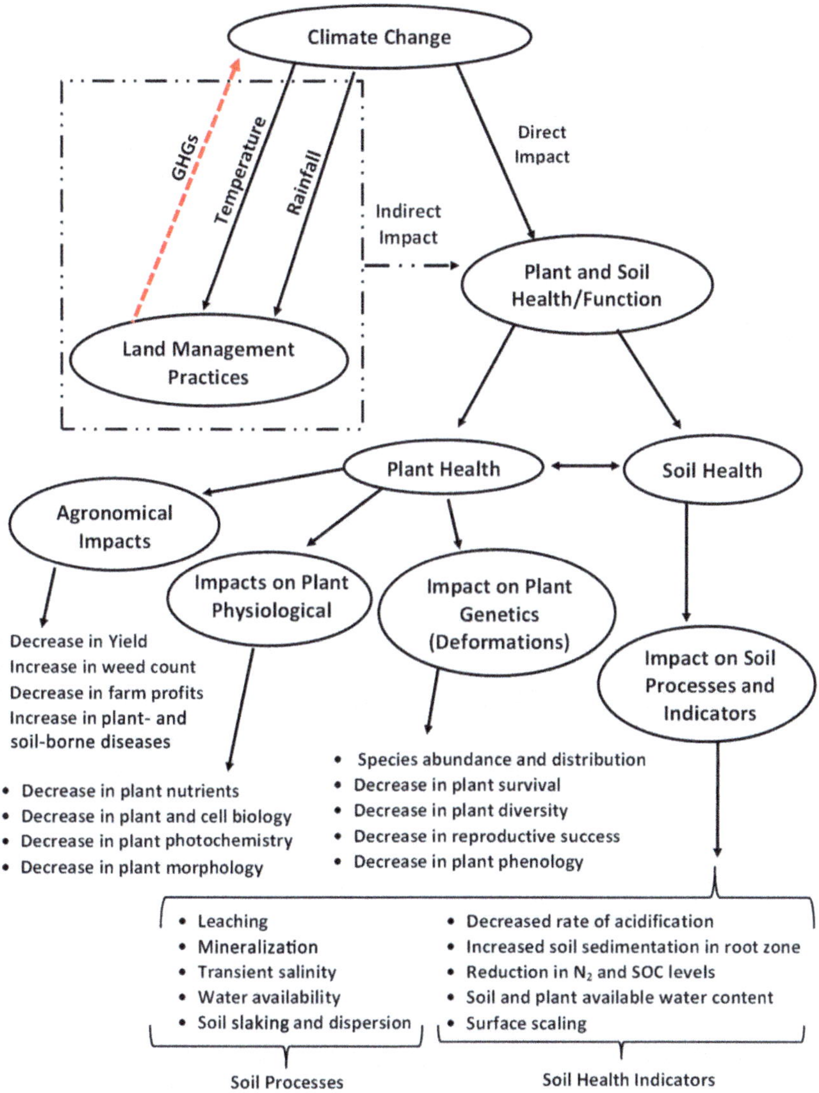

Figure 3 A conceptual figure showing soil–climate interactions and their impact (direct or indirect) on plant and soil properties/processes that are aligned with soil health.

change, for example via impacting plant C inputs and soil processes, can adversely influence soil physical health (Luo et al., 2017). Studies have demonstrated that poor soil structure (hardpans, surface sealing) due to anthropogenic activities such as continuous cultivation inhibits many plant processes (germination, root growth) (Turmel et al., 2015; Wang et al., 2002). Moreover, an increase in air and soil temperature inevitably creates an imbalance in the soil-thermal regime, which is governed by the processes

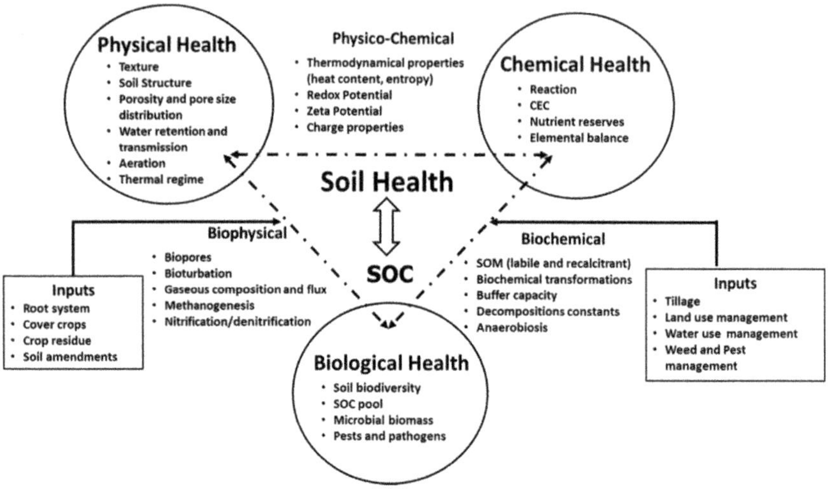

Figure 4 Components of soil health, influenced by management and organic matter inputs.

like evaporation, heat conduction and convective transfer *via* the movement of gas and water through the soil profile. This imbalance of soil-thermal regimes can impact key soil processes, such as rapid decomposition of SOM and nutrient loss *via* volatilization (DEFRA, 2005; Qian et al., 2011).

In agricultural systems, any changes in the balance of two dynamic soil properties, soil structure and SOM, can have a significant impact on soil fertility (Bauer and Black, 1994; Reeves, 1997; Sarker et al., 2018a), soil tilth (Karlen et al., 1990; Shepherd et al., 2001; Koehn et al., 2014; Hadas, 1997), water retention (Salem et al., 2015; Głab and Kulig, 2008) and susceptibility of soil to erosion (Sojka et al., 2007; Lal, 2003; Basher and Ross, 2002). It is a well-known fact that intensive tillage, especially using mouldboard or disc ploughs that invert the profile, disturbs the soil's physical stability and may impact root growth and belowground C allocation (Kay, 1990). For example, Mosaddeghi et al. (2009) demonstrated that a substantial increase in mechanical impediments by intensive tillage in the upper soil surface limits the distribution and downward progression of roots. In a 5-year study, Link et al. (2003) showed that increasing temperature by 5°C due to changing climate can rapidly decrease SOC in the top 5-cm layer by up to 32%, which would be further exacerbated if combined with intensive tillage operations (see Section 3.2). Numerous studies conducted after the Dust Bowl era of 1930s have reported that no-till improves water conservation and creates a more favourable soil-thermal regime in the top soil surface (Blevins et al., 1983; Ehlers and Claupein, 1994; Merrill et al., 1996). More recently, advanced tillage systems such as precision tillage and strategic tillage have been developed to optimize edaphic conditions (e.g. soil–water and soil–temperature regimes, soil aeration, seed–soil contact, nutrient availability and porosity) and influence pest, microbial and root activity, which depend on ecological niches (Mangalassery et al., 2015; Sarker et al., 2017, 2018a,b). For example, low-intensity tillage may create a favourable environment for root activity and belowground C allocation in some dryland cropping systems (e.g. canola *vs* wheat) with minimal or no negative impacts on soil structure (Fang et al., 2016; Sarker et al., 2017).

2.2 Soil chemical health

The most rapid degradation processes taking place in soil under current changing climatic conditions and through unsustainable management practices are the loss of SOM, salts and nutrient cations, with variation across soil types and land use regimes (Brinkman and Sombroek, 1996). These processes negatively impact chemical aspects of soil health as discussed in the following section.

2.2.1 Impact of climate change on soil acidification and soil salinity

Agricultural soils are commonly affected by acidification through a series of chemical processes that are influenced by interacting effects of management (fertilizer, biomass removal) and climate (Rengel, 2011). Bolan et al. (2003) outlined pathways of soil acidification: (1) increase in soil acidity or decrease in pH, (2) decrease in base saturation, (3) unbalanced availability of elements in the root environment or (4) decrease in the acid neutralizing capacity of soil. Changes in precipitation under climate change may influence the extent of soil acidification (Wild, 1993; Charman and Murphy, 2007). For example, although acidification is a natural process, nitrate leaching becomes the main cause of soil acidification during high rainfall conditions (Bolan et al., 1991). Moreover, rain water, which is usually slightly acidic (pH 5–5.6), may release protons (H^+) in soil from the dissolution of CO_2 and dissociation of carbonic acid (H_2CO_3), though this process is unlikely to affect soil pH on its own (because H_2CO_3 is a weak acid). Acid soils are common in the wet tropics, because enhanced leaching of basic cations to subsoil horizons facilitates the process of soil acidification (Rengel, 2011).

On the other hand, where there is soil surface sealing, infiltration of rainfall is slow, and leaching is limited, which increases the concentration of salts in soil solution (Rengasamy and Olsson, 1993). Transient salinity increases with capillary rise, moving salts into the root zone (Pang and Letey, 1998) and reducing plant growth. Choice of crop species (as the primary factor) and ameliorating strategies for soil acidity can be useful to minimize the impacts of variable and changing climate on soil health. For example, continuous legume crops or pastures can cause soil acidification where soils are unable to buffer against changes in pH due to the release of protons (H^+) during decomposition of organic matter and transformation of N and sulphur (Bolan et al., 2003; Bolan and Hedley, 2003). Notably, crop species and cultivars differ in tolerance of soil acidity.

2.2.2 Impact of climate change on nutrients and soil organic carbon

Climate change impacts on soil properties may be permanent or transient, and favourable or unfavourable, and can disrupt the equilibrium of different soil processes, such as nutrient cycling and SOC dynamics, in different cropping systems (Guenet et al., 2018; Köchy et al., 2015; Wieder et al., 2015). Out of many soil environmental factors, soil moisture plays a key role in regulating soil temperature and biological activity (Andrade et al., 2010). In particular, high soil moisture maintains an even temperature while reducing the thermal conductivity of soil and activity of different biological processes, such as soil C and nutrient mineralization (Bonito et al., 2003; Liu et al., 2010) and root growth (Boone et al., 1998). In addition to rainfall, soil moisture content can be affected by the high specific heat and low diffusivity of soil water, and also depends upon other constituents of soil such as clay and SOM (Monteith and Unsworth, 2007). Furthermore, the rate and magnitude of various soil and plant processes can be impacted by climate variables, affecting plant nutrition (Table 2). Plant roots take up nutrients from the soil through soil water. Thus, when soils are dry, this decreases the supply of nutrients at short distances (diffusion of phosphate and zinc), as well as the mass flow of water-soluble nutrients like sulphates, nitrate, calcium, magnesium and silicon at larger distances (Mackay and Barber, 1985; Barber, 1996). Lal (2000) reported that crop yields decrease exponentially with increasing aridity as the shortage of available soil moisture can also influence the availability and transportation of soil nutrients to the plant. Under optimum moisture conditions, the increase of root zone temperature from 15–18°C to 25–29°C enhanced the nutrient (K and P) uptake up to 100–300%, by increasing the root surface area and thus the rate of nutrient diffusion and water influx (Brouder and Volenec, 2008; Ching and Barber, 1979; Mackay and Barber, 1984). However, sensitivity to temperature varies between different plant growth stages (Cleland et al., 2007), so climate change has potential to affect plant phenology, nutrient uptake by plants and ultimately the soil nutrient status *via* impacting the (im)mobilization processes (Nord and Lynch, 2009).

Table 2 Potential interactions of soil and plant processes, climate variables and plant nutrients or other soil minerals

Soil/plant processes	Climate change variables	Interaction with different nutrients and minerals
Physical		
Soil erosion (wind/water)	Heavy precipitation/winds	Erosion causes losses of soil nutrients, SOC and fertilizer
Transpiration-driven mass flow	Drought, temperature, relative humidity, CO_2	NO_3, SO_4, Ca, Mg and Si
Root growth and architecture	Soil temperature, CO_2	Uptake of all nutrients, especially P and Ca
Chemical		
Soil organic carbon status	Soil moisture, soil temperature, CO_2	All nutrients
Salinization	Precipitation, temperature	Na, K, Ca and Mg
Soil redox status	Flooding	Mn, Fe, Al and B
Soil leaching	Precipitation	NO_3, SO_4, Ca and Mg
Biological		
Plant phenology	Temperature	P, N and K
Soil microbes (N cycling)	Drought, soil temperature	N
Biological N fixation	Drought, soil temperature	N
Mycorrhizas	CO_2	P, Zn [vesicular-arbuscular mycorrhiza – (VAM)], N (ectomycorrhizas)

Adapted and modified from Clair and Lynch (2010).

Evidence from different studies confirms that climate warming contributes to the degradation of soil quality under dry conditions through SOC and nutrient losses *via* higher rates of wind erosion, with implications for decreased primary productivity (Lee et al., 1996; Gupta, 1993). Experiments looking at the impact of increasing temperature on soil bio-chemical and physical-chemical processes indicated that soil warming and drying scenarios would decrease SOC *via* increased mineralization of SOM and loss of SOC as CO_2 to the atmosphere or decreased organic C input in soils, thus adversely impacting soil health (Jat et al., 2018; Smith et al., 2016). Climatic drivers, particularly mean annual temperature and precipitation, are the dominant drivers of SOC dynamics, along with soil texture, while management factors such as tillage regime or cropping intensity have lesser impact (Rabbi et al., 2015; Morrow et al., 2017). A recent investigation by Luo et al. (2017) across the diverse Australian cropping systems further demonstrated the relative importance of climate (e.g. precipitation and temperature), soil properties (e.g. cation exchange capacity, clay content, bulk density), C inputs (quantity and quality, that is pasture frequency in crop rotations) and soil C pools (size and composition) and their complex interconnections in regulating and predicting changes in SOC stocks.

2.3 Soil biological health

Climate change alters soil biological properties, which play a prominent role in managing SOM dynamics, with implication for soil health and global climate change (Lal, 2011; Singh et al., 2011). Soil biodiversity consists of soil microflora (e.g. bacteria, fungi and archea), soil microfauna (e.g. nematodes and protozoa), mesofauna (e.g. enchytraea, collembola and acarids) and macrofauna (e.g. earthworms, arthropods and termites). These soil organisms are the main drivers for C and nutrient cycling in pedosystems. Plant roots also play an important role because of their symbiotic relationships with soil rhizobia and mycorrhiza, and microbes in the rhizosphere, and these relationships can be impacted by climate change (McNear, 2013). Changes in the diversity and distribution of different biotic species in soil impact ongoing functional interactions among different soil organisms and plants, with impacts on plant nutrition and plant resistance to biotic and abiotic stresses (Gottfried et al., 2012; Langley and Hungate, 2014; Putten, 2012). These biological interactions among different soil community members and plants act cross-functionally, which can be beneficial, minimizing adverse impacts to ecological functions and services from climatic abnormalities (Vandenkoornhuyse et al., 2015).

Soil respiration is one of the representative indicators of soil biological activity (Allen et al., 2011) and is the second largest source of the annual flux of C (after photosynthesis) between the atmosphere and terrestrial systems (Houghton, 2007). It is therefore highly relevant to assess the change in soil respiration under climate change. Soil respiration represents a logistically useful measure of soil biological functions as it integrates a range of metabolic processes involved in C and nutrient cycling, as well as other aspects of soil organism activity (Mills et al., 2014). This is particularly important given that the long-term sustainability of ecosystems depends on maintaining or improving soil functions. Soil respiration is a function of soil organism's activity, which is highly dependent on soil temperature and moisture. High temperature regimes resulting from climate change can increase soil microbial activities, whereas lower moisture regimes in pedosystems can negate the increase, although the interactive effects of climate factors are not fully understood (Singh et al., 2017). Due to its importance in the global C cycle, changes in soil respiration may have significant feedback effects on climate change that in turn alter the aboveground and belowground biological diversity. However, knowledge gaps remain in our understanding of the intricate relationships between the diverse array of soil microorganisms and their responses to climate change. Nevertheless, application of high-throughput sequencing, multiple soil functional enzymes and stable C isotope probing tools has been enhancing understanding of the links between soil microbial diversity and their ecological functions, such as soil C and nutrient cycling (Bradford et al., 2008, 2014; Delgado-Baquerizo et al., 2016; Luo et al., 2018; Radajewski et al., 2000).

2.3.1 Impact of climate change on soil health and functions: soil microorganism interactions

The response and adaptation of microorganisms towards climate change can be quicker than those of plants or macroorganisms, which make them the ideal bioindicators to assess environmental stresses. The climatic variables such as atmospheric CO_2 concentration, temperature and precipitation regimes can have both direct and indirect impacts on soil microbial communities, but the direction and magnitude of these responses are uncertain. For instance, the response of soil microbial communities to changes in CO_2 concentrations was found to be positive or negative in different studies (Balser et al., 2010; Austin et al., 2009; Lesaulnier et al., 2008; Janus et al., 2005). While Austin et al. (2009) noticed no detectable effect on soil microbial communities from elevated CO_2 after 10 years in a sweet gum plantation, Lesaulnier et al. (2008) observed significant alteration of soil microbial community composition from elevated CO_2 after 4 years in a trembling aspen plantation. This contradictory response illustrates the poorly understood relationships between the diversity of soil microorganisms, their responses to climate change drivers and subsequent levels of microbially mediated CO_2 production. Further, these relationships are complicated by interactions between concurrent climate change drivers, for example temperature regime and altered precipitation (Balser et al., 2010; Castro et al., 2010). For example, Castro et al. (2010) examined the impact of multiple climate change drivers such as elevated atmospheric CO_2, increased temperature and varying precipitation, and found that the changes in precipitation had a greater effect than the others on the diversity of bacterial and fungal communities. However, the biomass and species composition of the plant community were also greatly impacted by the level of precipitation; hence, the observed changes in microbial community may have been an indirect response to changes in plant community rather than a direct effect of the changing climate.

Microbial physiological responses to climate change are the processes through which soil microorganisms regulate CO_2 emissions. Rising temperature can not only increase soil microbial activity and soil processes, but also be responsible for a shift in microbial community structure (Bradford et al., 2008; Pettersson and Bååth, 2003; Zak et al., 1999; Zogg et al., 1997). Many studies have shown that climatic change factors such as warming (Balser and Wixon, 2009; Rinnan et al., 2007, 2009; Bardgett et al., 2008) and altered precipitation (Zhang et al., 2016; Cregger et al., 2012) can significantly affect measures of microbial physiological activity in soil, such as carbon use efficiency (CUE). This would regulate heterotrophic respiration (Bhupinderpal-Singh et al., 2003) and consequently may result in either gain or loss of soil C (Zhang et al., 2013). Warming can increase the maintenance and respiratory costs, and decrease the CUE of microorganisms (Gillooly et al., 2001; Allison et al., 2010). In general, fungi and bacteria have different CUEs, with CUEs of fungi reported to be higher than those of bacteria (Keiblinger et al., 2010; Six et al., 2006). Reduction in CUE under unsustainable systems limits the biomass of microbial decomposers (e.g. fungal communities) and favours respiration. However, increased CUE promotes microbial biomass and SOC storage under sustainable management systems, possibly from a shift in microbial community structure and its adaptation to climate change (Allison et al., 2010; Fang et al., 2018; Manzoni et al., 2012).

Climate change is expected to increase variability in precipitation, which will lead to periods of soil moisture deficit and surplus, which can have a direct or indirect effect on soil microbial activities. Studies have reported that variable soil moisture regimes *via* uneven precipitation may increase or decrease the ratio of bacteria to fungi, as well as shift their community composition (Chen et al., 2007; Schimel et al., 1999; Williams, 2007). In particular, drought and elevated temperature could have significant consequences for soil C and nutrient cycling (Fang et al., 2017; Raich and Schlesinger, 1992), due to impacts on soil microbial activity caused by limited substrate availability and reduced enzyme activity (Steinweg et al., 2013; Zhang et al., 2013).

2.3.2 Soil fauna and functions: impact of climate change on soil fauna

In recent years, the importance of various soil fauna to soil health has been recognized. Soil microfauna (protozoa and nematodes) may cause shifts in microbial activity, thereby impacting the turnover of microbial biomass. Mesofauna such as mites and collembola increase the ratio of substrate surface area to mass of organic substrates by fragmentation, thus increasing their accessibility to microfauna and microorganisms (Haynes, 2014; Mulder et al., 2013). Soil macrofauna (earthworms, ants, termites, centipedes, millipedes) redistribute organic residues, vertically and horizontally, in the soil profile. This increases the substrate contact surface, loosens the soil (biotillage) and improves soil structure through the formation of macropores and aggregates *via* burrowing activities and interaction with faecal pellets (Belnap, 2003; Kostina et al., 2011). Also, the mixing of soil particles by fauna produces pores, tunnels and other biological compartments that facilitate the air and water flow, which further stimulates the soil microbial activity (Lavelle et al., 2006). On the other hand, a soil with low faunal activity shows more compaction, making it difficult for the plant roots to penetrate (Drewry et al., 2008). Clearly, soil fauna play a critical role in terrestrial ecosystems, at the interface between plants and microorganisms, in equilibrating different soil processes, in establishing relationships between different soil depths, and thus in facilitating C and nutrient cycling. Therefore, soil fauna are good indicators of changes in soil biological health as impacted by climate change and anthropogenic stresses (Cardoso et al., 2013).

Soil fauna are a poorly understood group of soil biota that may be greatly affected by climate change (Lindberg et al., 2002; McKenzie et al., 2013). There is evidence that climatic factors play an important role in determining soil fauna contributions to soil processes (García-Palacios et al., 2013; Swift et al., 1979). Soil fauna are generally well adapted to elevated CO_2 concentrations (Coûteaux and Bolger, 2000). However, there could be different responses to elevated CO_2 depending on body size of soil fauna; for example, microfauna may respond positively, while mesofauna tend to respond negatively to elevated CO_2 (Blankinship et al., 2011). Studies also reveal geographic variations among soil invertebrates upon increased CO_2 concentrations. For example, Kardol et al. (2011) noticed that mite and collembolan abundance was not affected after 5 years of elevated CO_2 treatments in Tennessee, USA; however, increased CO_2 caused >70% decreases in abundance of mites and collembolans at the Aspen FACE (Free Air CO_2 Enrichment) site in northern Wisconsin, USA (Loranger et al., 2004). Moreover, Xu et al. (2013) noticed a significant decrease (by 55–57%) in collembolan abundance from increased CO_2 after 7 years in a Swiss FACE experiment; however, the collembolan community richness was unaffected.

Other climatic factors such as temperature and changes in the quality and quantity of litter could have significant effects on soil fauna (García-Palacios et al., 2013; Coûteaux and Bolger, 2000). Also, varying response to warming relating to the size of the organisms is notable. Smaller organisms [e.g. Acari (mites)] are more sensitive to changes in soil temperature than larger organisms (e.g. Collembola), although there can be species-specific differences within a group (Bokhorst et al., 2012). Moreover, earthworms have been found to be more abundant in wetter than dry conditions (Eggleton et al., 2009) and their abundance has been shown to decrease under an increase in soil temperature of ~1.2°C (Zaller et al., 2009). Also, Eisenhauer et al. (2014) demonstrated a decrease in density of various species of earthworm from elevated soil temperature, unless there was increased rainfall. Lindberg et al. (2002) observed that a long-term decrease in precipitation (drought) can lead to a decrease in the abundance and diversity of soil fauna in terrestrial ecosystems. Gómez et al. (2003) demonstrated that ant assemblages are very sensitive to anthropogenic activities and climate change. A recent review by Coyle et al. (2017) highlights the need for studies on the responses of soil fauna to management and climate change, particularly in the context of soil health and sustainability.

3 The nexus of climate change and agricultural management strategies

As described earlier, SOC is one of the critical soil health indicators, and is highly vulnerable to climate change/variability and human activities such as deforestation, intensive tillage and land use change, fallowing, stubble removal and stubble burning. These activities can turn soil from a C sink to a source of GHG emissions. In the context of implementing sustainable agricultural and forestry practices, soils have the potential to sequester C in stable SOM pools, with implications for offsetting GHG emissions. Mineralized SOC is generally released as $CO_{2,}$ and also as CH_4 where oxygen is limited, such as in saturated soils. These GHGs are the main contributors to global warming. Climate change will become more severe as it accelerates the rate of soil degradation [United Nations Environment Program (UNEP), 2013].

Higher temperature conditions resulting from extreme climate variability and climate change can increase the rates of soil C and N mineralization and plant nutrient availability through increased turnover of SOM. In many agroecosystems, increasing N availability can help increase C input and storage in soil by encouraging higher rates of plant growth (Melillo et al., 1993; Sarker et al., 2017). In some instances, excess N increases mineralization of SOM, such as in Southern California, the Rocky Mountains and the eastern United States, where this priming effect on SOM turnover could increase C and N losses from ecosystems (Galloway et al., 2003; Saxe et al., 2001). Such losses could create other ecological problems in these ecosystems, such as eutrophication in nearby streams or lower pH in soils (Galloway et al., 2003; Saxe et al., 2001). Monitoring soil C and N status of an ecosystem is important for assessing how soil C and N storage and plant N availability may be affected by climate change.

To mitigate and adapt to anticipated effects of climate change, recommendations have been developed for agricultural and forestry systems to improve soil C sequestration and soil productivity while considering key principles of soil health (see Section 1). For example, no-till/reduced tillage (Sarker et al., 2018c) and stubble retention systems (Powlson et al., 2016), and reforestation or reducing deforestation (Poorter et al., 2016; Schwartz et al., 2017) would minimize soil disturbance, protect soil against erosion and improve soil C storage. Incorporating diverse rotations and cover crop mixes in farming systems can improve soil cover, biodiversity and ecosystem functions, such as plant available nutrients, water storage, nutrient use efficiency and crop yield (Poeplau and Don, 2015; Lorenz and Lal, 2014; Sarker et al., 2018a,b). Application of soil amendments, for example lime to neutralize acidity (Kunhikrishnan et al., 2016) and manures, composts and biochar to increase soil C (Singh et al., 2011, 2018; Wall et al., 2018; Xu et al., 2018), and livestock grazing in cropping systems to improve soil functions (Allan et al., 2016; Rakkar and Blanco-Canqui, 2018) are other key practices aiming to address soil productivity constraints.

3.1 Land use, land use change and land management strategies

More than 40% of total organic C in terrestrial ecosystems is stored in forest soils (IPCC, 2007). The conversion of forest to agricultural land commonly destroys the soil's macro- and micro-aggregate structure, which enhances SOC mineralization and CO_2 emissions (Don et al., 2011). While deforestation provides wood products to meet the needs of a growing population, there are negative effects of this land use change on ecosystem services such as rising soil erosion (wind and water), the decline of water regulation against floods and landslides, the loss of plant biodiversity and a decline in aesthetic and recreational values. Due to unsustainable land use and management, ~24% of global land had already suffered from declines in soil health and plant productivity over the past quarter of a century (Rhodes, 2014). Around 60% of C stored in soils and vegetation has been lost as a result of land use changes, such as clearing land for agriculture and human habitation, since the nineteenth century. According to Paustian et al. (2016), 10–14% of total global anthropogenic GHG emissions are derived from agricultural production (mainly *via* emissions from soils and livestock management) and 12–17% from land cover change. Conversions from natural grasslands to croplands, for example, can disturb the soil and vegetation and lead to losses of SOC and nutrients (Smith et al., 2016). Muñoz-Rojas et al. (2012) studied the impacts of land use changes on SOC stocks for different soil types for a period of 51 years (1956–2007) at a regional scale in Andalusia in Southern Spain and found a general decrease of SOC stocks due to the intensification of agriculture. In south-eastern Australia, there has been a decrease in SOC in areas where native vegetation was cleared and cultivated for crop production, with the greatest decline in SOC in the first 10 years and equilibrium reached after 50 years (Luo et al., 2010). From the Australian context, the study of Luo et al. (2010) highlighted that land use change has contributed substantially to Australia's GHG emissions. Nevertheless, this provides an opportunity through implementation of sustainable land uses and management practices in agroecosystems to sequester C and improve soil health, thus contributing to climate change mitigation and adaptation (Lal, 2011; Mehra et al., 2018; Singh et al., 2011, 2018; Stockmann et al., 2013).

3.2 Minimal soil disturbance strategy

A major factor that affects soil health in agricultural systems is tillage intensity. Intensive tillage disrupts soil structure, accelerates the decomposition of SOM and makes the bared topsoil vulnerable to erosion by rain and wind. Repeated tilling destroys the soil pores and tunnels and reduces the soil's infiltration capacity, which can lead to polluting aquatic systems with excess nutrients and pesticides through surface runoff (Mehra, 2017). Cultivation can also lead to soil compaction, thereby contributing to increased nitrous oxide emission due to reduced aeration (Bhandral et al., 2007). However, over the last five decades, there is evidence from field and on-farm research that the adoption of no-till farming has improved sustainability of cropping. Out of many tangible benefits, no-till not only helps in improving soil quality/health by increasing SOM (Mishra et al., 2010; Kahlon et al., 2013; Singh et al., 2018), soil biota and nutrient cycling (Chowdhury et al., 2015; Olson et al., 2014; Sarker et al., 2018a), but also helps in improving crop yield (Sá et al., 2001) and reducing GHG emissions (Zhang et al., 2014; Six et al., 2000). Sørensen et al. (2014) quantified the total GHG emission as 915 g CO_2 equivalents per kg product from conventional tillage systems, which was higher than reduced tillage (817 g CO_2 equivalents) and no-till systems (855 g CO_2 equivalents). This study also emphasized that SOC mineralization during the tillage process contributed the most (50–60%) to GHG emissions.

Recently, Mehra (2017) reviewed studies including Forte et al. (2017), Tellez-Rio et al. (2015), Álvaro-Fuentes and Martínez (2010), Lee et al. (2009) and Johnson et al. (2007) on tillage–carbon–climate change interaction and found strong evidence that minimum disturbance through no-till or reduced tillage minimizes soil GHG emissions relative to conventional tillage. These conservation tillage practices, especially when combined with stubble retention, may increase SOC (Petersen et al., 2008; Mangalassery et al., 2014), water retention (via increasing infiltration and minimizing runoff), nutrient use efficiency (Triplett and Dick, 2008) and herbicide efficacy (Bajwa et al., 2017; Preston, 2017). Mehra et al. (2018) noted that in addition to many other tangible soil health benefits, no-till contributes to mitigation of climate change through reducing the oxidation rate of SOM and fuel consumption per hectare (Lal, 2004; Singh et al., 2018), thus attracting farmers' interest, which is why no-till farming is becoming a mainstream approach in agricultural systems globally. Furthermore, no-till systems may assist in adapting to varying temperature and precipitation conditions (as a result of climate change) which would otherwise impair soil physical, chemical and biological properties in agricultural systems and diminish agricultural production. While it is generally accepted that conservation tillage practices provide a range of benefits to soil health and climate change adaptation, as summarized earlier, the magnitude of the climate change mitigation benefit through soil C sequestration is debated. Several meta-analyses have concluded that the potential to increase soil C stocks through adoption of no-till has been overstated (Luo et al., 2010; Powlson et al., 2014). However, this conclusion is questioned by Paustian et al. (2016), on the basis that variability in SOC inhibits capacity to detect statistically significant change (Kravchenko and Robertson, 2011), and recent reviews express confidence that conservation tillage, combined with stubble retention and cover crops, can contribute to substantial soil C sequestration, delivering significant climate benefits (Smith, 2016; Minasny et al., 2017).

3.3 Crop rotation and nitrogen management strategies

Diversified crop rotations in dryland farming systems can improve the utilization of agricultural lands, increase plant C inputs in soils and improve soil health and functions (Sarker et al., 2018a,b). Some cropping systems such as rice-based cropping systems may contribute substantially to global GHG emissions (Turmel et al., 2015). Since an increase in soil productivity is obtained in a relatively short time, many growers use inorganic fertilizers and pesticides, with benefits to crop yields. For example, nitrogen fertilization is one of the classical agronomic practices used in conventional cropping systems with the aim of increasing crop yield. However, overuse of N fertilizers contributes to N_2O emissions as well as surface and ground water pollution. Some studies have shown that N fertilization can negatively affect SOC stocks and decrease soil health in the long run (Al-Kaisi et al., 2008; Mulvaney et al., 2009). On the other hand, inclusion of cover crops in rotations reduces N leaching and controls the growth of weeds (Al-Kaisi et al., 2008; Mulvaney et al., 2009). Moreover, where legume crop species are incorporated in rotations, biological N_2 fixation may reduce N fertilization costs and improve soil functions (Blesh, 2018). Furthermore, SOC storage could be increased through improvement of crop rotations by changing from monoculture to more diverse cropping systems, integrating forages with/as cover crops in rotations, and perennial cover cropping, thus increasing the resilience of cropping systems to climate extremes (King and Blesh, 2018; Lal, 2016, 2004). Differences in the quantity and quality (chemical composition) of crop residue biomass and/or root exudates among crop species may also affect SOC stock change (Blanco-Canqui and Lal, 2009; Chivenge et al., 2007; Lal, 2004). Congreves et al. (2015) assessed the impact of crop rotations in temperate rainfed long-term experimental sites using new Ontario Soil Health Assessment (OSHA) overall scores. Crop rotations consisting of winter wheat or alfalfa had higher OSHA scores, while monoculture corn or soybean–corn showed lowest scores.

3.4 Crop residue management strategy

Surface crop residue offers protective mulch and shields the soil against the direct impact of rainfall during the wet season as well as against extreme desiccation and deflation by the wind during dry periods. Crop residue incorporation or retention on the surface of the soil improves soil physical properties such as aggregate stability, enhances SOM and soil fertility and increases nutrient retention and availability, thus maintaining or improving soil health (Sarker et al., 2018a,b). Another relevant practice often followed in mixed cropping/livestock enterprises is the on-farm utilization of crop residue for livestock grazing (Allan et al., 2016). Residue grazing in a no-till system generally has a positive impact on soil nutrients and may increase SOM compared to no grazing (Rakkar and Blanco-Canqui, 2018). However, negative impacts on soil health could arise from overgrazing of crop residues, causing soil compaction and uneven nutrient distribution through trampling actions (similar to mechanical baling), inhibiting soil microbial activity and increasing GHG emissions from over fertilization with faeces and urine (Blanco-Canqui and Lal, 2009; Bell et al., 2011; Rakkar and Blanco-Canqui, 2018).

While some meta-analyses and literature review studies have reported an increase in CO_2 and N_2O emissions in upland soils from straw incorporation, a decrease in CH_4 emission was also observed, and when all fluxes were considered, the practice of straw incorporation led to a net C sink (Chowdhury et al., 2015; Jiang et al., 2017; Lehtinen et al., 2014; Liu et al., 2014). For example, Yan et al. (2005) found that straw returned in the previous season (e.g. upland crop or fallow) decreased CH_4 emissions by 42%, relative to on-season straw return just before rice transplanting. Burning harvested crop residues is commonly practised especially in the tropics as a cheapest option for the residue management, but this causes nutrient loss, air pollution and reduction in SOM, and can also be harmful to beneficial soil organisms (Singh et al., 2018).

4 Soil health and climate change: strategic implications

Soils harbour the most biodiverse ecosystems on earth and are a foundation resource supporting natural and managed ecosystems (Delmont et al., 2011; Thiele-Bruhn et al., 2012). If, as anticipated, climate change becomes more severe, and unsustainable agricultural practices accelerate the rate of soil degradation, leading to more GHG emissions and soil erosion that depletes soil C and nutrients, soil health will continue to decline. However, agricultural soils present a significant opportunity for climate change mitigation through identifying practices and systems that minimize GHG emissions while enhancing soil C sequestration and also building soil health. Implementation of improved management practices based on soil health principles can help in mitigating and adapting to climate change *via* improving soil properties/functions, crop productivity and systems' resilience, known as *Climate-Smart Agriculture* (CSA). There are, however, knowledge gaps in understanding and overcoming the barriers to the adoption of CSA practices to achieve uptake at the scale required. The adoption of CSA is currently relatively low (Arslan et al., 2014; McCarthy et al., 2011). Implementing CSA practices may require investment by farmers in additional inputs, or may incur a yield reduction in the short term, and this will constrain uptake especially by smallholders with limited financial resources. Support from governments and non-government organizations is required to facilitate adoption, as discussed further. Impetus to implement CSA and other sustainable land management approaches may come from the new policy initiative for land management, 'Land degradation neutrality' (LDN). LDN promotes a dual-pronged approach encouraging sustainable land management to reduce land degradation, combined with land restoration and rehabilitation to reverse impacts of land degradation, in order to maintain and enhance the land resource base (Cowie et al., 2018). LDN has been adopted as a policy goal through Sustainable Development Goal 15, Target 15.3 'By 2030, combat desertification, restore degraded land and soil, including land affected by desertification, drought and floods, and strive to achieve a land degradation-neutral world'.

The evidence from empirical studies suggests that there are four main areas that require support to facilitate adoption of CSA: (1) dissemination/extension of information and particularly to adapt practices to local conditions; (2) coordination of efforts for objectives best managed at landscape scale, such as reducing the risks of flood, managing pest outbreaks or preserving biodiversity; (3) comprehensive risk assessment strategies to improve climate change resilience; and (4) reliable and timely access to specific inputs in farming systems to support resource-use efficiency. Recognizing the cross-cutting nature of these requirements, this chapter, therefore, highlights the need to focus on the nexus between soil health and climate change to address challenges of meeting global needs for food, fresh water and biomass-based products.

The first Green Revolution promoted adoption of chemical fertilizers and high-yielding cereal cultivars and utilization of pesticides in agroecosystems. It is now recognized that farming systems based on cereal monocultures and high inputs of chemical fertilizers and agrichemicals are less sustainable and have negative impacts on soil health. Considering the rising needs of the increasing population and anticipated impacts of climate change on yields, farmers need another revolution focused on enhancing the functional ecology of agroecosystems, to improve agricultural sustainability. The conceptual framework of this revolution should focus on improving soil health and mitigating climate impacts *via* the adoption of CSA practices, which is one of the central pillars to address the climate–food–water nexus.

The conundrum of climate change faced by agricultural researchers and policymakers is how to address the problem of food security to meet the requirements of the rising population without further impacting the climate and degrading the natural resource base. Furthermore, agricultural policies cannot focus solely on this biophysical challenge but need to incorporate social and economic policies that address the challenges to adoption (e.g. Brown et al., 2016; Mitter et al., 2018). Policies should be designed and developed while considering the needs of farmers (large- and smallholders) and consumers at the centre of these policies, to meet the Sustainable Development Goals of eradicating the hunger/malnutrition, improving rural livelihoods and achieving LDN. New developments in plant breeding have enhanced the resilience of crop varieties to drought and other climate extremes such as high temperature and waterlogging (Kole et al., 2015; Misra, 2014). The use of mixed cropping (polycultures) with several crops growing at one time can increase resilience during periods of water or heat stress. As discussed earlier, other practices that enhance soil health include use of cover crops, mulches and organic amendments such as manure and compost. These practices are often used by smallholders but rarely applied in industrial-scale agriculture.

From the perspective of global food security, it is not possible to simply revert to more sustainable traditional practices, when there is an increasing pressure on the natural resources, for example arable land ha/person that is 13.04% from 2000 to 2014 and 37.84% from 1961 to 2000 (World Bank, 2016). Therefore, closing the sustainability gaps will require a focus on sustainable intensification, to develop high-yielding agricultural systems that maintain ecological integrity of agroecosystems and minimize off-site environmental impacts. Promising systems include integrated soil fertility management (utilizing organic amendments such as manure and compost supplemented with judicious use of chemical fertilizers where appropriate; incorporating legumes and pasture phase in crop rotations), integrated weed and pest management (minimizing use of agrichemicals and encouraging beneficial organisms), integrated natural resources management (integrated management of land, water and biological resources) and integrated landscape management (that integrates production, conservation and urban land uses, and determines appropriate agricultural practices based on land potential) (e.g. Altieri, 2002; Thiele-Bruhn et al., 2012; Lal, 2015; LaCanne and Lundgren, 2018). Effort is required to develop and promote sustainable systems that are suited across the range of scales, from smallholders to large-scale mechanized production systems. This will require multidisciplinary collaborations between a variety of technical experts, ranging from agricultural and ecological experts to statisticians and social scientists, working with local landholders. These collaborations will enable policymakers, entrepreneurs and farming communities to foster new strategies and re-engineer the conventional agricultural practice guidelines. New technologies can be applied to disseminate information and enable participatory approaches to land management. For example, the Land-Potential Knowledge System (Herrick et al., 2016) uses mobile apps, cloud computing and crowdsourcing to support evidence-based, site-specific land management by guiding the user in land potential assessment based on easily observed site properties, and provides site-specific information about degradation risk under different management scenarios.

If sustainable strategies and management practices that focus on improving soil health and maintaining the land resource base are adopted and implemented efficiently and consistently, then such approaches can become powerful entities in mitigating GHG emissions, adapting to climate change and maintaining biodiversity while enhancing global food security.

5 Future trends and conclusion

Climate is a major factor in shaping the plant and animal life on and in the soil. However, the climate extreme events and unsustainable land use and management, such as conversions from forestlands or grasslands to croplands and intensive tillage in croplands, have been threatening soil health and the health of local and global environment. These results, rather than turning soil into a C sink, make soil a source of GHG emissions, thus degrading the physical, chemical and biological components of soil quality/health. Holistically, under the nexus of climate change and agricultural systems, this chapter provides an overview of sustainable management practices that build SOM and improve soil resilience to management and environmental changes *via* meeting the soil health principles. The key soil health principles described in this chapter are: minimal soil disturbance, enhanced plant diversity and soil biodiversity, long-lasting soil cover through cover crops and stubble management, and careful integration of livestock into farming systems.

This chapter also highlights that implementation of sustainable farming practices can improve soil functionality and plant performance while delivering climate mitigation and adaptation benefits. Collectively, climate change impacts on soil properties may be permanent or temporary, but factors affecting adoption rates for sustainable practices in croplands are more specific to the techniques, knowledge and growers' skills to reduce the climate change impact. Currently, farmers have started using precision agriculture for site-specific farm management decisions and contributions to climate mitigation while employing their ground-truth knowledge and the data collected through precision agriculture technologies. Therefore, it is imperative to fortify the nexus between soil health and climate change by tailoring new strategic and CSA approaches while considering the practicality of soil health knowledge at the farmers' paddocks. Nevertheless, devising novel management strategies can set new directions to address soil productivity constraints under the changing climate and land use scenarios, while the knowledge gaps relating to the realisms of CSA strategies require multidisciplinary collaboration and communications.

6 Where to look for further information

For further insights into the nexus of soil health, agricultural management systems and/or climate change, the authors recommend reading the following scholarly books and articles:

1. Books: Al-Kaisi and Lowery (2017), Duiker et al. (2016), Muñoz and Zornoza (2018), Singh et al. (2011) and Stirling et al. (2016).
2. Articles: Al-Kaisi et al. (2017), Al-Kaisi and Lal (2017), Byrnes et al. (2018), Dalal et al. (2017), Seneviratne et al. (2018) and Ward et al. (2018).

7 References

Al-Kaisi, M. M. and Lal, R. (2017). Conservation agriculture systems to mitigate climate variability effects on soil health. In *Soil Health and Intensification of Agroecosytems* (Eds., Al-Kaisi M. M. and Lowery, B.):79–107. Elsevier Academic Press, London.

Al-Kaisi, M. M. and Lowery, B. (2017). *Soil Health and Intensification of Agroecosystems*. Academic Press, London, 418.

Al-Kaisi, M. M., Kruse, M. L. and Sawyer, J. E. (2008). Effect of nitrogen fertilizer application on growing season soil carbon dioxide emission in a corn–soybean rotation. *Journal of Environmental Quality* 37(2):325–32.

Al-Kaisi, M. M., Lal, R., Olson, K. R. and Lowery, B. (2017). Fundamentals and functions of soil environment. In *Soil Health and Intensification of Agroecosystems* (Eds., Al-Kaisi M. M. and Lowery, B.):1–20. Elsevier Academic Press, London.

Allan, C. J., Jones, B., Falkiner, S., Nicholson, C., Hyde, S., Mauchline, S., Ferrier, D.-A., Ward, P., Siddique, K. H. M. and Flower, K. C. (2016). Light grazing of crop residues by sheep in a Mediterranean-type environment has little impact on following no-tillage crops. *European Journal of Agronomy* 77:70–80.

Allen, D. E., Singh, B. P. and Dalal, R. C. (2011). Soil health indicators under climate change : A review of current knowledge. In *Soil Health and Climate Change* (Eds., Singh, B. P., Cowie, A. L. and Chan, K. Y.), Soil Biology Series (29):25–46. Springer-Verlag, Berlin.

Allison, S. D., Wallenstein, M. D. and Bradford, M. A. (2010). Soil-carbon response to warming dependent on microbial physiology. *Nature Geoscience* 3:336–40.

Altieri, M. A. (2002). Agroecology: The science of natural resource management for poor farmers in marginal environments. *Agriculture, Ecosystems and Environment* 93(1–3):1–24.

Álvaro-Fuentes, J. and Martínez, C. C. (2010). Short communication. Potential to mitigate anthropogenic CO_2 emissions by tillage reduction in dryland soils of Spain. *Spanish Journal of Agricultural Research* 8(4):1271–6.

Andrade, V. J. A., Abreu, F. M. G. and Madeira M. A. V. (2010). Influence of litter layer removal on the soil thermal regime of a pine forest in a mediterranean climate. *Revista Brasileira de Ciência Do Solo* 34(5):1481–90.

Arslan, A., McCarthy, N., Lipper, L., Asfaw, S. and Cattaneo, A. (2014). Adoption and intensity of adoption of conservation farming practices in Zambia. *Agricultural Ecosystems and Environment* 187:72–86.

Austin, E. E., Castro, H. F., Sides, K. E., Schadt, C. W. and Classen, A. T. (2009). Assessment of 10 years of CO_2 fumigation on soil microbial communities and function in a sweetgum plantation. *Soil Biology and Biochemistry* 41(3):514–20.

Bajwa, A. A., Walsh, M. and Chauhan, B. S. (2017). Weed management using crop competition in Australia. *Crop Protection* 95:8–13.

Balser, T. C. and Wixon, D. L. (2009). Investigating biological control over soil carbon temperature sensitivity. *Global Change Biology* 15(12):2935–49.

Balser, T. C, Gutknecht, J. L. M. and Liang, C. (2010). How will climate change impact soil microbial communities? In *Soil Microbiology and Sustainable Crop Production* (Eds., Dixon, G. R. and Tilston, E. L.):373–97. Springer Science Business Media B.V, Dordrecht, the Netherlands.

Barber, S. A. (1996). *Soil Nutrient Bioavailability: A Mechanistic Approach*. John Wiley and Sons, New York.

Bardgett, R. D., Freeman, C. and Ostle, N. J. (2008). Microbial contributions to climate change through carbon cycle feedbacks. *The ISME Journal* 2:805–14.

Basher, L. R. and Ross, C. W. (2002). Soil erosion rates under intensive vegetable production on clay loam, strongly structured soils at Pukekohe, New Zealand. *Australian Journal of Soil Research* 40(6):947–61.

Bastos, A., Gouveia, C. M., Trigo, R. M. and Running, S. W. (2013). Comparing the Impacts of 2003 and 2010 heatwaves in NPP over Europe. *Biogeosciences Discussions* 10(10):15879–911.

Bauer, A. and Black, A. L. (1994). Quantification of the effect of soil organic matter content on soil productivity. *Soil Science Society of America Journal* 58(1):185–93.

Bell, L. W., Kirkegaard, J. A., Swan, A., Hunt, J. R., Huth, N. I. and Fettell, N. A. (2011). Impacts of soil damage by grazing livestock on crop productivity. *Soil and Tillage Research* 113:19–29.

Belnap, J. (2003). The world at your feet: Desert biological soil crusts. *Frontiers in Ecology and the Environment* 1(4):181–9.

Bhandral, R., Saggar, S., Bolan, N. S. and Hedley, M. J. (2007). Transformation of nitrogen and nitrous oxide emission from grassland soils as affected by compaction. *Soil and Tillage Research* 94(2):482–92.
Bhupinderpal-Singh, Nordgren, A., Ottosson Löfvenius, M., Högberg, M. N., Mellander, P. E. and Högberg, P. (2003). Tree root and soil heterotrophic respiration as revealed by girdling of boreal scots pine forest: Extending observations beyond the first year. *Plant, Cell and Environment* 26(8):1287–96.
Blanco-Canqui, H. and Lal, R. (2009). Crop residue removal impacts on soil productivity and environmental quality. *Critical Reviews in Plant Sciences* 28:139–63.
Blankinship, J. C., Niklaus, P. A. and Hungate, B. A. (2011). A meta-analysis of responses of soil biota to global change. *Oecologia* 165(3):553–65.
Blesh, J. (2018). Functional traits in cover crop mixtures: Biological nitrogen fixation and multifunctionality. *Journal of Applied Ecology* 55(1):38–48.
Blevins, R. L., Smith, M. S., Thomas, G. W. and Frye, W. W. (1983). Influence of conservation tillage on soil properties. *Journal of Soil and Water Conservation* 38(3):301–5.
Bokhorst, S., Phoenix, G. K., Bjerke, J. W., Callaghan, T. V., Huyer-Brugman, F. and Berg, M. P. (2012). Extreme winter warming events more negatively impact small rather than large soil fauna: Shift in community composition explained by traits not taxa. *Global Change Biology* 18:1152–62.
Bolan, N. S. and Hedley, M. J. (2003). Role of carbon, nitrogen, and sulfur cycles in soil acidification. In *Handbook of Soil Acidity* (Ed., Rengel, Z.):43–70. CRC Press, Marcel Dekker, NY.
Bolan, N. S., Hedley, M. J. and White, R. E. (1991). Processes of soil acidification during nitrogen cycling with emphasis on legume based pastures. *Plant and Soil* 134(1):53–63.
Bolan, N. S., Adriano, D. C. and Curtin, D. (2003). Soil acidification and liming interactions with nutrient and heavy metal transformation and bioavailability. *Advances in Agronomy* 78:215–72.
Bonito, G. M., Coleman, D. C., Haines, B. L. and Cabrera, M. L. (2003). Can nitrogen budgets explain differences in soil nitrogen mineralization rates of forest stands along an elevation gradient? *Forest Ecology and Management* 176(1–3):563–74.
Boone, R. D., Nadelhoffer, K. J., Canary, J. D. and Kaye, J. P. (1998). Roots exert a strong influence on the temperature sensitivity of soil respiration. *Nature* 396(6711):570.
Bradford, M. A., Davies, C. A., Frey, S. D., Maddox, T. R., Melillo, J. M., Mohan, J. E., Reynolds, J. F., Treseder, K. K. and Wallenstein, M. D. (2008). Thermal adaptation of soil microbial respiration to elevated temperature. *Ecology Letters* 11(12):1316–27.
Bradford, M. A., Wood, S. A., Bardgett, R. D., Black, H. I., Bonkowski, M., Eggers, T., Grayston, S. J., Kandeler, E., Manning, P., Setälä, H. and Jones, T. H. (2014). Discontinuity in the responses of ecosystem processes and multi-functionality to altered soil community composition. *Proceedings of the National Academy of Sciences of the United States of America* 111:14478–83.
Brinkman, R. and Sombroek, W. G. (1996). The effects of global change on soil conditions in relation to plant growth and food production. In *Global Climate Change and Agricultural Production* (Eds., Bazzaz, F. and Sombroek, W.):49–63. John Wiley and Sons, Chichester, England.
Brouder, S. M. and Volenec, J. J. (2008). Impact of climate change on crop nutrient and water use efficiencies. *Physiologia Plantarum* 133:705–24.
Brown, P. R., Bridle, K. L. and Crimp, S. J. (2016). Assessing the capacity of Australian broadacre mixed farmers to adapt to climate change: Identifying constraints and opportunities. *Agricultural Systems* 146:129–41.
Brussaard, L., de Ruiter, P. C. and Brown, G. G. (2007). Soil biodiversity for agricultural sustainability. *Agriculture, Ecosystems and Environment* 121(3):233–44.
Byrnes, R. C., Eastburn, D. J., Tate, K. W. and Roche, L. M. (2018). A global meta-analysis of grazing impacts on soil health indicators. *Journal of Environmental Quality*. doi:10.2134/jeq2017.08.0313.
Cardoso, E. J. B. N., Vasconcellos, R. L. F., Bini, D., Miyauchi, M. Y. H., dos Santos, C. A., Alves, P. R. L., de Paula, A. M., Nakatani, A. S., de Moraes Pereira, J. and Nogueira, M. A. (2013). Soil health: Looking for suitable indicators. What should be considered to assess the effects of use and management on soil health? *Scientia Agricola* 70 (4):274–89

Castro, H. F., Classen, A. T., Austin, E. E., Norby, R. J. and Schadt, C. W. (2010). Soil microbial community responses to multiple experimental climate change drivers. *Applied and Environmental Microbiology* 76:999–1007.

Charman, P. E. V. and Murphy, B. W. (2007). *Soils: Their Properties and Management*, 3rd ed. Oxford University Press, South Melbourne, Vic.

Chen, M. M., Zhu, Y. G., Su, Y. H., Chen, B. D., Fu, B. J. and Marschner, P. (2007). Effects of soil moisture and plant interactions on the soil microbial community structure. *European Journal of Soil Biology* 43(1):31–8.

Ching, P. C. and Barber, S. A. (1979). Evaluation of temperature effects on K uptake by corn. *Agronomy Journal* 71(6):1040–4.

Chivenge, P. P., Murwira, H. K., Giller, K. E., Mapfumo, P. and Six, J. (2007). Long-term impact of reduced tillage and residue management on soil carbon stabilization: Implications for conservation agriculture on contrasting soils. *Soil Tillage Research* 94:328–37.

Chowdhury, S., Farrell, M., Butler, M. and Bolan, N. (2015). Assessing the effect of crop residue removal on soil organic carbon storage and microbial activity in a no-till cropping system. *Soil Use and Management* 31(4):450–60.

Christensen, N. L., Bartuska, A. M., Brown, J. H., Carpenter, S., D'Antonio, C., Francis, R., Franklin, J. F., MacMahon, J. A., Noss, R. F. and Parsons, D. J. (1996). The report of the Ecological Society of America Committee on the scientific basis for ecosystem management. *Ecological Applications* 6(3):665–91.

Clair, S. B. and Lynch, J. P. (2010). The opening of Pandora's Box: Climate change impacts on soil fertility and crop nutrition in developing countries. *Plant and Soil* 335:101–15.

Cleland, E. E., Chuine, I., Menzel, A., Mooney, H. A. and Schwartz, M. D. (2007). Shifting plant phenology in response to global change. *Trends in Ecology and Evolution* 22(7):357–65.

Congreves, K. A., Hayes, A., Verhallen, E. A. and Van Eerd, L. L. (2015). Long-term impact of tillage and crop rotation on soil health at four temperate agroecosystems. *Soil and Tillage Research* 152:17–28.

Coûteaux, M.-M. and Bolger, T. (2000). Interactions between atmospheric CO_2 enrichment and soil fauna. *Plant and Soil* 224(1):123–34.

Cowie, A. L., Orr, B. J., Sanchez, V. M. C., Chasek, P., Crossman, N. D., Erlewein, A., Louwagie, G., Maron, M., Metternicht, G. I., Minelli, S. and Tengberg, A. E. (2018). Land in balance: The scientific conceptual framework for Land Degradation Neutrality. *Environmental Science and Policy* 79:25–35.

Coyle, D. R, Nagendra, U. J., Taylor, M. K., Campbell, J. H., Cunard, C. E., Joslin, A. H., Mundepi, A., Phillips, C. A. and Callaham, Mac A. (2017). Soil fauna responses to natural disturbances, invasive species, and global climate change: Current state of the science and a call to action. *Soil Biology and Biochemistry* 110:116–33.

Cregger, M. A., Schadt, C. W., McDowell, N. G., Pockman, W. T. and Classen, A. T. (2012). Response of the soil microbial community to changes in precipitation in a semiarid ecosystem. *Applied and Environmental Microbiology* 78(24):8587–94.

Dalal, R. C., Kopittke, P. M. and Menzies, N. W. (2017). Impact of climate change on soil carbon storage. In *The Nexus of Soils, Plants, Animals and Human Health* (Eds., Singh, B. R., McLaughlin, M. J. and Brevik, E.):156–63. Schweizerbart Science Publishers, Stuttgart, Germany.

Datiri, B. C. and Lowery, B. (1991). Effects of conservation tillage on hydraulic properties of a Griswold silt loam soil. *Soil and Tillage Research* 21(3–4):257–71.

Davidson, E. A. and Janssens, I. A. (2006). Temperature sensitivity of soil carbon decomposition and feedbacks to climate change. *Nature* 440:165–73.

DEFRA (2005). Impact of climate change on soil functions. Final project report, Research and Development, London, UK.

Delgado-Baquerizo, M., Maestre, F. T., Reich, P. B., Jeffries, T. C., Gaitan, J. J., Encinar, D., Berdugo, M., Campbell, C. D. and Singh, B. K. (2016). Microbial diversity drives multifunctionality in terrestrial ecosystems. *Nature Communications* 7:10541.

Delmont, T. O., Robe, P., Cecillon, S., Clark, I. M., Constancias, F., Simonet, P., Hirsch, P. R. and Vogel, T. M. (2011). Accessing the soil metagenome for studies of microbial diversity. *Applied and Environmental Microbiology* 77(4):1315–24.

Don, A., Schumacher, J. and Freibauer, A. (2011). Impact of tropical land-use change on soil organic carbon stocks - A meta-analysis. *Global Change Biology* 17(4):1658–70.

Drewry, J. J., Cameron, K. C. and Buchan, G. D. (2008). Pasture yield and soil physical property responses to soil compaction from treading and grazing – A review. *Soil Research* 46(3):237–56.

Duiker, S. W., Myers, J. C. and Blazure, L. C. (2016). *Soil Health in Field and Forage Crop production.* USDA Natural Resources Conservation Service, Penn State University Extension, Capital Resource Conservation & Development, and Clinton County Conservation District. Desktop publishing by Molly McDonough, USDA NRCS, Washington DC.

Eggleton, P., Inward, K., Smith, J., Jones, D. T. and Sherlock, E. (2009). A six year study of earthworm (*Lumbricidae*) populations in pasture woodland in southern England shows their responses to soil temperature and soil moisture. *Soil Biology and Biochemistry* 41:1857–65.

Ehlers, W. and Claupein, W. (1994). Approaches toward conservation tillage in Germany. In *Conservation Tillage in Temperate Agroecosystems* (Ed., Carter, M. R.):141–65. Lewis Publishers, Boca Raton, FL.

Eisenhauer, N., Stefanski, A., Fisichelli, N. A., Rice, K., Rich, R. and Reich, P. B. (2014). Warming shifts 'worming': Effects of experimental warming on invasive earthworms in northern North America. *Scientific Reports* 4:6890.

Fang, Y., Singh B. P., Badgery W. and He, X. (2016). *In situ* assessment of new carbon and nitrogen assimilation and allocation in contrastingly managed dryland wheat crop–soil systems. *Agriculture, Ecosystems and Environment* 235:80–90.

Fang, Y., Singh B. P., Matta, P., Cowie, A. L. and Van Zwieten, L. (2017). Temperature sensitivity and priming of organic matter with different stabilities in a vertisol with aged biochar. *Soil Biology and Biochemistry* 115:346–56.

Fang, Y., Singh, B. P., Yu, L., Boersma, M. and Van Zwieten, L. (2018). Biochar carbon dynamics in physically separated fractions and microbial use efficiency in contrasting soils under temperate pastures. *Soil Biology and Biochemistry* 116:399–409.

FAO (2016). *FAO Statistical Year Book 2015.* Food and Agriculture Organization of the United Nations, Rome, Italy.

FAO and ITPS (2015). *Status of the World's Soil Resources (SWSR) – Main Report.* Food and Agriculture Organization of the United Nations and Intergovernmental Technical Panel on Soils, Rome, Italy.

Forte, A., Fiorentino, N., Fagnano, M. and Fierro, A. (2017). Mitigation impact of minimum tillage on CO_2 and N_2O emissions from a mediterranean maize cropped soil under low-water input management. *Soil and Tillage Research* 166:167–78.

Galloway, J. N., Aber, J. D., Erisman, J. W., Seitzinger, S. P., Howarth, R. W., Cowling, E. B. and Cosby, J. B. (2003). The nitrogen cascade. *BioScience* 53(4):341–56.

García-Palacios, P., Maestre, F. T., Kattge, J. and Wall, D. H. (2013). Climate and litter quality differently modulate the effects of soil fauna on litter decomposition across biomes. *Ecology Letters* 16(8):1045–53.

Ghee, C., Neilson, R., Hallett, P. D., Robinson, D. and Paterson, E. (2013). Priming of soil organic matter mineralisation is intrinsically insensitive to temperature. *Soil Biology and Biochemistry* 66:20–8.

Gillooly, J. F., Brown, J. H., West, G. B., Savage, V. M. and Charnov, E. L. (2001). Effects of size and temperature on metabolic rate. *Science* 293(5538):2248–51.

GISTEMP Team. (2018). GISS Surface Temperature Analysis (GISTEMP). NASA Goddard Institute for Space Studies. https://data.giss.nasa.gov/gistemp/graphs_v3/Fig.A2.txt (last accessed on 25 February 2018).

Głab, T. and Kulig, B. (2008). Effect of mulch and tillage system on soil porosity under wheat (*Triticum aestivum*). *Soil and Tillage Research* 99(2):169–78.

Gómez, C., Casellas, D., Oliveras, J. and Bas, J. M. (2003). Structure of ground-foraging ant assemblages in relation to land-use change in the Northwestern Mediterranean Region. *Biodiversity and Conservation* 12(10):2135–46.

Gottfried, M., Pauli, M., Futschik, A., Akhalkatsi, M., Barančok, P., Alonso, J. L. B., Coldea, G., Dick, J., Erschbamer, B., Calzado, M. R. F., Kazakis, G., Krajči, J., Larsson, P., Mallaun, M., Michelsen, O., Moiseev, D., Moiseev, P., Molau, U., Merzouki, A., Nagy, L., Nakhutsrishvili, G., Pedersen, B., Pelino, G., Puscas, M., Rossi, G., Stanisci, A., Theurillat, J. P., Tomaselli, M., Villar, L., Vittoz, P., Vogiatzakis, L. and Grabherr, G. (2012). Continent-wide response of mountain vegetation to climate change. *Nature Climate Change* 2(2):111–15.

Guenet, B., Camino-Serrano, M., Ciais, P., Tifafi, M., Maignan, F., Soong, J. L. and Janssens, I. A. (2018). Impact of priming on global soil carbon stocks. *Global Change Biology* 24(5):1–11.

Gupta, J. P. (1993). Wind erosion of soil in drought-prone areas. In *Desertification and Its Control in the Thar, Sahara and Sahel Regions* (Eds. Kar, A., Garg, B. K., Singh, M. P. and Kathju, S.):91–105. Scientific Publishers, Jodhpur.

Hadas, A. (1997). Soil tilth – the desired soil structural state obtained through proper soil fragmentation and reorientation processes. *Soil and Tillage Research* 43:7–40.

Haynes, R. J. (2014). Nature of the belowground ecosystem and its development during pedogenesis. *Advances in Agronomy* 127:43–109.

Herrick, J. E., Beh, A., Barrios, E., Bouvier, I., Coetzee, M., Dent, D., Elias, E., Hengl, T., Karl, J. W., Liniger, H., Matuszak, J., Neff, J. C., Ndungu, L. W., Obersteiner, M., Shepherd, K. D., Urama, K. C., van den Bosch, R. and Webb. N. P. (2016). The Land-Potential Knowledge System (LandPKS): Mobile apps and collaboration for optimizing climate change investments. *Ecosystem Health and Sustainability* 2(3):e01209.

Houghton, R. A. (2007). Balancing the global carbon budget. *Annual Review Earth Planetary Science* 35:313–47.

IPCC (2007). IPCC fourth assessment report. *IPCC* 1:976.

IPCC (2012). Managing the risks of extreme events and disasters to advance climate change adaptation. In *A Special Report of Working Groups I and II of the Intergovernmental Panel on Climate Change* (Eds., Stocker, T. F., Field, B. and Barros, V.). IPCC and Cambridge University Press, Cambridge, NY.

Izaurralde, R. C., Williams, J. R., Post, W. M., Thomson, A. M., McGill, W. B., Owens, L. B. and Lal, R. (2007). Long-term modelling of soil C erosion and sequestration at the small watershed scale. *Climatic Change* 80(1–2):73–90.

Janus, L. R., Angeloni, N. L., McCormack, J., Rier, S. T., Tuchman, N. C. and Kelly, J. J. (2005). Elevated atmospheric CO_2 alters soil microbial communities associated with trembling aspen (*Populus tremuloides*) roots. *Microbial Ecology* 50(1):102–9.

Jat, M. L., Stirling, C. M., Jat, H. S., Tetarwal, J. P., Jat, R. K., Singh, R., Lopez-Ridaura, S. and Shirsath, P. B. (2018). Soil processes and wheat cropping under emerging climate change scenarios in South Asia. *Advances in Agronomy* 148:111–71.

Jiang, C. M., Yu, W. T., Ma, Q., Xu, Y. G. and Zou, H. (2017). Alleviating global warming potential by soil carbon sequestration: A multi-level straw incorporation experiment from a maize cropping system in Northeast China. *Soil and Tillage Research* 170:77–84.

Johnson, J. M. F., Franzluebbers, A. J., Weyers, S. L. and Reicosky, D. C. (2007). Agricultural opportunities to mitigate greenhouse gas emissions. *Environmental Pollution* 150(1):107–24.

Kahlon, M. S., Lal, R. and Ann-Varughese, M. A. (2013). Twenty two years of tillage and mulching impacts on soil physical characteristics and carbon sequestration in Central Ohio. *Soil and Tillage Research* 126:151–8.

Kardol, P., Reynolds, W. N., Norby, R. J. and Classen, A. T. (2011). Climate change effects on soil microarthropod abundance and community structure. *Applied Soil Ecology* 47:37–44.

Karlen, D. L., Erbach, D. C., Kaspar, T. C., Colvin, T. C., Berry, E. C. and Timmons, D. R. (1990). Soil tilth: A review of past perceptions and future needs. *Soil Science Society of America Journal* 54(1):153–61.

Karmakar, R., Das, I., Dutta, D. and Rakshit, A. (2016). Potential effects of climate change on soil properties: A review. *Science International* 4:51–73.

Kay, B. D. (1990). Rates of change of soil structure under different cropping systems. *Advances in Soil Science* 12:1–52.

Keesstra, S. D., Geissen, V., Mosse, K., Piiranen, S., Scudiero, E., Leistra, M. and van Schaik, L. (2012). Soil as a filter for groundwater quality. *Current Opinion in Environmental Sustainability* 4(5):507–16.

Keiblinger, K. M., Hall, E. K., Wanek, W., Szukics, U., Hämmerle, I., Ellersdorfer, G., Böck, S., Strauss, S., Sterflinger, K. and Richter, A. (2010). The effect of resource quantity and resource stoichiometry on microbial carbon-use-efficiency. *FEMS Microbiology Ecology* 73(3):430–40.

Kibblewhite, M. G., Ritz, K. and Swift, M. J. (2008). Soil health in agricultural systems. *Philosophical Transactions of the Royal Society B: Biological Sciences* 363(1492):685–701.

King, A. E. and Blesh, J. (2018). Crop rotations for increased soil carbon: Perenniality as a guiding principle. *Ecological Applications* 28(1):249–61.

Kladivko, E. J. (2001). Tillage systems and soil ecology. *Soil and Tillage Research* 61:61–76.

Köchy, M., Don, A., Molen, M. K. V. and Freibauer, A. (2015). Global distribution of soil organic carbon-part 2: Certainty of changes related to land use and climate. *Soil* 1(1):367–80.

Koehn, A. C., Lehrsch, G. A., Busscher, W. J., Evans, D. E., King, B. A., Stieneke, D. L. and Sojka, R, E. (2014). Aggregate tensile strength and friability characteristics of furrow and sprinkler irrigated fields in Southern Idaho. *Communications in Soil Science and Plant Analysis* 45(20):2712–20.

Kole, C., Muthamilarasan, M., Henry, R., Edwards, R., Sharma, R., Abberton, M., Batley, J., Bentley, A., Blakeney, M., Bryant, J., Cai, H., Cakir, M., Cseke, L. J., Cockram, J., de Oliveira, A. C., De Pace, C., Dempewolf, H., Ellison, S., Gepts, P., Greenland, A., Hall, A., Hori, K., Hughes, S., Humphreys, M. W., Iorizzo, M., Ismail, A. M., Marshall, A., Mayes, S., Nguyen, H. T., Ogbonnaya, F. C., Ortiz, R., Paterson, A. H., Simon, P. W., Tohme, J., Tuberosa, R., Valliyodan, B., Varshney, R. K., Wullschleger, S. D. and Yano, M. and Prasad, M. (2015). Application of genomics-assisted breeding for generation of climate resilient crops: Progress and prospects. *Frontiers in Plant Science* 6:563.

Kostina, N. V., Bogdanova, T. V. and Umarov, M. M. (2011). Biological activity of the coprolites of earthworms. *Moscow University Soil Science Bulletin* 66(1):18–23.

Kravchenko, A. N. and Robertson, G. P. (2011). Whole-profile soil carbon stocks: The danger of assuming too much from analyses of too little. *Soil Science Society of America Journal* 75(1):235.

Kunhikrishnan, A., Thangarajan, R., Bolan, N. S., Xu, Y., Mandal, S., Gleeson, D. B., Seshadri, B., Zaman, M., Barton, L., Tang, C., Luo, J., Dalal, R., Ding, W., Kirkham, M. B. and Naidu, R. (2016). Functional relationships of soil acidification, liming, and greenhouse gas flux. *Advances in Agronomy* 139:1–71.

LaCanne, C. E. and Lundgren, J. G. (2018). Regenerative agriculture: Merging farming and natural resource conservation profitably. *Peer Journal* 6:e4428.

Lal, R. (2000). Mulching effects on soil physical quality of an alfisol in western Nigeria. *Land Degradation and Development* 11(4):383–92.

Lal, R. (2003). Soil erosion and the global carbon budget. *Environment International* 29(4):437–50.

Lal, R. (2004). Soil carbon sequestration impacts on global climate change and food security. *Science* 304(5677):1623–7.

Lal, R. (2011). Soil health and climate change: An overview. In *Soil Health and Climate Change* (Eds., Singh, B. P., Cowie, A. L. and Chan, K. Y.), Soil Biology Series (29):3–24. Springer-Verlag, Berlin.

Lal, R. (2015). Restoring soil quality to mitigate soil degradation. *Sustainability* 7:5875–95.

Lal, R. (2016). Soil health and carbon management. *Food and Energy Security* 5(4):212–22.

Langley, J. A. and Hungate, B. A. (2014). Plant community feedbacks and long-term ecosystem responses to multi-factored global change. *AoB Plants* 6:plu035.

Lavelle, P., Decaëns, T., Aubert, M., Barot, S., Blouin, M., Bureau, F., Margerie, P., Mora, P. and Rossi, J. P. (2006). Soil invertebrates and ecosystem services. *European Journal of Soil Biology* 42:S3–S15.

Lee, J. J., Phillips, D. L. and Dodson, R. F. (1996). Sensitivity of the US corn belt to climate change and elevated CO_2: II. Soil erosion and organic carbon. *Agricultural Systems* 52(4):503–21.

Lee, J., Hopmans, J. W., Kessel, C. V., King, A. P., Evatt, K. J., Louie, D., Rolston, D. E. and Six, J. (2009). Tillage and seasonal emissions of CO_2, N_2O and NO across a seed bed and at the field scale in a mediterranean climate. *Agriculture, Ecosystems and Environment* 129(4):378–90.

Lehtinen, T., Schlatter, N., Baumgarten, A., Bechini, L., Krüger, J., Grignani, C., Zavattaro, L., Costamagna, C. and Spiegel, H. (2014). Effect of crop residue incorporation on soil organic

carbon and greenhouse gas emissions in European agricultural soils. *Soil Use and Management* 30(4):524–38.
Lesaulnier, C., Papamichail, D., McCorkle, S., Ollivier, B., Skiena, S., Taghavi, S., Zak, D. and Lelie, D. V. D. (2008). Elevated atmospheric CO_2 affects soil microbial diversity associated with trembling aspen. *Environmental Microbiology* 10(4):926–41.
Lindberg, N., Engtsson, J. B. and Persson, T. (2002). Effects of experimental irrigation and drought on the composition and diversity of soil fauna in a coniferous stand. *Journal of Applied Ecology* 39(6):924–36.
Link, S. O., Smith, J. L., Halvorson, J. J. and Bolton, H. (2003). A reciprocal transplant experiment within a climatic gradient in a semiarid shrub-steppe ecosystem: Effects on bunchgrass growth and reproduction, soil carbon, and soil nitrogen. *Global Change Biology* 9(7):1097–105.
Liu, X., Wang, G., Li, J. and Wang, Q. (2010). Nitrogen isotope composition characteristics of modern plants and their variations along an altitudinal gradient in Dongling Mountain in Beijing. *Science in China Series D: Earth Sciences* 53(1):128–40.
Liu, C., Lu, M., Cui, J., Li, B. and Fang, C. (2014). Effects of straw carbon input on carbon dynamics in agricultural soils: A meta-analysis. *Global Change Biology* 20(5):1366–81.
Loranger, G. I., Pregitzer, K. S. and King, J. S. (2004). Elevated CO_2 and O_3 concentrations differentially affect selected groups of the fauna in temperate forest soils. *Soil Biology and Biochemistry* 36:1521–4.
Lorenz, K. and Lal, R. (2014). Soil organic carbon sequestration in agroforestry systems. A review. *Agronomy for Sustainable Development* 34(2):443–54.
Luo, Z., Wang, E. and Sun, O. J. (2010). Can no-tillage stimulate carbon sequestration in agricultural soils? A meta-analysis of paired experiments. *Agriculture, Ecosystems and Environment* 139(1):224–31.
Luo, Z., Feng, W., Luo, Y., Baldock, J. and Wang, E. (2017). Soil organic carbon dynamics jointly controlled by climate, carbon inputs, soil properties and soil carbon fractions. *Global Change Biology* 23(10):4430–9.
Luo, G., Rensing C., Chen, H., Liu, M., Wang, M., Guo. S., Ling, N. and Shen, Q. (2018). Deciphering the associations between soil microbial diversity and ecosystem multifunctionality driven by long-term fertilization management. *Functional Ecology* 32(4):1103–16.
Mackay, A. D. and Barber, S. A. (1984). Soil temperature effects on root growth and phosphorus uptake by corn. *Soil Science Society of America Journal* 48(4):818–23.
Mackay, A. D. and Barber, S. A. (1985). Soil moisture effect on potassium uptake by corn. *Agronomy Journal* 77(4):524–7.
Mangalassery, S., Sjögersten, S., Sparkes, D. L., Sturrock, C. J., Craigon, J. and Mooney, S. J. (2014). To what extent can zero tillage lead to a reduction in greenhouse gas emissions from temperate soils? *Scientific Reports* 4:4586.
Mangalassery, S., Mooney, S. J., Sparkes, D. L., Fraser, D. L. and Sjögersten, S. (2015). Impacts of zero tillage on soil enzyme activities, microbial characteristics and organic matter functional chemistry in temperate soils. *European Journal of Soil Biology* 68:9–17.
Manzoni, S., Taylor, P., Richter, A., Porporato, A. and Ågren, G. I. (2012). Environmental and stoichiometric controls on microbial carbon-use efficiency in soils. *New Phytologist* 196(1):79–91.
McCarthy, N., Lipper, L. and Branca, G. (2011). Climate-smart agriculture: Smallholder adoption and implications for climate change adaptation and mitigation. *Mitigation of Climate Change in Agriculture Working Paper* 3:1–37.
McKenzie, S. W., Hentley, W. T., Hails, R. S., Jones, T. H., Vanbergen, A. J. and Johnson, S. N. (2013). Global climate change and above-belowground insect herbivore interactions. *Frontiers in Plant Science* 4:412.
McNear Jr., D. H. (2013). The rhizosphere - roots, soil and everything in between. *Nature Education Knowledge* 4(30):1.
Mehra, P. (2017). Return to tillage under no-till system: Impacts on soil carbon dynamics. PhD thesis, University of South Australia, Future Industries Institute, Adelaide, Australia. http://researchoutputs.unisa.edu.au/11541.2/130686 (accessed 09 April 2018).

Mehra, P., Desbiolles, J., Baker, J., Sojka, R. E., Bolan, N., Kirkham, M. B., Ross, C. W. and Gupta, R. (2018). A review of tillage practices and their potential to impact the soil carbon dynamics. *Advances in Agronomy*. doi:10.1016/bs.agron.2018.03.002.

Melillo, J. M., McGuire, A. D., Kicklighter, D. W., Moore, B., Vorosmarty, C. J. and Schloss, A. L. (1993). Global climate change and terrestrial net primary production. *Nature* 363(6426):234–40.

Merrill, S. D., Black, A. L. and Bauer, A. (1996). Conservation tillage affects root growth of dryland spring wheat under drought. *Soil Science Society of America Journal* 60:575–83.

Mills, R. T. E., Gavazov, K. S., Spiegelberger, T., Johnson, D. and Buttler, A. (2014). Diminished soil functions occur under simulated climate change in a sup-alpine pasture, but heterotrophic temperature sensitivity indicates microbial resilience. *Science of the Total Environment* 473–474:465–72.

Minasny, B., Malone, B. P., McBratney, A. B., Angers, D. A., Arrouays, D., Chambers, A., Chaplot, V., Chen, Z. S., Cheng, K., Das, B. S. and Field, D. J. (2017). Soil carbon 4 per mille. *Geoderma* 292:59–86.

Mishra, U., Ussiri, D. A. N. and Lal, R. (2010). Tillage effects on soil organic carbon storage and dynamics in corn belt of Ohio USA. *Soil and Tillage Research* 107(2):88–96.

Misra, A. K. (2014). Climate change and challenges of water and food security. *International Journal of Sustainable Built Environment* 3(1):153–65.

Mitchell, J. K. and Soga, K. (2005). *Fundamentals of Soil Behaviour*, 3rd ed. John Wiley and Sons, New York, 592.

Mitter, H., Schönhart, M., Larcher, M. and Schmid, E. (2018). The Stimuli-Actions-Effects-Responses (SAER)-framework for exploring perceived relationships between private and public climate change adaptation in agriculture. *Journal of Environmental Management* 209:286–300.

Monteith, J. and Unsworth, M. (2007). *Principles of Environmental Physics. Plant, Animals and the Animals*, 4th ed. Academic Press, Boston, MA, iii.

Morrow, J. G., Huggins, D. R. and Reganold, J. P. (2017). Climate change predicted to negatively influence surface soil organic matter of dryland cropping systems in the Inland Pacific Northwest, USA. *Frontiers in Ecology and Evolution* 5:10.

Mosaddeghi, M. R., Mahboubi, A. A. and Safadoust, A. (2009). Short-term effects of tillage and manure on some soil physical properties and maize root growth in a sandy loam soil in Western Iran. *Soil and Tillage Research* 104(1):173–9.

Mu, L., Liang Y. and Han R. (2014). Assessment of the soil organic carbon sink in a project for the conversion of farmland to forestland: A case study in Zichang County, Shaanxi, China. *PLoS ONE* 9(4):e94770.

Mulder, C., Ahrestani, F. S., Bahn, M., Bohan, D. A., Bonkowski, M., Griffiths, B. S. Guicharnaud, R., Kattge, J., Krogh, P., Lavorel, S., Lewis, O., Mancinelli, G., Naeem, S., Peñuelas, J., Poorter, H., Reich, P., Rossi, L., Rusch, G., Sardans, J. and Wright, I. J. (2013). Connecting the green and brown worlds: Allometric and stoichiometric predictability of above- and below-ground networks. *Advances in Ecological Research* 49:69–175.

Mulvaney, R. L., Khan, S. A. and Ellsworth, T. R. (2009). Synthetic nitrogen fertilizers deplete soil nitrogen: A global dilemma for sustainable cereal production. *Journal of Environmental Quality* 38(6):2295–314.

Muñoz, M. Á. and Zornoza, R. (2018). *Soil Management and Climate Change, First Edition. Effects on Organic Carbon, Nitrogen Dynamics, and Greenhouse Gas Emissions*. Academic Press, London, p. 396.

Nannipieri, P., Ascher, J., Ceccherini, M., Landi, L., Pietramellara, G. and Renella, G. (2003). Microbial diversity and soil functions. *European Journal of Soil Science* 54:655–70.

Nord, E. A. and Lynch, J. P. (2009). Plant phenology: A critical controller of soil resource acquisition. *Journal of Experimental Botany* 60(7):1927–37.

Olson, K. R., Al-Kaisi, M. M., Lal, R. and Lowery, B. (2014). Experimental consideration, treatments, and methods in determining soil organic carbon sequestration rates. *Soil Science Society of America Journal* 78(2):348–60.

Pang, X. P. and Letey, J. (1998). Development and evaluation of ENVIRO-GRO, an integrated water, salinity, and nitrogen model. *Soil Science Society of America Journal* 62(5):1418–27.

Paustian, K., Lehmann, J., Ogle, S., Reay, D., Robertson, G. P. and Smith, P. (2016). Climate-smart soils. *Nature* 532:49–57.

Petersen, S. O., Schjønning, P., Thomsen, I. K. and Christensen, B. T. (2008). Nitrous oxide evolution from structurally intact soil as influenced by tillage and soil water content. *Soil Biology and Biochemistry* 40(4):967–77.

Pettersson, M. and Bååth, E. (2003). Temperature-dependent changes in the soil bacterial community in limed and unlimed soil. *FEMS Microbiology Ecology* 45(1):13–21.

Poeplau, C. and Don, A. (2015). Carbon sequestration in agricultural soils via cultivation of cover crops – A meta-analysis. *Agriculture, Ecosystems and Environment* 200:33–41.

Poorter, L., Bongers, F., Aide, M. T., Zambrano, A. M., Balvanera, R., Becknell, J. M., Boukili, V., Brancalion, P. H. S., Broadbent, E. N. and Chazdon, R. L. (2016). Biomass resilience of Neotropical secondary forests. *Nature* 530:211–14.

Potter, S., Cabbage, M. and McCarthy, L. (2017). NASA, NOAA data show 2016 warmest year on record globally. https://www.nasa.gov/press-release/nasa-noaa-data-show-2016-warmest-year-on-record-globally (accessed 20 February 2018).

Poulter, B., Frank, D., Ciais, P., Myneni, R. B., Andela, N., Bi J., Broquet, G., Canadell, J. G., Chevallier, F., Liu, Y. Y., Running, S. W., Sitch, S. and van der Werf, G. R. (2014). Contribution of semi-arid ecosystems to interannual variability of the global carbon cycle. *Nature* 509:600–3.

Powlson, D. S., Stirling, C. M., Jat, M. L., Gerard, B. G., Palm, C. A., Sanchez, P. A. and Cassman, K. G. (2014). Limited potential of no-till agriculture for climate change mitigation. *Nature Climate Change* 4:678.

Powlson, D. S., Stirling, C. M., Thierfelder, C., White, R. P. and Jat, M. L. (2016). Does conservation agriculture deliver climate change mitigation through soil carbon sequestration in tropical agro-ecosystems? *Agriculture, Ecosystems and Environment* 220:164–74.

Preston, C. (2017). With the grain: Right conditions needed to maximise post-em effectiveness, 2017. https://grdc.com.au/news-and-media/news-and-media-releases/south/2017/07/withthegrain-right-conditions-needed-to-maximise-post-em-effectiveness (accessed 09 April 2018).

Putten, W. H. Van der. (2012). Climate change, aboveground-belowground interactions, and species' range shifts. *Annual Review of Ecology, Evolution, and Systematics* 43(1):365–83.

Qian, B., Gregorich, E. G., Gameda, S., Hopkins, D. W. and Wang, X. L. (2011). Observed soil temperature trends associated with climate change in Canada. *Journal of Geophysical Research* 116(D2). doi:10.1029/2010JD015012.

Rabbi, S. M. F., Tighe, M., Delgado-Baquerizo, M., Cowie, A., Robertson, F., Dalal, R., Page K., Crawford, D., Wilson, B. R., Schwenke, G., McLeod, M., Badgery, W., Dang, Y. P., Bell, M., O'Leary, G., Liu, D. L. and Baldock, J. (2015). Climate and soil properties limit the positive effects of land use reversion on carbon storage in Eastern Australia. *Scientific Reports* 5: 17866.

Radajewski, S., Ineson, P., Parekh, N. R. and Murrell, J. C. (2000). Stable-isotope probing as a tool in microbial ecology. *Nature* 403:646–9.

Rakkar, M. K. and Blanco-Canqui, H. (2018). Grazing of crop residues: Impacts on soils and crop production. *Agricultural Ecosystems and Environment* 258:71–90.

Raich, J. W. and Schlesinger, W. H. (1992). The global carbon dioxide flux in soil respiration and its relationship to vegetation and climate. *Tellus B* 44(2):81–99.

Reeves, D. W. (1997). The role of soil organic matter in maintaining soil quality in continuous cropping systems. *Soil and Tillage Research* 43(1–2):131–67.

Rengasamy, P. and Olsson, K. A. (1993). Irrigation and sodicity. *Soil Research* 31(6):821–37.

Rengel, Z. (2011). Soil pH, soil health and climate change. In *Soil Health and Climate Change* (Eds., Singh, B. P., Cowie, A. L. and Chan, K. Y.), Soil Biology Series (29):69–85. Springer-Verlag, Berlin.

Rhodes, C. J. (2014). Soil erosion, climate change and global food security: Challenges and strategies. *Science Progress* 97(Pt 2):97–153.

Rinnan, R., Michelsen A., Bååth E. and Jonasson S. (2007). Fifteen years of climate change manipulations alter soil microbial communities in a subarctic heath ecosystem. *Global Change Biology* 13(1):28–39.

Rinnan, R., Rousk, J., Yergeau, E., Kowalchuk, G. A. and Bååth, E. (2009). Temperature adaptation of soil bacterial communities along an Antarctic climate gradient: Predicting responses to climate warming. *Global Change Biology* 15(11):2615–25.

Rogers, C. E. and McCarty, J. P. (2000). Climate change and ecosystems of the mid-Atlantic Region. *Climate Research* 14(3):235–44.

Sá, J. C. de M., Cerri, C. C., Dick, W. A., Lal, R., Venske Filho, S. P., Piccolo, M. C. and Feigl, B. E. (2001). Organic matter dynamics and carbon sequestration rates for a tillage chronosequence in a Brazilian Oxisol. *Soil Science Society of America Journal* 65(5):1486–99.

Salem, H. M., Valero, C., Muñoz, M. Á., Rodríguez, M. G. and Silva, L. L. (2015). Short-term effects of four tillage practices on soil physical properties, soil water potential, and maize yield. *Geoderma* 237–238:60–70.

Sarker, J. R., Singh, B. P., He, X., Fang, Y., Li, G. D., Collins, D. and Cowie, A. L. (2017). Tillage and nitrogen fertilization enhanced belowground carbon allocation and plant nitrogen uptake in a semi-arid canola crop–soil system. *Scientific Reports* 7:10726.

Sarker, J. R., Singh, B. P., Cowie, A, L., Fang, Y., Collins, D., Dougherty, W. J. and Singh, B. K. (2018a). Carbon and nutrient mineralisation dynamics in aggregate-size classes from different tillage systems after input of canola and wheat residues. *Soil Biology and Biochemistry* 116:22–38.

Sarker, J. R., Singh, B. P. Dougherty, W. J., Fang, Y., Badgery, W., Hoyle, F. C., Dalal, R. C. and Cowie A. L. (2018b). Impact of agricultural management practices on the nutrient supply potential of soil organic matter under long-term farming systems. *Soil and Tillage Research* 175:71–81.

Sarker, J. R., Singh, B. P., Cowie, A. L., Fang, Y., Collins, D., Badgery, W. and Dalal, R. C. (2018c) Agricultural management practices impacted carbon and nutrient concentrations in soil aggregates, with minimal influence on aggregate stability and total carbon and nutrient stocks in contrasting soils. *Soil and Tillage Research* 178:209–23.

Saxe, H., Cannell, M. G. R., Johnsen, Ø., Ryan M. G. and Vourlitis, G. (2001). Tree and forest functioning in response to global warming. *New Phytologist* 149(3):369–99.

Schimel, J. P., Gulledge, J. M., Clein-Curley, J. S., Lindstrom, J. E. and Braddock, J. F. (1999). Moisture effects on microbial activity and community structure in decomposing birch litter in the Alaskan Taiga. *Soil Biology and Biochemistry* 31(6):831–8.

Schwartz, N., Uriarte, M., DeFries, R., Gutierrez-Velez, V. H. and Pinedo-Vasquez, M. A. (2017). Land-use dynamics influence estimates of carbon sequestration potential in Tropical Second-Growth Forest. *Environmental Research Letters* 12(7):74023.

Seneviratne, S. I., Phipps, S. J., Pitman, A. J., Hirsch, A. L., Davin, E. L., Donat, M. G., Hirschi, M., Lenton, A., Wilhelm, M. and Kravitz, B. (2018). Land radiative management as contributor to regional-scale climate adaptation and mitigation. *Nature Geoscience* 11:88–96.

Shepherd, T. G., Saggar, S., Newman, R. H., Ross, C. W. and Dando, J. L. (2001). Tillage-induced changes to soil structure and organic carbon fractions in New Zealand soils. *Australian Journal of Soil Research* 39(3):465–89.

Singh, B. P., Cowie, A. L. and Chan, K. Y. (2011). *Soil Health and Climate Change*, Soil Biology Series (29). Springer-Verlag, Berlin.

Singh, M., Sarkar, B., Biswas, B., Bolan, N. and Churchman, G. J. (2017). Relationship between soil clay mineralogy and carbon protection capacity as influenced by temperature and moisture. *Soil Biology and Biochemistry* 109:95–106.

Singh, B. P., Setia, R., Wiesmeier, M. and Kunhikrishnan, A. (2018). Agricultural management practices and soil organic carbon storage. In *Soil Carbon Storage, 1st Edition: Modulators, Mechanisms and Modeling* (Ed., Singh B.):207–44. Academic Press, London, UK.

Six, J., Elliott, E. T. and Paustian, K. (2000). Soil macroaggregate turnover and microaggregate formation: A mechanism for C sequestration under no-tillage agriculture. *Soil Biology and Biochemistry* 32(14):2099–103.

Six, J., Frey, S. D., Thiet, R. K. and Batten K. M. (2006). Bacterial and fungal contributions to carbon sequestration in agroecosystems. *Soil Science Society of America Journal* 70(2):555–69.
Smith, P. (2016). Soil carbon sequestration and biochar as negative emission technologies. *Global Change Biology* 22(3):1315–24.
Smith, P., House, J. I., Bustamante, M., Sobocká, J., Harper, R., Pan, G., West, P. C., Clark, J. M., Adhya, T. and Rumpel, C. (2016). Global change pressures on soils from land use and management. *Global Change Biology* 22(3):1008–28.
Sojka, R. E., Bjorneberg, D. L., Trout, T. J., Strelkoff, T. S. and Nearing, M. A. (2007). The importance and challenge of modelling irrigation-induced erosion. *Journal of Soil and Water Conservation* 62(3):153–62.
Sørensen, C. G., Halberg, N., Oudshoorn, F. W., Petersen, B. M. and Dalgaard, R. (2014). Energy inputs and GHG emissions of tillage systems. *Biosystems Engineering* 120:2–14.
Steinweg, J. M., Dukes, J. S., Paul, E. A. and Wallenstein, M. D. (2013). Microbial responses to multi-factor climate change: Effects on soil enzymes. *Frontiers in Microbiology* 4:146.
Stirling, G., Hayden, H., Pattison, T. and Stirling, M. (2016). *Soil Health, Soil Biology, Soil Borne Diseases and Sustainable Agriculture: A Guide*. CSIRO Publishing, Melbourne, 280.
Stockmann, U., Adams, M. A., Crawford, J. W., Field, D. J., Henakaarchchi, N., Jenkins, M., Minasny, B., McBratney, A. B., Courcelles, V. de R. de, Singh, K., Wheeler, I., Abbott, L., Angers, D. A., Baldock, J., Bird, M., Brookes, P. C., Chenu, C., Jastrow, J. D., Lal, R., Lehmann, J., O'Donnell, A. G., Parton, W. J., Whitehead, D. and Zimmermann, M. (2013). The knowns, known unknowns and unknowns of sequestration of soil organic carbon. *Agriculture Ecosystems and Environment* 164:80–99.
Swift, M. J., Heal, O. W. and Anderson, J. M. (1979). *Decomposition in Terrestrial Ecosystems*, vol. 5. University of California Press, Berkeley, CA.
Tellez-Rio, A., García-Marco, S., Navas, M., López-Solanilla, E., Tenorio, J. L. and Vallejo, A. (2015). N_2O and CH_4 emissions from a fallow-wheat rotation with low N input in conservation and conventional tillage under a mediterranean agroecosystem. *Science of the Total Environment* 508:85–94.
Thiele-Bruhn, S., Bloem, J., de Vries, F. T., Kalbitz, K. and Wagg, C. (2012). Linking soil biodiversity and agricultural soil management. *Current Opinion in Environmental Sustainability* 4(5):523–8.
Thiessen, S., Gleixner, G., Wutzler, T. and Reichstein, M. (2013). Both priming and temperature sensitivity of soil organic matter decomposition depend on microbial biomass – an incubation study. *Soil Biology and Biochemistry* 57:739–48.
Thornton, P. K., Ericksen, P. J., Herrero, M. and Challinor, A. J. (2014). Climate variability and vulnerability to climate change: A review. *Global Change Biology* 20(11):3313–28.
Titeux, N., Henle, K., Jean-Baptiste, M., Regos, A., Geijzendorffer, I. R., Cramer, W., Verburg, P. H. and Brotons, L. (2016). Biodiversity scenarios neglect future land-use changes. *Global Change Biology* 22(7):2505–15.
Trenberth, K. E. and Fasullo, J. T. (2012). Climate extremes and climate change: The Russian heat wave and other climate extremes of 2010. *Journal of Geophysical Research* 117:D17103.
Triplett, G. B. and Dick, W. A. (2008). No-tillage crop production: A revolution in agriculture! *Agronomy Journal* (100):S-153–65.
Turmel, M.-S., Speratti, A., Baudron, F., Verhulst, N. and Govaerts, B. (2015). Crop residue management and soil health: A systems analysis. *Agricultural Systems* 134:6–16.
UNEP (2013). The Emissions Gap Report. Report no. DEW/1742/NA. United Nations Environment Programme (UNEP), Nairobi. http://web.unep.org/sites/default/files/EGR2013/EmissionsGapReport_2013_high-res.pdf (accessed February 2018).
Ussiri, D. A. N., Lal, R. and Jarecki, M. K. (2009). Nitrous oxide and methane emissions from long-term tillage under a continuous corn cropping system in Ohio. *Soil and Tillage Research* 104(2):247–55.
Vandenkoornhuyse, P., Quaiser, A., Duhamel, M., Le Van, A. and Dufresne, A. (2015). The importance of the microbiome of the plant holobiont. *New Phytologist* 206(4):1196–206.

Wall, D. P., O'Sullivan, L., Creamer, R. and McLaughlin, M. J. (2018). Soil fertility and nutrient cycling. In *The Soils of Ireland* (Eds., Creamer, R. and O'Sullivan, L.):223–34. Springer, Cham.
Wang, J., Liu, X., Zhang, F., Lu, S., Zeng, X. and Cao, Y. (2002). The effect of different soil mulch materials on the growth and yield of rice. *Acta Ecologica Sinica* 22(6):922–9.
Ward, A., Sharpley, A., Miller, K., Dick, W., Hoorman, J., Fulton, J. and LaBarge, G. A. (2018). An assessment of in-field nutrient best management practices for agricultural crop systems with subsurface drainage. *Journal of Soil and Water Conservation*, 73(1):5A–10A.
Wieder, W. R. Cleveland, C. C., Kolby S. W. and Todd-Brown, K. (2015). Future productivity and carbon storage limited by terrestrial nutrient availability. *Nature Geoscience* 8:441–4.
Wild, A. (1993). Soils in relation to the environment. In *Soils and the Environment* (Ed., Wild, A.):107–8. Cambridge University Press, Cambridge.
Williams, M. A. (2007). Response of microbial communities to water stress in irrigated and drought-prone tallgrass prairie soils. *Soil Biology and Biochemistry* 39(11):2750–7.
World Bank (2016). The World Bank DataBank [WWW Document]. Data Bank. URL https://data.worldbank.org/indicator/AG.LND.ARBL.HA.PC?end=2014&start=2014&type=points&view=map (last accessed on 30 Nov 2017).
Xu, G.-L., Fu, S.-L., Schleppi, P. and Li, M.-H. (2013). Responses of soil Collembola to long-term atmospheric CO_2 enrichment in a mature temperate forest. *Environmental Pollution* 173:23–8.
Xu, Y., Seshadri, B., Sarkar, B., Wang, H., Rumpel, C., Sparks, D., Farrell, M., Hall, T., Yang, X. and Bolan, N. (2018). Biochar modulates heavy metal toxicity and improves microbial carbon use efficiency in soil. *Science of the Total Environment* 621:148–59.
Yan, X., Yagi, K., Akiyama, H. and Akimoto, H. (2005). Statistical analysis of the major variables controlling methane emission from rice fields. *Global Change Biology* 11(7):1131–41.
Zak, D. R., Holmes, W. E., MacDonald, N. W. and Pregitzer, K. S. (1999). Soil temperature, matric potential, and the kinetics of microbial respiration and nitrogen mineralization. *Soil Science Society of America Journal* 63(3):575–84.
Zaller, J. G., Caldwell, M. M., Flint, S. D., Ballaré, C. L., Scopel, A. L. and Sala, O. E. (2009). Solar UVB and warming affect decomposition and earthworms in a fen ecosystem in Tierra del Fuego, Argentina. *Global Change Biology* 15:2493–502.
Zhang, N., Liu, W., Yang, H., Yu, X., Gutknecht, J., Zhang, Z., Wan, S. and Ma, K. (2013). Soil microbial responses to warming and increased precipitation and their implications for ecosystem C cycling. *Oecologia* 173(3):1125–42.
Zhang, H. L., Lal, R., Zhao, X., Xue, J. F. and Chen, F. (2014). Opportunities and challenges of soil carbon sequestration by conservation agriculture in China. *Advances in Agronomy* 124:1–36.
Zhang, K., Shi, Y., Jing, X., He, J., Sun, R., Yang, Y., Shade, A. and Chu, H. (2016). Effects of short-term warming and altered precipitation on soil microbial communities in alpine grassland of the Tibetan Plateau. *Frontiers in Microbiology* 7:1032.
Zogg, G. P., Zak, D. R., Ringelberg, D. B., White, D. C., MacDonald, N. W. and Pregitzer, K. S. (1997). Compositional and functional shifts in microbial communities due to soil warming. *Soil Science Society of America Journal* 61(2):475–81.

Chapter 4

Integrated soil health management: a framework for soil conservation and regeneration

Daniel K. Manter and Jorge A. Delgado, USDA-ARS, USA; and Jennifer Moore-Kucera, USDA-NRCS, USA

1 Introduction

Soil degradation threatens global food production and security, clean water and air, and animal and human health. Degradation occurs in different forms and at different rates and can be categorized into three types – chemical, physical and biological – each of which does not exist individually but is the product of multiple interactions and feedback loops. Chemical deterioration primarily involves loss of nutrients and/or soil organic matter (SOM), salinization, acidification, pollution and eutrophication. Physical deterioration includes loss of soil, nutrients and SOM from wind and water erosion, compaction, sealing, crusting, water logging and loss of rooting depth. Biological deterioration is likely the least understood but critical to overall soil function and crop productivity and involves the loss of biodiversity, keystone taxa or consortia, and an overall reduction in efficient biological processes. Soil microbes play a direct role in driving multiple soil chemical and physical processes important for overall ecosystem function, but also have direct and indirect effects on plant productivity and quality. As a result, we suggest that soil conservation (i.e. protection and minimization of soil loss) and regeneration (i.e. rejuvenation of degraded soils) should focus not only on erosion and soil loss but also on the status of the soil

http://dx.doi.org/10.19103/AS.2017.0033.05
Published by Burleigh Dodds Science Publishing Limited, 2018.

biological community and its function. In other words, both are different forms of soil health management and although minimizing erosion is a worthy goal, optimum soil function and health remains paramount to plant productivity and sustainability.

2 The role of soil microbes in chemical and physical processes

Soil health and the critical role that soil biology plays in maintaining soil health are not new concepts (e.g. Kibblewhite et al., 2008; Magdoff, 2001) with two key concepts often emerging in the definition of soil health – its capacity to perform functions; and soil is a complex, vital living ecosystem harbouring a functionally and taxonomically diverse array of soil organisms (Kibblewhite, 2018). Healthy, high-functioning soils are characterized by a diverse and active community of organisms that maintain and drive key ecosystem functions. Important soil physical and chemical processes include the ability of the soil to form stable soil aggregates and SOM, as well as, cycle and recycle nutrients. The formation of soil aggregates is possible via the synergistic interactions between plant roots, fungal hyphae and microbial extracellular polysaccharides and proteins that help 'glue' soil into various sized aggregates (Lynch and Bragg, 1985; Six et al., 2004) important for erosion resistance, microbial habitat and air and water flow. SOM (plant residues and rhizodeposits, living/dead microbial biomass, detritus and humus) formation, and decomposition, and the flow of C through soil are key drivers of soil function that influence, and are influenced by, the soil biological community (Paterson et al., 2009; Kallenbach et al., 2016). Like SOM, soil microbes are a critical driver of the continuous supply of nutrients through both direct and indirect processes. Direct processes include the decomposition of SOM and the subsequent release of excess nutrients. Microbial biomass, both dead and living, is a source of nutrients for plants (Paungfoo-Lonhienne et al., 2010). Indirect processes include the chelation and solubility of macro- (e.g. Alori et al., 2017) and micro- (Ahmed and Holmström, 2014) nutrients for plant availability.

3 The role of soil microbes in plant processes

Soil microbes have a wide-ranging impact on plant health through a variety of mechanisms, and several reviews have been written on this subject (e.g. Babalola, 2010; Cummings, 2009; Esitken, 2011; Lugtenberg and Kamilova, 2009; Maheshwari, 2011). Briefly, these processes include: increasing resistance and resilience to pests, pathogens and weed pressure; forming beneficial relationships to aid plant productivity; recycling essential plant nutrients; storing and cycling carbon, CO_2, and other atmospheric gases important in climate change; and increasing resilience to environmental disturbances such as drought, fire, floods and temperature extremes. Together, these plant–microbial interactions help improve crop productivity. Much of our knowledge regarding the mechanisms of plant–microbial interactions has come from studies involving individual microbial strains; however, the mechanistic basis (i.e. key genes and their functions) continues to be elucidated (e.g. Saharan and Nehra, 2011). Further, these fundamental roles of soil organisms, along with the high costs of agricultural inputs, have resulted in a strong focus on using soil microbes to remediate degraded or contaminated soils (Cai et al., 2013; Patil, 2014; Vaxevanidou et al., 2015) and increase crop productivity.

4 Soil conservation: moving beyond tolerable soil loss (T) to integrated soil health management (ISHM)

Historically, erosion control has been the primary goal of conservation efforts in the USA. The Soil Conservation Service (now called the Natural Resource Conservation Service, NRCS) was formed in 1932 to address the deleterious impacts of the Dust Bowl in the Great Plains of the USA. Erosion contributes to the removal of soil particles, SOM and key nutrients, and can transport soil microbes off-site, negatively impacting soil productivity (Acosta-Martinez et al., 2015). It also reduces soil productivity between 4.3% and 29.6% per every 10 cm of soil loss, thus decreasing crop yields and soil quality (Bakker et al., 2004) in part due to reductions in soil organic carbon content, nutrient levels, soil water storage and soil aggregates. Although major achievements have been made through the adoption of various conservation practices to address soil erosion, it is still a major contributor to soil degradation and declining soil health (Adams, 1949; Andraski and Lowery, 1992; Bakker et al., 2004; Bramble-Brodahl et al., 1984; Cruse and Herndl, 2009; Follett and Stewart, 1985; Lal, 1995, 1998; Langdale et al., 1979; Mokma and Sietz, 1992; Quine and Zhang, 2002). An approach to assess acceptable erosion levels was proposed by Smith (1941) and Smith and Whitt (1948). This approach uses tolerable soil loss (T value), which estimates acceptable levels of erosion that maintain sustainable systems. At a 1956 conference (USDA, 1956), it was recommended that T values consider economic and physical information to estimate a value of erosion that is acceptable, considering the value of nitrogen and phosphorus (25 dollars per ha) and rate of soil loss versus rate of soil formation. Johnson (1987) reported that the use of the T value to allow annual erosion losses to be higher than the rate of formation is not sustainable in the long term due to the loss or degradation of the soil resource. Montgomery (2007) reported that with conventional agricultural practices, the use of T will never be sustainable as current rates of soil formation cannot replenish soil erosion losses (T values > 0); furthermore, T values based solely on fertility considerations (i.e. nutrient loss only) can never be maintained as long as soil is being lost from the system. Recently, Cox (2008) called for a more comprehensive assessment of management and conservation systems based on multiple production and environmental endpoints that do not consider only T values.

Delgado et al. (2013) reported on the need for a new framework that considers the impact on chemical, physical, biological and ecological degradation. Furthermore, Delgado et al. (2013) suggested that such a framework should consider '(1) the effects of soil erosion rates on the potential to lower soil productivity and the corresponding short- and long-term crop yield economic losses; (2) effects of soil quality, soil health, and ecology that are also closely related to soil productivity and thus maintaining long-term economic returns; (3) climatic feedback implications of soil carbon balance and the effects of soil erosion on off-site losses of soil organic matter as well as the effects of climate change and how conservation practices can contribute to maintain soil organic matter levels needed for long-term sustainability and soil quality; and (4) off-site sediment pollution potential relative to recreation, water supplies, wildlife, associated agrochemicals, costs of dredging, water quality'. In other words, soil regeneration and conservation should focus on promoting not only soil profile depths, but also soil biology and overall health. However, to our knowledge, no framework and/or assessment tools quantifying soil microbial diversity and/or functions are currently being used for soil conservation, restoration and/or management.

Since soil health is comprised of physical, chemical and biological features and each component is intricately linked and contributes to soil functions, we suggest all three soil health components be considered when designing plans for soil conservation and/or regeneration. With this concept in mind, we propose an integrated soil health management (ISHM, Fig. 1) framework for selecting appropriate management strategies. Similar to an integrated pest management programme, ISHM is an integrated approach that involves a potentially wide array of management strategies to not only achieve soil conservation and regeneration, but also provide a problem-driven approach to soil management. ISHM consists of the following five key steps (described in detail).

5 Integrated soil health management

5.1 ISHM step 1: knowledge

The likelihood for success in any management planning process will depend on a detailed knowledge of the agroecosystem being managed. Knowledge is comprised of two key components: the desired objectives for the management area and typical problems encountered on a given piece of land. For example, while erosion may be the primary

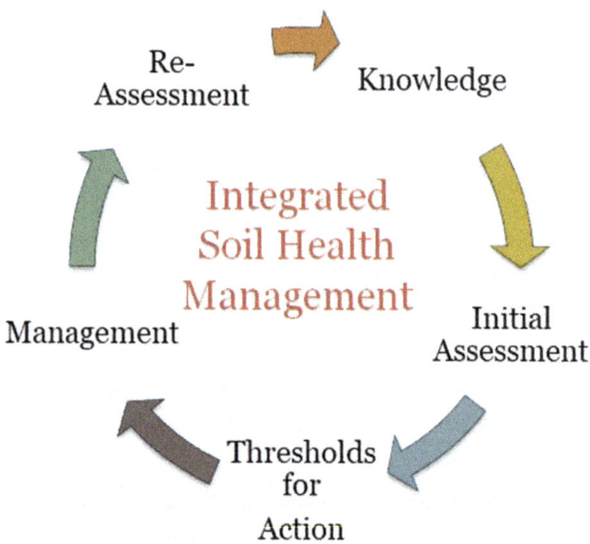

Figure 1 Flow diagram of integrated soil health management.

Published by Burleigh Dodds Science Publishing Limited, 2018.

concern on steep slopes or fine sandy soils, it may not be equally important in other locations where conservation agriculture is being implemented and/or the risk of soil erosion is low. Additionally, in dryland agriculture of the semi-arid western USA, soil management strategies should be designed placing critical importance on promotion of soil water holding capacity and beneficial bacteria and/or reduction of wind erosion.

As part of the knowledge gathering step, management should also gather information on how soils are currently managed and what, if any, best management practices are currently available. NRCS proposes four basic principles of managing for soil health: manage soils by disturbing them less, keep the soil covered as much as possible, use plant and animal diversity to increase soil biodiversity and keep plants growing throughout the year to feed the soil biology (USDA-NRCS 2016). In terms of soil regeneration and conservation, these principles, at a minimum, have the following management goals: (1) protecting the soil; (2) feeding the microbes (i.e. add carbon to the soil) and (3) promoting microbial biodiversity and functions.

5.1.1 Protecting the soil

As outlined above, early concepts of soil management have largely focused on protecting soil by minimizing erosion, as soil formation rates are quite low (global mean = 0.058 mm year^{-1}, Wakatsuki and Rasyidin, 1992). The relationship between soil productivity and erosion is not a simple one; however, Stocking (2003) showed a significant negative exponential relationship between crop yields and soil loss. Simplistically then, any measure that can minimize erosion should have a beneficial impact on crop productivity. Management strategies to prevent erosion have been extensively studied and reviewed (e.g. Montgomery, 2007) and can best be grouped into strategies that provide soil cover and/or minimize disturbance. For example, through the establishment of grasslands, forests or wetlands on marginal lands, CRP (Conservation Reserve Program) enrolment increased surface cover and by 2009 reduced soil erosion nationwide due to water and wind by 215 and 230 Tg, respectively (USDA-FSA, 2009). In a recent report (USDA, 2015), soil erosion on US croplands has decreased by 44% since 1982. Soil protection extends beyond reduction in erosion potential as soil aggregates also function as habitat for soil microorganisms and protection of organic carbon that can become entrapped within microaggregates (Six et al., 2000; McCarthy et al., 2008). Furthermore, protection of the soil surface aids in buffering against swings in temperature and reducing evaporation losses that increases water-use efficiency.

5.1.2 Feeding the microbes

Organic matter and soil carbon can not only directly influence soil physical properties but are also the food source for all heterotrophic organisms in the soil. SOM is a critical component of all aspects of soil health (Johnston, 1986) and is a complex mix of plant residues, microorganisms (living and dead) and the substrates produced by these organisms during decomposition. SOM influences the chemical, biological and physical properties of soil in ways that are almost universally beneficial to crop production, such as promoting soil water infiltration and retention (Franzluebbers, 2002), nutrient availability (Fontaine et al., 2003) and microbial biomass (Manter et al., 2016a). Plants are the carbon producers of the world converting CO_2 to sugars via photosynthesis. While we are usually focused on the aboveground conversion of these sugars to biomass and crop yield, much of this carbon will remain in the system after harvest as crop residues and rhizodeposition.

Recent research has demonstrated that root exudation is critical to mediating interactions between plants and other organisms, including insects, neighbouring plants and microorganisms (Badri et al., 2009; Badri and Vivanco, 2009; Broeckling et al., 2008). Various factors affect the quantity and quality of root exudates, including plant species, age and biotic and abiotic stressors, and N fertilizer (Zhu et al., 2016). Up to 50% of the carbon fixed by plants may be transferred to the soil rhizosphere (Lynch and Whipps, 1990) with root exudates primarily consisting of relatively labile carbon-based compounds (Bais et al., 2006; Uren, 2007), but released ions (such as H^+), inorganic acids, oxygen and water are also present. The organic compounds include low-molecular weight compounds, like amino acids, organic acids, sugars, phenolics and secondary metabolites, and higher-molecular weight compounds like mucilage and proteins. Given this complexity in both the quantity and quality of root exudation and its importance in regulating soil microbial communities, this is an area of research and management focus that can have great potential to modify soil microbial communities and soil health in a cost-effective and sustainable way.

5.1.3 Promoting microbial functions

As outlined above, soil microbe and microbial functions are the key drivers in determining both soil chemical and physical characteristics and plant structure and function. Key recent studies have shown the importance of understanding microbial functional diversity along with microbial species diversity in the rhizosphere (Nannipieri et al., 2003, 2008). Classic examples of beneficial associations between plant roots and soil microbes include nodulation by nitrogen-fixing bacteria (e.g. Phillips, 1980; Shridhar, 2012) and arbuscular mycorrhizae (e.g. Parniske, 2008). In addition, a wide variety of plant-growth-promoting rhizobacteria (PGPR) strains can promote plant growth and yield, disease resistance, drought tolerance and heavy metal remediation (Chen et al., 2013; Joo and Chang 2005; Karlidag et al., 2007; Khalid et al., 2004; Mayak et al., 2004; Sgroy et al., 2009). Despite the large number of studies showing the potential efficacy of PGPRs, field application of PGPRs is often hindered by poor survival rates (Acea et al., 1988) and inconsistent efficacy (Labuschagne et al., 2011). However, it is increasingly clear that these microbial strains are highly specific to plant species, cultivars and genotypes (Figueiredo et al., 2011; Gupta et al., 2000; Lucy et al., 2004; Siddiqui and Shaukat, 2003) and susceptible to the presence of competing microbes and soil factors such as temperature, water content, oxygen and pH (Dutta and Podile, 2010; Brimecombe et al., 2007; Hrynkiewicz et al., 2010). Furthermore, Maherali and Klironomos (2007) showed that plant productivity increased only when arbuscular mycorrhizal diversity spanned a range of functional groups, not taxonomic groups, based on spatial niches. Due to these difficulties, we suggest that not only additional studies are needed to evaluate the factors limiting successful PGPR colonization and survival, but also management strategies that promote specific functions within the natural microbial community to increase plant productivity and stress tolerance.

Identifying the appropriate functions, and methodology (e.g. proteomic, genomic, etc.) to assess these functions, is currently a very active area of research with, arguably, little consensus. However, several soil microbial enzymes (e.g. amylase, arylsulphatase, β-glucosidase, cellulase, chitinase, dehydrogenase, phosphatase and urease) have been proposed as useful indicators (e.g. Das and Varma, 2011). Further, a suite of common genes (nitrogenase, *nifHDK*; pyrroloquinoline quinone-encoding genes, *pqqBCDEFG*; indole-3-pyruvate decarboxylase/phenylpyruvate decarboxylase gene, *ipdC/ppdC*; nitrite reductase *nirK*, 1-aminocyclopropane-1-carboxylate deaminase, *acdS*; acetoin genes *budAB*; hydrogen cyanide genes, *hcnABC*; 2,4-diacetylphloroglucinol, *phlABCD*) associated with the positive effects of root-colonizing bacteria has recently emerged; while many PGPR strains may contain more than one gene, none are common to a single isolate (Bruto et al., 2014). This redundancy of these genes across divergent taxa, and often unique gene combinations within taxa, suggests that PGPRs work in cooperation, and a diversity of taxonomic groups contribute to overall optimum soil health.

5.2 ISHM step 2: initial assessment

An important consideration for any management strategy is to determine the current state of the ecosystem to be managed. Since soil health is defined by both its continued capacity to function and the cumulative interaction of chemical, physical and biological processes, the selection of appropriate indicators is not trivial. There is a general consensus in the scientific community that a single indicator may not be sufficient to quantify soil health. For example, the Comprehensive Assessment of Soil Health or CASH (Moebius-Clune et al., 2016) and the Soil Management Assessment Framework or SMAF (Andrews et al., 2004) include a variety of chemical, physical and biological indicators. In addition, depending upon the management objectives and differences in regional problems – due to climate, crop selection and soil types – differing indicators may be desirable. Furthermore, the use of multiple indicators may be best suitable for identifying individual soil health characteristics that are especially low, most easily targeted for regeneration or address specific problems.

The assessment, and understanding, of the contribution of soil biology to soil health has lagged behind soil chemical and physical processes, in large part, due to inadequate tools to identify biological members and their functions (e.g. Lehman et al., 2015). However, it is becoming increasingly clear that all of the attributes of soil quality, as defined by Doran and Parkin (1994), are influenced by soil biology. This includes multiple chemical and physical attributes commonly used in soil health assessments (organic matter, nutrient release, water flow, storage and purity, resistance to degradation, etc.). Thus, to truly reverse soil degradation and continue to produce enough food, fibre and fuel for our growing population, we must enhance and understand soil biology.

Key challenges to identification and adoption of methods for microbial community composition and structure include a shift of how soil is sampled, transported and stored and a paucity of interpretable or actionable information. Thresholds or ranges to define relative poor/good functioning could be established if standard procedures are followed; however, they currently have not been widely adopted, although some efforts have been initiated as part of the Earth Microbiome Project (e.g. Caporaso et al., 2011, 2012). Furthermore, beyond method (Manter and Vivanco, 2007; Manter et al., 2010b) and analytical tool (Manter and Bakker, 2015; Manter et al., 2016b) development, we need to continue to work on quantifying those indicators that capture appropriate microbial functions, current baselines in agricultural systems, the natural bounds of indicators and the management factors that influence each indicator. Manter et al. (2017) documented the need for standardized sample handling, storage and analysis of soil samples for current and future agricultural needs, particularly, in light, of ever-changing technology.

5.3 ISHM step 3: thresholds for action

The concept of thresholds for action is a natural extension of the initial assessment step of ISHM. This concept is ingrained in such practices as nutrient management, where fertilizer regimes are often based on the status of nutrients currently in the soil. By extension, a similar strategy should be applied to managing for soil regeneration, conservation and health. For many soil characteristics, a variety of indicator curves [i.e. scoring functions ranging from 0 (low) to 1 (optimum)] have been previously identified (Wienhold et al., 2009). However, due to potential upper limits on many soil characteristics (e.g. soil carbon, Stewart et al., 2007; Bernal et al., 2016) and/or significant thresholds before changes in soil functions are observed (e.g. Fig. 1b in Wienhold et al., 2009), significant changes in soil health may not be easily achievable or cost-effective. It is also important to consider thresholds for action based on expected variability in the system. For example, what happens when new biotic and/or abiotic stresses are introduced into the system?

5.4 ISHM step 4: management

Once appropriate soil characteristics to be targeted are identified, the next step is to identify appropriate strategies to target the desired changes. For a variety of indicators, several options may be available; however, in some cases, such as stimulating specific microbial groups, little information is currently available. Like any management decision, the strategies selected must coincide with the overall management objectives, may be location and/or regionally specific; in some cases, multiple actions may need to be considered.

In the next few sections, we discuss various management options and their potential use in soil conservation and regeneration. In general, management strategies should be selected using the previous ISHM principles outlined above and may differ depending upon whether the goal is to conserve soil (minimize soil erosion and maximize soil depth), regenerate soil, maximize yield or restore optimum soil health and function. Where possible we provide examples of those soil health characteristics known to be associated with each management strategy; however, in the case of some soil characteristics (i.e. soil microbial communities), information is often lacking and may require additional assessments and/or learning as part of the soil management process.

5.4.1 Crop residue management

Crop residue management has been reported by several scientists to have the potential to improve soil health, carbon sequestration and soil fertility, and to have the potential for climate change adaptation (Cruse and Herndl, 2009; Delgado, 2010; Johnson et al., 2010; Lal, 2004; Newman et al., 2010). As reviewed by Turmel et al. (2015), crop residue has been shown to benefit nearly all aspects of soil health, including chemical (soil organic carbon, pH and cation exchange capacity), physical (soil structure, run-off, erosion, compaction, soil temperature and moisture content) and biological properties (biodiversity and biomass) presumably, due to increased soil organic carbon. The retention of organic carbon in the system, and potential benefit on soil health, is driven by both total residue input and decomposition rates (Paustian et al., 1997), which are also influenced by other management factors (e.g. tillage), soil type and climate (Govaerts et al., 2009). As a result, soil organic carbon levels may have regional limits that should be considered when determining if crop residue management will be effective in promoting soil health.

5.4.2 Conservation tillage

Conservation tillage, such as no-till or minimum tillage, is a practice aimed to minimize physical disturbance to soil and thus reduce soil erosion losses by maintaining crop residues at the surface. Organic matter levels and thus carbon content increase in surface soils, providing food and energy sources to soil organisms, which will also facilitate the formation of soil aggregates that are important in erosion control. Several researchers over the last decade have reported on the importance of considering deep carbon sequestration when evaluating the effects of conservation tillage and conducting carbon balances (Baker et al., 2007; Blanco-Canqui and Lal, 2008; Six et al., 2002; VandenBygaart et al., 2011; West and Post, 2002). VandenBygaart et al. (2011) found in western Canada that no-till had higher soil organic carbon in the soil profile compared to conventional tillage, while in eastern Canada, there was no difference in the soil profile. Independent of these differences that could be found for some specific sites, conservation agriculture contributes to increased soil carbon sequestration at the soil surface, reduced erosion and off-site transport of SOM, and improved soil health.

5.4.3 Crop rotations/cover crops

Crop rotations/cover crops can be used to manage agroecosystems and reduce soil erosion, and they can even result in higher yields and improved soil health, especially if cover crops are included in the rotation. Like many of the management options covered in this chapter, much of the benefit of cover crops can be related to adding a carbon source to the soil, leading to increased soil biology activity and improved soil health. Furthermore, summer cover crops in Colorado can contribute to increased nitrogen use efficiencies and reduced losses of reactive nitrogen while increasing yields (Delgado et al., 2007; Essah et al., 2012). They can also improve water quality by reducing nitrate leaching or mining nitrate from groundwater (Meisinger and Delgado, 2002; Delgado, 1998; Delgado et al., 1999, 2001; Dabney et al., 2001). Improving grazing management of rangeland also aids in carbon sequestration (Eagle et al., 2010). Crop conversion to natural systems can contribute to carbon sequestration when native C inputs exceed crop productivity (Eagle et al., 2010). Other practices that can lead to reduced erosion are grass waterways, buffers and riparian buffers, and should contribute to increased carbon sequestration, but additional information about their effects on soil health is needed.

As suggested above, the strong link between plants and microbial community structure and function represents a unique opportunity for selecting and promoting microbial functions in the soil. Perhaps the best known symbiotic relationship is that between legumes and nitrogen-fixing bacteria. Individual cover crops have also been shown to promote particular functional groups, such as oats for arbuscular mycorrhizal fungi (Finney et al., 2017) and *Brassicaceae* spp. for biocontrol (Delgado et al., 2007; Bryant et al., 2014; O'Neill, 2016). Additionally, differences associated with varieties and clones within a crop species also promote different microbial taxa and functions (Barnett et al., 2015; Manter et al., 2010a; Stromberger et al., 2017).

5.4.4 Nutrient applications

Nutrient applications as they relate to crop productivity are perhaps the most common management decision related to plant productivity; however, it is becoming increasingly clear that such nutrient additions influence soil health and microbial community structure and function. For example, long-term fertilization enhanced microbial biomass from 8% to 15% primarily due to increased soil C content (Geisseler and Scow, 2014); however, most studies have not evaluated the impact on community composition or functional capacity of the community. Furthermore, many studies have shown limited responses of soil microbial biomass to increasing N rate (Lupwayi et al., 2011, 2012) or even declines (Biederbeck et al., 1996; Liebig et al., 2002; Stewart et al., 2015) due to a significant decrease in soil pH with N fertilizer. Further, numerous studies have shown the negative impact of fertilizer concentrations on symbiotic associations between plants and mycorrhizal fungi (Treseder, 2004), nitrogen-fixing bacteria abundance and activity (Weese et al., 2015; Klinger et al., 2016), and nutrient availability due to microbe–plant competition (Zhu et al., 2016).

5.4.5 Soil amendments

Soil amendments such as biochar or manure can be used as direct replacement of both carbon and nutrients to the soil (Wright et al., 1998). For example, Mikkha et al. (2017) reported that manure and tillage can be used to remediate eroded soils and improve the quality of eroded soils. Biochar increases soil carbon stocks and microbial communities while mitigating greenhouse gas (GHG) emissions (Lone et al., 2015). Biochar amendment to soils is proposed as a means to increase soil organic carbon stocks with a recalcitrant form of carbon, but impacts on soil microbes and GHGs are often site- and biochar-specific, making responses difficult to predict (Lehmann et al., 2011). Compost additions have largely been related to increases in microbial biomass and diversity (Garcia-Gil et al., 2000); however, it can also be used to affect the community and/or function. For example, Carrera et al. (2007) showed that the resultant microbial communities will differ depending upon whether the compost is manure- or plant-based.

5.5 ISHM step 5: reassessment

The reassessment phase is essentially the learning phase of the management framework proposed. In this step, monitoring of the current status of the soil, and identification of any changes associated with management, leads to a better understanding of soil processes and helps define uncertainties in the system, and can be used to design future management directions. In some adaptive management models, a double-loop structure has been suggested (e.g. Williams and Brown, 2014), whereby the initial steps (i.e. steps 1–3 as outlined here) may be skipped, and new management strategies effectively implemented. However, as management objectives may change over time (step 1) and the threshold for action (step 3) may no longer apply after the first round of management, the entire ISHM framework should be considered for any management decisions being considered. It is also important to note that reassessment may also need to be considered over multiple time periods. For example, ecosystem resilience encompasses both resistance – the ability to buffer a system against negative changes in structure or function – and recovery – the speed of return following disturbance (Côté and Darling, 2010). And in either case, temporal changes associated with disturbance (i.e. management decisions or natural) may require multiple time points to reflect the true long-term changes and resilience of the system.

6 Future trends and conclusion

In summary, as Cox (2008) reported, there is a need to go 'beyond T' to develop sustainable systems. Delgado et al. (2013) reported that we need to integrate T values with more complex assessments to prevent chemical, physical and biological degradation, so that we could develop sustainable systems. If we are to achieve soil conservation and regeneration, there is a need to move forward in applying a comprehensive analysis with an ISHM framework that not only aims to eliminate soil erosion, but also integrates all physical, chemical and biological aspects of soil health.

The concepts, and the adaptive management-based framework, contained within ISHM are an extension of current soil health assessment frameworks, such as the CASH (Moebius-Clune et al., 2016). However, we suggest that soil management, conservation and/or regeneration will only be truly successful by careful knowledge, assessment and management of the soil biological community and its structure and function that link soil health and plant productivity. For example, recent ecological system restoration efforts have shown that the soil microbial community is a key component in plant survival and successful restoration efforts (Harris, 2009). Furthermore, soil microbes not only promote ecosystem restoration through their continued function, but can also have broader ecosystem impacts, such as shifting plant community structure (Wubs et al., 2016). Microbial communities can also significantly mediate biotic (herbivory, Badri et al., 2013) and abiotic (drought, Zolla et al., 2013) stresses. By assessing microbial functions, a targeted, or problem-driven, management approach becomes more readily achievable. This proposed approach is not intended to replace the chemical and physical centric approach that currently dominates many soil assessment approaches. Rather, we propose this in addition to assessments, such as CASH (Moebius-Clune et al., 2016) or SMAF (Andrews et al., 2004) and a detailed assessment of the biological community, how to manage for C quality, root exudation and specific soil functions. This opens up the possibility for problem-specific management of soils based on the underlying biology.

7 Abbreviations

ISHM	Integrated soil health management.
SOM	Soil organic matter.
T value (T factor)	Tolerable soil loss or soil loss tolerance (tons per acre) is the maximum amount of soil loss that can be tolerated and still permit a high level of crop productivity.

8 Where to look for further information

The field of soil health, particularly as it relates to soil conservation and regeneration, is a rapidly growing field, with public and private entities conducting a variety of soil censuses and long-term studies to better explore the role of soil biology in soil health. The belowground environment contains a tremendous number and assortment of microscopic and macroscopic organisms that interact within a varied and dynamic matrix – that together exert influences on agriculture and rangeland productivity. The task of understanding what organisms are present and active in soils is a daunting challenge, but developing this understanding will help producers better manage their lands for maintaining or improving productivity, resiliency and sustainability. Such a gain in understanding will provide more information regarding 'who' is present, 'what' they are doing, 'how' they are doing it, as well as how they are interacting with each other. This understanding of a census will provide better information and knowledge as to how the belowground community interacts positively and negatively (or unremarkably) with the abiotic components of the belowground environment as well as components of the larger ecosystem, and will provide clues for how to manage soil ecosystems to meet agricultural/rangeland goals. Such efforts at understanding the diversity of soil biology and its contribution to soil structure and function are paramount to our future ability to not only keep soil in place but also maintain the functions necessary for a healthy and vibrant soil that maintains ecosystem services and plant productivity.

Selected groups performing soil-health-related research include:

1. African Soil Microbiology Project: http://blog.globalsoilbiodiversity.org/article/2016/10/20/african-soil-microbiology-project (summary)
2. AgBiome: http://agbiome.com/whoisagbiome/
3. E. O. Wilson Biodiversity Laboratory at Gorongosa National Park, Mozambique: https://eowilsonfoundation.org/the-biodiversity-collection/
4. Earth Microbiome Project: http://www.earthmicrobiome.org/
5. Forest Inventory and Analysis: https://www.fia.fs.fed.us/
6. Foundation for Food and Agriculture Research: https://foundationfar.org/
7. Indigo Agriculture: https://www.wired.com/2016/10/farmers-manipulating-microbiomes-help-grow-crops/
8. Microbiome Initiative: https://obamawhitehouse.archives.gov/the-press-office/2016/05/12/fact-sheet-announcing-national-microbiome-initiative
9. National Ecological Observatory Network (NEON): http://www.neonscience.org/observatory
10. Noble Research Institute: https://www.noble.org/
11. Soil Heath Institute: https://soilhealthinstitute.org/
12. TerraGenome: http://www.terragenome.org/
13. The Agricultural Microbiome Research Coordination Network: https://twin-cities.umn.edu/news-events/unlocking-secrets-plant-microbes
14. Trace Genomics: https://www.tracegenomics.com/#/
15. USDA-ARS: https://www.ars.usda.gov
16. USDA-NRCS: https://www.nrcs.usda.gov/wps/portal/nrcs/site/national/home/

9 References

Acea, M. J., C. R. Moore and M. Alexander. 1988. Survival and growth of bacteria introduced into soil. *Soil Biol. Biochem.* 20:509–15.

Acosta-Martinez, V., S. Van Pelt, J. Moore-Kucera, M. C. Baddock and T. M. Zobeck. 2015. Microbiology of wind-eroded sediments: Current knowledge and future research directions. *Aeolian Res.* 18:99–113.

Adams, W. E. 1949. Loss of topsoil reduced crop yield. *J. Soil Water Conserv.* 4:130.

Ahmed, E. and S. J. M. Holmström. 2014. Siderophores in environmental research: Roles and applications. *Microb. Biotechnol.* 7(3):196–208.

Alori, E. T., B. R. Glick and O. O. Babalola. 2017. Microbial phosphorus solubilization and its potential for use in sustainable agriculture. *Front. Microbiol.* 8:971.

Andraski, B. J. and B. Lowery. 1992. Erosion effects on soil water storage, plant water uptake, and corn growth. *Soil Sci. Soc. Am. J.* 56:1911–19.

Andrews, S. S., D. L. Karlen and C. A. Cambardella. 2004. The soil management assessment framework: A quantitative soil quality evaluation method. *Soil Sci. Soc. Am. J.* 68:1945–62.

Babalola, O. O. 2010. Beneficial bacteria of agricultural importance. *Biotechnol. Lett.* 32:1559–70.

Badri, D. V. and J. M. Vivanco. 2009. Regulation and function of root exudates. *Plant Cell Environ.* 32:666–81.

Badri, D. V., N. Quintana, E. G. El Kassis, H. K. Kim, Y. H. Choi, A. Sugiyama, R. Verpoorte, E. Marinoia, D. K. Manter and J. M. Vivanco. 2009. An aBC transporter mutation alters root exudation of phytochemicals that provoke and overhaul of natural soil microbiota. *Plant Phys.* 151:2006–17.

Badri, D. V., G. Zolla, M. G. Bakker, D. K. Manter and J. M. Vivanco. 2013. Potential impact of soil microbiomes on the leaf metabolome and on herbivore feeding behavior. *New Phytol.* 198:264–73.

Bais, H. P., T. L. Weir, L. G. Perry, S. Gilroy and J. M. Vivanco. 2006. The role of root exudates in rhizosphere interactions with plants and other organisms. *Annu. Rev. Plant Biol.* 57:233–66.

Baker, J. M., T. E. Oschner, R. T. Venterea and T. J. Griffis. 2007. Tillage and soil carbon sequestration – what do we really know? *Agric. Ecosys. Environ.* 118:1–5.

Bakker, M. M., G. Govers and M. D. A. Rounsevell. 2004. The crop productivity-erosion relationship: An analysis based on experimental work. *Catena* 57:55–76.

Barnett, B. A., D. G. Holm, J. W. Koym, R. G. Wilson and D. K. Manter. 2015. Site and clone effects on the potato root-associated core microbiome and its relationship to tuber yield and nutrients. *Am. J. Potato Res.* 92(1):1–9.

Bernal, B., D. C. McKinley, B. A. Hungate, P. M. White, T. J. Mozdzer and J. P. Megonigal. 2016. Limits to soil carbon stability; deep, ancient soil carbon decomposition stimulated by new labile organic inputs. *Soil Biol. Biochem.* 98:85–94.

Biederbeck, V. O., C. A. Campbell, H. Ukrainetz, D. Curtin and O. T. Bouman. 1996. Soil microbial and biochemical properties after ten years of fertilization with urea and anhydrous ammonia. *Can. J. Soil Sci.* 76:7–14.

Blanco-Canqui, H. and R. Lal. 2008. No-tillage and soil-profile carbon sequestration: An on-farm-assessment. *Soil Sci. Soc. Am. J.* 72:693–701.

Bramble-Brodahl, M., M. A. Fosberg, D. J. Walker and A. L. Falen. 1984. Changes in soil productivity related to changing topsoil depth on two Idaho Palouse soils. *Erosion and Soil Productivity: Proceedings of the National Symposium on Erosion and Soil Productivity.* 10–11 December 1984, Hyatt Regency, New Orleans, LA and American Society of Agricultural Engineering, New Orleans, LA.

Brimecombe, M. J., F. A. A. M. De Leij and J. M. Lynch. 2007. Rhizodeposition and microbial populations. In: R. Pinton, Z. Varanini and P. Nannipieri (Eds), *The Rhizosphere: Biochemistry and Organic Substances at the Soil-plant Interface.* CRC Press, Boca Raton, FL, pp. 73–100.

Broeckling, C. D., A. K. Broz, J. Bergelson, D. K. Manter and J. M. Vivanco. 2008. Root exudates regulate soil fungal community composition and diversity. *Appl. Environ. Microbiol.* 74:738–44.

Bruto, M., C. Prigent-Combaret, D. Muller and Y. Moenne-Loccoz. 2014. Analysis of genes contributing to plant-beneficial functions in plant-growth-promoting rhizobacteria and related Proteobacteria. *Sci. Rep.* 4:6261.

Bryant, A., T. Coudron, D. Brainard and Z. Szendrel. 2014. Cover crop mulches influence biological control of the imported cabbageworm (*Pieris rapae* L., Lepidoptera: Pieridae) in cabbage. *Biol. Con.* 73:75–83.

Cai, M., J. Yao, H. Yang, R. Wang and K. Massakorala. 2013. Aerobic biodegradation process of petroleum and pathway of main compounds in water flooding well of Dagang oil field. *Bioresour. Technol.* 144:100–6.

Caporaso, J. G., C. L. Lauber, W. A. Walters, D. Berg-Lyons, C. A. Lozupone, P. J. Turnbaugh, N. Fierer and R. Knight. 2011. Global patterns of 16S rRNA diversity at a depth of millions of sequences per sample. *Proc. Natl. Acad. Sci. USA* 108:4516–22.

Caporaso, J. G., C. L. Lauber, W. A. Walters, D. Berg-Lyons, J. Huntley, N. Fierer, S. M. Owens, J. Betley, L. Fraser, M. Bauer, N. Gormley, J. A. Gilbert, G. Smith and R. Knight. 2012. Ultra-high throughput microbial community analysis on the Illumina HiSeq and MiSeq platforms. *ISME J.* 6(8):1621–4.

Carrera, L. M., J. S. Buyer, B. Vinyard, A. A. Abdul-Baki, L. J. Sikora and J. R. Teasdale. 2007. Effects of cover crops, compost, and manure amendments on soil microbial community structure in tomato production systems. *Appl. Soil Ecol.* 37(3):247–55.

Chen, Y., F. Yan, Y. Chai, H. Liu, R. Kolter, R. Losick and J. H. Guo. 2013. Biocontrol of tomato wilt disease by *Bacillus subtilis* isolates from natural environments depends on conserved genes mediating biofilm formation. *Environ. Microbiol.* 15:848–64.

Côté, I. M. and E. S. Darling. 2010. Rethinking ecosystem resilience in the face of climate change. *PLoS Biol.* 8(7):e1000438.

Cox, C. A. 2008. Beyond T: Guiding sustainable soil management. *J. Soil Water Conserv.* 63:162A–4A.

Cruse, R. M. and C. G. Herndl. 2009. Balancing corn stover harvest for biofuels with soil and water conservation. *J. Soil Water Conserv.* 64:286–91.

Cummings, S. P. 2009. The application of plant growth promoting rhizobacteria (PGPR) in low input and organic cultivation of graminaceous crops; potential and problems. *Environ. Biotechnol.* 5:43–50.

Dabney, S. M., J. A. Delgado and D. W. Reeves. 2001. Using winter cover crops to improve soil and water quality. *Comm. Soil Sci. Plant Anal.* 32:1221–50.

Das, S. K. and A. Varma. 2011. Role of enzymes in maintaining soil health. In: G. Shukla and A. Varma (Eds), *Soil Enzymology, Soil Biology*, vol. 22. Springer-Verlag, Berlin, pp. 25–42.

Delgado, J. A. 1998. Sequential NLEAP simulations to examine effect of early and late planted winter cover crops on nitrogen dynamics. *J. Soil Water Conserv.* 53:241–4.

Delgado, J. A. 2010. Crop residue is a key for sustaining maximum food productivity and for conservation of our biosphere. *J. Soil Water Conserv.* 65:111A–16A.

Delgado, J. A., R. T. Sparks, R. F. Follett, J. L. Sharkoff and R. R. Riggenbach. 1999. Use of winter cover crops to conserve soil and water quality in the San Luis Valley of South Central Colorado. In: R. Lal (Ed.), *Soil Quality and Soil Erosion*. CRC Press, Boca Raton, FL, pp. 125–42.

Delgado, J. A., R. R. Riggenbach, R. T. Sparks, M. A. Dillon, L. M. Kawanabe and R. J. Ristau. 2001. Evaluation of nitrate-nitrogen transport in a potato-barley rotation. *Soil Sci. Soc. Am. J.* 65:878–83.

Delgado, J. A., M. A. Dillon, R. T. Sparks and S. Y. C. Essah. 2007. A decade of advances in cover crops: Cover crops with limited irrigation can increase yields, crop quality, and nutrient and water use efficiencies while protecting the environment. *J. Soil Water Conserv.* 62:110A–17A.

Delgado, J. A., M. A. Nearing and C. W. Rice. 2013. Conservation practices for climate change adaptation. *Adv. Agron.* 121:47–115.

Doran, J. W. and T. B. Parkin. 1994. Defining and assessing soil quality. In: J. W. Doran, D. C. Coleman, D. F. Bezdicek and B. A. Stewart (Eds), *Defining Soil Quality for a Sustainable Environment*. Soil Science Society of America, Madison, WI, pp. 1–21.

Dutta, S. and A. R. Podile. 2010. Plant growth promoting rhizobacteria (PGPR): The bugs to debug the root zone. *Crit. Rev. Microbiol.* 36:232–44.

Eagle, A. J., L. R. Henry, L. P. Olander, K. Haugen-Kozyra, N. Millar and G. P. Robertson. 2010. Greenhouse gas mitigation potential of agricultural land management in the United States: A synthesis of the literature. Technical Working Group on Agricultural Greenhouse Gases (T-AGG) Report. Nicholas Institute for Environmental Policy Solutions, Duke University, Durham, NC.

Esitken, A. 2011. Use of plant growth promoting rhizobacteria in horticultural plants. In: D. K. Maheshwari (Ed.), *Bacteria in Agrobiology: Crop Ecosystems*. Springer, Berlin, pp. 189–235.

Essah, S. Y. C., J. A. Delgado, M. Dillon and R. Sparks. 2012. Cover crops can improve potato tuber yield and quality. *HortTechnology* 22:185–90.

Figueiredo M., L. Seldin, F. Araujo and R. Mariano. 2011. Plant growth promoting rhizobacteria: Fundamentals and applications. In: D. K. Maheshwari (Ed.), *Plant Growth and Health Promoting Bacteria*. Springer, Berlin Heidelberg, pp. 21–43.

Finney, D. M., J. S. Buyer and J. P. Kaye. 2017. Living cover crops have immediate impacts on soil microbial community structure and function. *J. Soil Water Conserv.* 72(4):361–73.

Follett, R. F. and B. A. Stewart. (Eds). 1985. *Soil Erosion and Crop Productivity*. ASA, Madison, WI.

Fontaine, S., A. Mariotti and L. Abbadie. 2003. The priming effect of organic matter: A question of microbial competition? *Soil Biol. Biochem.* 35:837–43.

Franzluebbers, A. J. 2002. Water infiltration and soil structure related to organic matter and its stratification with depth. *Soil Till. Res.* 66(2):197–205.

Garcia-Gil, J. C., C. Plaza, P. Soler-Rovira and A. Polo. 2000. Long-term effects of municipal solid waste compost application on soil enzyme activities and microbial biomass. *Soil Biol. Biochem.* 32(13):1907–13.

Geisseler, D. and K. M. Scow. 2014. Does long-term use of mineral fertilizers affect the soil microbial biomass? *Better Crops* 98(4):13–15.

Govaerts, B., N. Verhulst, A. Castellanos-Nevarrete, K. D. Sayre, J. Dixon and L. Dendooven. 2009. Conservation agriculture and soil carbon sequestration: Between myth and famer reality. *Crit. Rev. Plant Sci.* 28:97–122.

Gupta, A., M. Gopal and K. V. Tilak. 2000. Mechanism of plant growth promotion by rhizobacteria. *Indian J. Exp. Biol.* 38:856–62.

Harris, J. 2009. Soil microbial communities and restoration ecology: Facilitators or followers? *Science* 325:573–4.

Hrynkiewicz, K., C. Baum and P. Leinweber. 2010. Density, metabolic activity, and identity of cultivable rhizosphere bacteria on *Salix viminalis* in disturbed arable and landfill soils. *J. Plant Nutr. Soil Sci.* 173:747–56.

Johnson, L. C. May-June 1987. Soil loss tolerance: Fact or myth? *J. Soil Water Conserv.* 42(3):155–60.

Johnson, J. M. F., D. L. Karlen and S. S. Andrews. 2010. Conservation considerations for sustainable bioenergy feedstock production: If, what, where, and how much? *J. Soil Water Conserv.* 65:88A–91A.

Johnston, A. E. 1986. Soil organic matter, effects on soils and crops. *Soil Use Manage.* 2(3):97–105.

Joo, H. S. and C. S. Chang. 2005. Production of protease from a new alkalophilic *Bacillus* sp. I-312 grown on soybean meal: Optimization and some properties. *Process Biochem.* 40:1263–70.

Kallenbach, C. M., S. D. Frey and A. S. Grandy. 2016. Direct evidence for microbial-derived soil organic matter formation and its ecophysiological controls. *Nat. Commun.* 7:13630.

Karlidag, H., A. Esitken, M. Turan and F. Sahin. 2007. Effects of root inoculation of plant growth promoting rhizobacteria (PGPR) on yield, growth and nutrient element contents of leaves of apple. *Sci. Hort.* 114:16–20.

Khalid, A., M. Arshad and Z. A. Zahir. 2004. Screening plant growth promoting rhizobacteria for improving growth and yield of wheat. *J. Appl. Microbiol.* 96:473–80.

Kibblewhite, M. G. 2018. Soil and soil health: An overview (evolution and definitions of soil health concepts). In: D. Reicosky (Ed.), *Managing Soil Health for Sustainable Agriculture*. Burleigh Dodds, Cambridge, UK.

Kibblewhite, M. G., K. Ritz and M. J. Swift. 2008. Soil health in agricultural systems. *Phil. Trans. R. Soc. B* 363:685–701.

Klinger, C. R., J. A. Lau and K. D. Heath. 2016. Ecological genomics of mutualism decline in nitrogen-fixing bacteria. *Proc. R. Soc. B* 283:20152563.

Labuschagne, N., T. Pretorius, A. H. Idris. 2011. Plant growth promoting rhizobacteria as biocontrol agents against soil-borne plant diseases. In: D. K. Maheshwari (Ed.), *Plant Growth and Health Promoting Bacteria*. Springer, Berlin, Heidelberg, pp. 211–30.

Lal, R. 1995. Global soil erosion by water and carbon dynamics. In: R. Lal, J. M. Kimle, E. Levine and B. A. Stewart (Eds), *Soils and Global Change*. Lewis Publishers, Boca Raton, FL, pp. 131–40.

Lal, R. 1998. Soil erosion impact on agronomic productivity and environmental quality. *Crit. Rev. Plant Sci.* 17:319–464.

Lal, R. 2004. Soil carbon sequestration impacts on global climate change and food security. *Science* 304:1623–7.

Langdale, G. W., J. E. Box, Jr., R. A. Leonard, A. P. Barnett and W. G. Fleming. 1979. Corn yield reduction on eroded Southern Piedmont soils. *J. Soil Water Conserv.* 34:226–8.

Lehman, R. M., C. A. Cambardella, D. E. Stott, V. Acosta-Martinez, D. K. Manter, J. S. Buyer, J. E. Maul, J. L. Smith, H. P. Collins, J. J. Halvorson, R. J. Kremer, J. G. Lundgren, T. F. Ducey, V. L. Jin and D. L. Karlen. 2015. Understanding and enhancing soil biological health: The solution for reversing soil degradation. *Sustainability* 7:988–1027.

Lehmann, J., M. C. Rillig, J. Thies, C. A. Masiello, W. C. Hockaday and D. Crowley. 2011. Biochar effects on soil biota – A review. *Soil Biol. Biochem.* 43:1812–36.

Liebig, M. A., G. E. Varvel, J. W. Doran and B. J. Wienhold. 2002. Crop sequence and nitrogen fertilization effects on soil properties in the Western Corn Belt. *Soil Sci. Soc. Am. J.* 66:596–601.

Lone, A. H., G. R. Najar, M. A. Ganie, J. A. Sofi and T. Ali. 2015. Biochar for sustainable soil health: A review of prospects and concerns. *Pedosphere* 25(5):639–53.

Lucy, M., E. Reed and B. Glick. 2004. Applications of free living plant growth-promoting rhizobacteria. *Antonie Van Leeuwenhoek* 86:1–25.

Lugtenberg, B. and F. Kamilova. 2009. Plant-growth-promoting rhizobacteria. *Annu. Rev. Microbiol.* 63:541–56.

Lupwayi, N. Z., G. W. Clayton, J. T. O'Donovan and C. A. Grant. 2011. Soil microbial response to nitrogen rate and placement and barley seeding rate under no till. *Agron. J.* 103:1064–71.

Lupwayi, N. Z., G. P. Lafond, N. Ziadi and C. A. Grant. 2012. Soil microbial response to nitrogen fertilizer and tillage in barley and corn. *Soil Till. Res.* 118:139–46.

Lynch, J. M. and E. Bragg. 1985. Microorganisms and soil aggregate stability. *Adv. Soil Sci.* 2:133–71.

Lynch, J. M. and J. M. Whipps. 1990. Substrate flow in the rhizosphere. *Plant Soil* 129:1–10.

Magdoff, F. 2001. Concept, components, and strategies of soil health in agroecosystems. *J. Nematol.* 33:169–72.

Maherali, H. and J. N. Klironomos. 2007. Influence of phylogeny on fungal community assembly and ecosystem functioning. *Science* 316:1746–8.

Maheshwari, D. K. 2011. *Plant Growth and Health Promoting Bacteria*, vol. 18. Microbiology monographs. Springer, Munster, Germany.

Manter, D. K. and M. G. Bakker. 2015. Estimating beta diversity for under-sampled communities using the variably weighted Odum dissimilarity index and OTUshuff. *Bioinformatics* 31(21):3451–9.

Manter, D. K. and J. M. Vivanco. 2007. Use of the ITS primers, ITS1F and ITS4, to characterize fungal abundance and diversity in mixed-template samples by qPCR and length heterogeneity analysis. *J. Microbiol. Meth.* 71:7–14.

Manter, D. K., J. A. Delgado, D. G. Holm, R. A. Stong. 2010a. Pyrosequencing reveals a highly diverse and cultivar-specific bacterial endophyte community in potato roots. *Microb. Ecol.* 60:157–66.

Manter, D. K., T. L. Weir and J. M. Vivanco. 2010b. Negative effects of sample pooling on PCR-based estimates of soil microbial richness and community structure. *Appl. Environ. Microbiol.* 76(7):2086–90.

Manter, D. K., L. Sherrod and G. Peterson. 2016a. Soil biology and carbon in dryland agriculture. *Proceedings of the Cover Your Acres Winter Conference*, vol. 13, pp. 15–21.

Manter, D. K., M. Korsa, C. Tebbe, J. A. Delgado. 2016b. myPhyloDB: A local web server for the storage and analysis of metagenomic data. *Database (Oxford)*. pii: baw037.

Manter, D. K., J. A. Delgado, H. D. Blackburn, D. Harmel, A. A. Perez de Leon and C. W. Honeycutt. 2017. A new concept: National living soil repository. *Proc. Nat. Acad. Sci. USA* 114:13587–90.

Mayak, S., T. Tirosh and B. R. Glick. 2004. Plant growth-promoting bacteria that confer resistance to water stress in tomatoes and peppers. *Plant Sci.* 166:525–30.

McCarthy, J. F., J. Ilavsky, J. D. Jastrow, L. M. Mayer, E. Perfect and J. Zhuang. 2008. Protection of organic carbon in soil microaggregates via restructuring of aggregate porosity and filling of pores with accumulating organic matter. *Geochim. Cosmochim. Acta* 72:4725–44.

Meisinger, J. J. and J. A. Delgado. 2002. Principles for managing nitrogen leaching. *J. Soil Water Conserv.* 57:485–98.

Mikha, M. M., J. G. Benjamin, M. F. Vigil and D. J. Poss. 2017. Manure and tillage use in remediation of eroded land and impacts on soil chemical properties. *PLoS ONE* 12(4):e0175533.

Moebius-Clune, B. N., D. J. Moebius-Clune, B. K. Gugino, O. J. Idowu, R. R. Schindelbeck, A. J. Ristow, H. M van Es, J. E. Thies, H. A. Shayler, M. B. McBride, K. S. M. Kurtz, D. W. Wolfe and G. S. Abawi. 2016. *Comprehensive Assessment of Soil Health – The Cornell Framework*, Edition 3.2. Cornell University, Geneva, NY.

Mokma, D. L. and M. A. Sietz. 1992. Effects of soil erosion on corn yields on Marlette soils in south-central Michigan. *J. Soil Water Conserv.* 47:325–7.

Montgomery, D. R. 2007. Soil erosion and agricultural sustainability. *Proc. Natl. Acad. Sci. USA* 104:13268–72.

Nannipieri, P., J. Ascher, M. T. Ceccherini, L. Landi, G. Pietramellara and G. Renella. 2003. Microbial diversity and soil functions. *Eur. J. Soil Sci.* 54:655–70.

Nannipieri, P., J. Ascher, M. T. Ceccherini, L. Landi, G. Pietramellara, G. Renella and F. Valori. 2008. Effects of root exudates in microbial diversity and activity in rhizosphere soils. In: Nautiyal, C. S. and P. Dion (Eds), *Molecular Mechanisms of Plant and Microbe Coexistence*, vol. 15, Soil Biology. Springer-Verlag, Berlin, pp. 339–65.

Newman, J. K., A. L. Kaleita and J. M. Laflen. 2010. Soil erosion hazard maps for corn stover management using National Resources Inventory data and the Water Erosion Prediction Project. *J. Soil Water Conserv.* 65:211–22.

O'Neill, K. P. J. 2016. Cover cropping for control of Columbia root knot nematodes in short season potato production. M.S. Thesis, Washing State University, Pullman, WA, 73p.

Parniske, M. 2008. Arbuscular mycorrhiza: The mother of plant root endosymbioses. *Nat. Rev. Microbiol.* 6:763–75.

Paterson, E., A. J. Midwood and P. Millard. 2009. Through the eye of the needle: A review of isotope approaches to quantify microbial processes mediating soil carbon balance. *New Phytol.* 184(1):19–33.

Patil, S. S., E. M. Adetutu, J. Rochow, J. G. Mitchell and A. S. Ball. 2014. Sustainable remediation: Electrochemically assisted microbial dechlorination of tetrachloroethene-contaminated groundwater. *Microb. Biotechnol.* 7(1):54–63.

Paungfoo-Lonhienne, C., D. Rentch, S. Robatzek, R. I. Webb, E. Sagulenko, T. Näsholm, S. Schmidt and T. G. A. Lonhienne. 2010. Turning the table: Plants consume microbes as a source of nutrients. *PLOS One* 5(7):e11915.

Paustian, K., O. Andren, H. H. Janzen, R. Lal, P. Smith, G. Tian, H. Tiessen, M. Van Noordwijk and P. L. Woomer. 1997. Agricultural soils as a sink to mitigate CO_2 emissions. *Soil Use Manage.* 13:240–4.

Phillips, D. A. 1980. Efficiency of symbiotic nitrogen fixation in legumes. *Annu. Rev. Plant Physiol.* 31:29–49.

Quine, T. A. and Y. Zhang. 2002. An investigation of spatial variation in soil erosion, soil properties, and crop production within an agricultural field in Devon, United Kingdom. *J. Soil Water Conserv.* 57:55–64.

Saharan, B. S. and V. Nehra. 2011. Plant growth promoting rhizobacteria: A critical review. *Life Sci. Med. Res.* 21:1–30.

Sgroy, V., F. Cassán, O. Masciarelli, M. Papa, A. Lagares and V. Luna. 2009. Isolation and characterization of endophytic plant growth-promoting (PGPB) or stress homeostasis-regulating (PSHB) bacteria associated to the halophyte *Prosopis strombulifera*. *Appl. Microbiol. Biotechnol.* 85:371–81.

Shridhar, B. S. 2012. Review: Nitrogen fixing microorganisms. *Int. J. Microbiol. Res.* 3(1):46–52.

Siddiqui, I. A. and S. S. Shaukat. 2003. Plant species, host age and host genotype effects on *Meloidogyne incognita* biocontrol by *Pseudomonas fluorescens* strain CHA0 and its genetically modified derivatives. *J .Phytopathol.* 151:231–8.

Six, J., K. Paustian, E. T. Elliot, C. Combrink. 2000. Soil structure and organic matter: I. Distribution of aggregate-size classes and aggregate-associated carbon. *Soil Sci. Soc. Am. J.* 64:681–9.

Six, J., R. T. Conant, E. A. Paul and K. Paustian. 2002. Stabilization mechanisms of soil organic matter: Implications for C-saturation of soils. *Plant Soil* 241:155–76.

Six, J., H. Bossuyt, S. Degryze and K. A. Denef. 2004. A history of research on the link between (micro) aggregates, soil biota, and soil organic matter dynamics. *Soil Till. Res.* 79(1):7–31.

Smith, D. D. 1941. Interpretation of soil conservation data for field use. *Agric. Eng.* 22:173–5.

Smith, D. D. and D. M. Whitt. 1948. Evaluating soil losses from field area. *Agric. Eng.* 29:394–6.

Stewart, C. E., K. Paustian, R. T. Conant, A. F. Plante and J. Six. 2007. Soil carbon saturation: Concept, evidence and evaluation. *Biogeochemistry* 86(1):19–31.

Stewart, C. E., P. Moturi, R. F. Follett and A. D. Halvorson. 2015. Lignin biochemistry and soil N determine crop residue decomposition and soil priming. *Biogeochemistry* 124:335–51.

Stocking, M. A. 2003. Tropical soils and food security: The next 50 years. *Science* 302(5649):1356–0.

Stromberger, M. E., I. Abduelafez, P. Byrne, M. M. Canela, A. A. Elamari, D. K. Manter, T. Weir. 2017. Genotype-specific enrichment of 1-aminocyclopropane-1-carboxylic acid deaminase-positive bacteria in winter wheat rhizospheres. *Soil Biol. Biochem.* 81:796–805.

Treseder, K. K. 2004. A meta-analysis of mycorrhizal responses to nitrogen, phosphorus, and atmospheric CO_2 in field studies. *New Phytol.* 164(2):347–55.

Turmel, M.-S., A. Speratti, F. Baudron, N. Verhulst and B. Govaerts. 2015. Crop residue management and soil health: A systems analysis. *Agric. Sys.* 134:6–16.

Uren, N. C. 2007. Types, amounts, and possible functions of compounds released into the rhizosphere by soil-grown plants. In: R. Pinton, Z. Varanini and P. Nannipieri (Eds), *The Rhizosphere: Biochemistry and Organic Substances at the Soil-plant Interface*, 2nd edition. CRC, Boca Raton, FL, pp. 1–21.

USDA. 1956. Joint conference on slope-practice, Washington DC.

USDA. 2015. Summary Report: 2012 National Resources Inventory. Natural Resources Conservation Service, Washington DC and Center for Survey Statistics and Methodology, Iowa State University, Ames, IA. http://www.nrcs.usda.gov/technical/nri/12summary.

USDA-FSA. 2009. Conservation Reserve Program annual summary and enrollment statistics-FY 2009. Available at https://www.fsa.usda.gov/Internet/FSA_File/fyannual2009.pdf.

USDA-NRCS. 2016. Soil health key points. https://www.nrcs.usda.gov/wps/portal/nrcs/detailfull/national/soils/health/?cid=stelprdb1049236. (accessed on 12 December 2017).

VandenBygaart, A. J., E. Bremer, B. G. McConkey, B. H. Ellert, H. H. Janzen, D. A. Angers, M. R. Carter, C. F. Drury, G. P. Lafond and R. H. McKenzie. 2011. Impact of sampling depth on differences in soil carbon stocks in long-term agroecosystem experiments. *Soil Sci. Soc. Am. J.* 75:226–34.

Vaxevanidou, K., C. Christou, G. F. Kremmydas, D. G. Georgakopoulos and N. Papassiopi. 2015. Role of indigenous arsenate and iron(III) respiring microorganisms in controlling the mobilization of arsenic in a contaminated soil sample. *Bull. Environ. Contam. Toxicol.* 94(3):282–8.

Wakatsuki, T. and A. Rasyidin. 1992. Rates of weathering and soil formation. *Geoderma* 52:251–63.

Weese, D. J., K. D. Heath, B. T. M. Dentinger and J. A. Lau. 2015. Long-term nitrogen addition causes the evolution of less-cooperative mutualists. *Evolution* 69:631–42.

West, T. O. and W. M. Post. 2002. Soil organic carbon sequestration rates by tillage and crop rotation: A global data analysis. *Soil Sci. Soc. Am. J.* 66:1930–46.

Wienhold, B. J., D. L. Karlen, S. S. Andrews and D. E. Stott. 2009. Protocol for indicator scoring in the soil management assessment framework (SMAF). *Renew. Agric. Food Sys.* 24(4):260–6.

Williams, B. K. and E. D. Brown. 2014. Adaptive management: From more talk to real action. *Environ. Manage.* 53(2):465–79.

Wright, R. J., W. D. Kemper, P. D. Millner, J. F. Power and R. F. Korcak (Eds). 1998. Agricultural uses of municipal, animal, and industrial byproducts. Conservation Research Report 44. U.S. Department of Agriculture, Agricultural Research Service, Washington DC.

Wubs, E. R. J., W. H. van der Putten, T. M. Bosch and T. M. Bezemer. 2016. Soil inoculation steers restoration of terrestrial ecosystems. *Nat. Plants* 2:16107.

Zhu, S., J. M. Vivanco and D. K. Manter. 2016. Nitrogen fertilizer rate affects root exudation, the rhizosphere microbiome and nitrogen-use-efficiency of maize. *Appl. Soil Ecol.* 107:324–33.

Zolla, G., D. V. Badri, M. G. Bakker, D. K. Manter and J. M. Vivanco. 2013. Soil microbiomes vary in their ability to confer drought tolerance to Arabidopsis. *Appl. Soil Ecol.* 68:1–9.

Chapter 5

The economics of soil health

*Maria Bowman, USDA-ERS, USA**

1 Introduction

Much of the current and previous research on the economics of soil conservation and soil health has focused on the farmer's adoption decision as it relates to discrete soil health practices, such as no-till or reduced tillage, cover cropping or nutrient management (see USDA Natural Resources Conservation Service (2017) for a list of practices considered a part of a soil health management system). This research is data-intensive, as it requires data that link on-farm soil health variables or changes in soil health or environmental benefits to changes in practices *and environmental and economic outcomes*. In addition, economists have often focused on evaluating one practice or programme at a time (or

* The views expressed are those of the author and should not be attributed to the Economic Research Service or USDA.

http://dx.doi.org/10.19103/AS.2017.0033.06
Published by Burleigh Dodds Science Publishing Limited, 2018.

one outcome variable, such as yield or profit) in order to more easily identify the effect or change they care about. However, this historical focus on individual practices in research and in public policy design means that there is less economic research that examines whether soil health practices are substitutes or complements for other soil health practices in terms of the production of benefits for agricultural production, or of environmental benefits.

Economists' contributions to research on the economics of soil health go beyond efforts to describe the current patterns of adoption and quantify the costs and benefits of soil health and soil health practices to farmers and society. Using econometric techniques, socioeconomic surveys, spatially explicit modelling and behavioural experiments, economists are researching the patterns of adoption of soil health practices, the barriers to adoption of soil health practices, the role of cost-share or incentive programmes in farmer decisions to adopt soil health practices and the cost-effectiveness and optimal design of incentive programmes.

This chapter looks at basic patterns of adoption of a few key soil health practices with a focus on the United States, followed by a discussion of factors that affect the farmer's adoption decision about soil health practices. The bulk of the chapter discusses several active areas of research that relate to the economics of soil health and provide context for why these research areas are important for public policy.

2 Use of key soil health practices by farmers in the USA

In the United States, there are several farmer surveys administered by the U.S. Department of Agriculture (USDA) that are nationally representative and collect information about soil health practices including tillage practices, cover cropping, some nutrient management practices, soil health testing and rotational grazing. These data allow us to track farmer adoption rates in space and time. Data from the most recent Census of Agriculture, conducted in 2012, suggest that more than a third of farmland area is engaged in no-till agriculture, while less than 3% of cropland area is cover-cropped. However, approximately 9% of the *farms* with cropland report using cover crops on some portion of their farm. Although these are the national rates for no-till and cover crop adoption, the rates of adoption of these practices vary by region, crop grown and other variables (Wade et al., 2015). Wade et al. (2015) use several years of data from the USDA Agricultural Resource Management Survey (ARMS) to show how adoption rates of tillage practices and cover cropping vary over time, as well as regionally. The authors found that no-till/strip-till and cover cropping are used more frequently in the southern and eastern regions of the United States, and that no-till/strip-till use is the highest on soybean fields (between 40 and 45% of the cropland planted to soybeans used no-till in the 2006 and 2012 versions of the survey) when compared to cropland planted to corn, wheat and cotton. Wade et al. also report that 64% of the cropland planted to cotton in 2007 and 31% of the cropland planted to corn in 2010 reported splitting applications of nitrogen fertilizer so that parts of the crop's nitrogen requirements were applied after planting. In addition, 24% of the cropland planted to cotton and 6% of the cropland planted to corn reported combining split

Table 1 Adoption rates in the United States for several key soil health practices

		Adoption rate	Year	Data source
Cover cropping	Wade et al. (2015)	<2% of corn, soybean, wheat, and cotton planted area	United States, 2010–11	USDA Agricultural Resource Management Survey
	Bowman et al. (2016)	8.6% of farms with cropland and 2.9% of cropland	United States, 2012	USDA Census of Agriculture
No-till/strip-till	Bowman et al. (2016)	34.6% of cropland in no-till (for which tillage practices are reported)	United States, 2012	USDA Census of Agriculture
	Wade et al. (2015)	39% of corn, soybean, wheat, and cotton planted area in no-till/ strip-till	United States, 2010–11	USDA Agricultural Resource Management Survey
Combined nutrient management practices[1]	Wade et al. (2015)	24% of cotton planted area 6% of corn planted area	United States; 2007 (cotton) and 2010 (corn)	USDA Agricultural Resource Management Survey
Testing for soil organic matter	Bowman et al. (2016)	11% of oat fields and 15.8% of cotton fields surveyed tested at least once in last ten years	United States, 2015	USDA Agricultural Resource Management Survey
Rotational or management-intensive grazing		Approximately 23% of farms with pastureland[2]	United States, 2012	USDA Census of Agriculture

[1] No fall application of N, some N application after planting, N application at rates below a 'benchmark' and fertilizer incorporation or injection below the soil surface.
[2] 288 719 farms/1 239 085 farms with pastureland (excluding woodland pastured).

nitrogen application with no fall application, some nitrogen application after planting, applying less than the recommended amount of nitrogen and fertilizer incorporation or injection. Table 1 summarizes the most recent national adoption rates for soil health practices reported in USDA datasets.

Adoption rates of soil health practices vary regionally due to a number of factors (Wade et al. 2015). These include inherent differences in agronomic factors such as soil types and climatic factors, demographic and social factors and economic factors. State and federal incentive programmes, policies and regulations also impact adoption rates, as do regional differences in farmer outreach and extension activities. Figure 1 shows dramatic regional differences in cover crop adoption patterns in the United States.

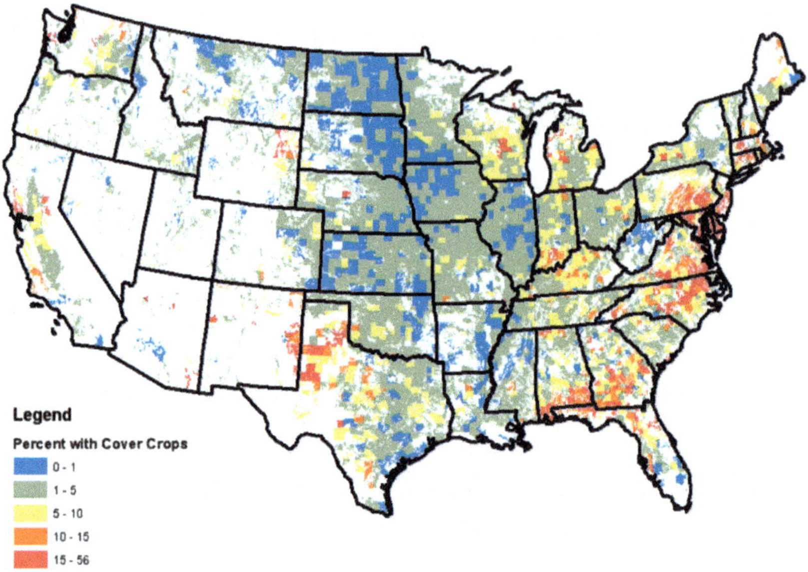

Figure 1 Share of harvested cropland area with cover crops by county in the United States as reported in the 2012 USDA Census of Agriculture. Source: Bowman et al. (2016), using data from the 2012 USDA Census of Agriculture.

3 Costs and benefits of soil health practices

When farmers make decisions about adopting soil health practices, they are usually weighing whether the benefits they receive from adopting a practice or suite of practices exceed the costs of implementing the practice(s), and how these costs and benefits might compare to other types of production systems or practices over time. From an economic perspective, these types of costs and benefits that accrue to or are borne by the farmer are referred to as *private* costs and benefits. Soil health practices can also have benefits for society, or *public benefits*. Table 2 summarizes the potential private costs and benefits associated with cover cropping, a key soil health practice. Snapp et al. (2005) summarize some of the key private costs and benefits associated with cover cropping in Michigan. In general, private benefits of adopting one or more soil health practices might be improved crop yields, reduced agricultural input costs due to reduced need for synthetic fertilizer or other inputs, greater resilience of crops to extreme weather (droughts and floods) as a result of improved soil water-holding capacity, and an opportunity to engage in environmental stewardship. Private costs could include investments in time or resources as the farmer learns about a new practice, increased labour associated with monitoring or managing (some studies cite this as a benefit – that is, decreased labour requirements associated with using soil health practices relative to conventional management practices) and increased costs of certain other inputs (such as cover crop seeds or additional herbicides).

One body of literature related to soil health practice adoption aims to simply estimate when and where it makes economic sense to the farmer to adopt soil health or conservation practices, and what the production outcomes or expected financial returns are to the

Published by Burleigh Dodds Science Publishing Limited, 2018.

Table 2 Potential private costs and benefits of planting a cover crop

Costs	Benefits
• Investment in learning about new practice • Cover crop seed and other costs associated with establishment • Increased herbicide use, if chemically terminated • Increased labour associated with seeding, managing, and terminating cover crop • Decreased yield (yield penalty) in cash crop • Opportunity cost of not planting cash crop, if there is a cash crop option • Potential loss of flexibility due to timing of planting and termination of cover crop with harvest and planting of cash crops and/or poor establishment of cash crop • Difficult to estimate N benefit and timing of N release • Competition for water in water-scarce regions	• Improved yields (yield boost) • Weed or pest suppression can result in reduced input costs for herbicides or pesticides • Provides nutrients (e.g. N from legumes) which can reduce need for fertilizer inputs • Increased water use efficiency might reduce irrigation costs or improve yields • Greater resilience to extreme weather events such as droughts and floods can mitigate yield losses • Improved nutrient cycling • Reduced erosion and/or improved soil stability/soil structure • Pride in farm and/or in environmental stewardship • Grazing (forage) benefit (if grazed)

Note: Some items listed could fall in either column depending upon the direction of the effect on yield, planting time, labour requirements or other variables.

farmer compared to other practices or combinations of practices. These types of studies are sometimes based on the data from field/farmer trials, models, or farmer surveys. For example, Karlen et al. (2013a,b) evaluate the impacts of different types of tillage over 30 years in Iowa on yield and soil fertility, as well as the production costs associated with each. Corn and soybean rotations boosted corn yield by 17% when compared to continuous corn, and although tillage produced the highest yields, no-till had the highest net returns over the study period due to reduced costs. They also confirm that mouldboard ploughing over the long run has a significant and negative effect on soil quality. Al-Kaisi et al. (2016) also found that the net return is higher for no-till corn and soybean rotations in a study in Iowa due to reduced input costs and comparable yields, and Cook and Trlica (2016) show that yield penalties associated with no-till over the long term in Southern Illinois disappear with adequate NPK.

Hanson et al. (1993) and Lichtenberg et al. (1994) evaluated the profitability of cover crops for no-till corn systems in the Mid-Atlantic region. Hanson et al. (1993) found that no-till corn after a hairy vetch cover crop was more profitable than either after winter fallow or a winter wheat cover crop in the Piedmont and Coastal Plain regions of Maryland. Lichtenberg et al. also found hairy vetch to be more profitable than fallow and cover crop alternatives, and that this was primarily due to a yield advantage for hairy vetch. Watkins et al. (2004) found that no-till may be more profitable than conventional tillage in rice systems in Arkansas in some years but not in others.

These types of studies highlight an important point: though agronomic studies often focus on what practices have the highest yield, high-yielding practices are not inherently the most profitable practices. Where this is the case, it might be due to reduced input use – and, in turn, input costs – associated with the lower yielding practice relative to conventional systems that may not engage in soil-health-related practices. A farmer may also care about yield variability in addition to just average yield; so to the extent that soil health practices impact average yields and yield variability, this will also affect farmer decision-making about soil health practices (see Box 1 'Risk and farmers' decision making').

Box 1 Risk and farmers' decision-making

Farming is an inherently risky endeavour, and farmers face different types of risk. Within a growing season, these include price risks (i.e. the risk that input prices or commodity prices will change) and yield risks, or production uncertainty (i.e. the risk that a farmer's crop will not produce the expected yield at the expected time), among others (Chavas et al., 2010). There are also risks that farmers must consider over a longer time horizon, such as the risk associated with trying a new technology.

Farmers have many tools at their disposal to manage risks. These include using credit markets to borrow at times of low business cash flow, participating in government insurance programmes, engaging in contracts that guarantee a set purchase price or volume for inputs or for a final agricultural product and diversifying input or output mixes. Finally, farmers can use management practices that may help to mitigate risks.

Adopting soil health practices and systems has the potential to increase or decrease risk. Adopting soil health practices might increase the yield risk due to pest or weed pressure, particularly in the short term, if a farmer is trying out a practice for the first time. On the other hand, soil health practices may decrease risks by making a farmer more resilient to droughts or extreme weather, for example.

This potential benefit of soil health practices, risk mitigation, is an increasing focus of the agronomic and economic literatures. The risk of drought is perhaps the most important weather/climate-related risk a non-irrigated farmer faces, and the economic literature suggests that soil health practices become a tool that farmers use to mitigate and adapt to drought risk. Between 1948 and 2010, 38% of the crop insurance indemnity payments, at a value of $32.5 billion, were due to farm production losses from drought (followed by excess moisture, at 24% or $20.5 billion) (Wallander et al., 2013). Counterintuitively, when controlling for recent indemnity and ad hoc disaster payments, Schoengold et al. (2015) found that the number of recent flood years increased the use of no-till and that the number of recent drought years decreased the use of no-till. They also found that counties with high disaster and crop insurance indemnity payments were more likely to use no-till and less likely to use other types of conservation tillage. Their results suggest that there may be complex interactions between federal crop insurance and disaster payment programmes and farmer decision-making, and that future research might consider how such programmes impact farmer behaviour and may incentivize or disincentivize farmers to engage in on-farm risk management through the adoption of practices (such as conservation tillage).

A slightly different collection of the literature looks at the role of specific factors in influencing a farmer decision to adopt a practice, often while controlling for other factors known to influence farmer decision-making (such as input prices, crop prices, location, agri-environmental variables, government programme participation, farmer characteristics, etc.). We already discussed how location-specific biophysical variables might affect farmer adoption decisions and drive differences in regional adoption rates. But, what about other variables like farm/field size and farmer characteristics, and economic variables like prices? Prokopy et al. (2008) and Baumgart-Getz et al. (2012) conducted a review and a meta-analysis (respectively) of the published literature in order to determine the factors affecting the adoption of best management practices (BMPs) by farmers. Some of the factors that have a positive relationship with BMP adoption across the studies reviewed in Prokopy et al. (2008) and Baumgart-Getz et al. (2012) include capital, farm size, specialization in farming, access to information and environmental attitudes/awareness.

Characterizing patterns of practice use over time and space is also important in order to understand public and private benefits, as well as farmer decision-making. Understanding these dynamics is especially important for practices that farmers can make choices about every year or planting season. For example, the adoption of many practices, such as no-till and cover crops, can occur in some fields but not others, or in one year but not the next. This may have something to do with changes under annual weather conditions, or with crop choice and crop residue management. There may also be on-farm or off-farm spillover effects. Farmers that adopt a practice on a field may be more likely to adopt it on others if, for example, they learn how to successfully implement the practice on a small portion of the farm and later expand, or if they have to purchase new farm machinery (capital) in order to adopt a practice. Off the farm, farmers share knowledge and information within their social and professional networks about practices and about the economics of practice adoption (Foster and Rosenzweig, 1995). If a practice is observed to be profitable in a region or if the learning costs are reduced by transmission of information, other farmers may be more likely to adopt, and social norms and attitudes about particular practices may also change as more farmers adopt. Sunding and Zilberman (2001) discuss different models of how patterns of technology adoption can be described over space/time, such as the diffusion and threshold models of adoption.

In the case of soil health and conservation technologies, social and professional groups in the United States provide important opportunities for farmers and researchers to network with peers and learn about technologies through outlets, such as publications, radio broadcasts and conferences. For example, the National No-Till Conference has been occurring annually for 25 years, and the newly established National Conference on Cover Crops and Soil Health is organized by the Soil and Water Conservation Society (Meitner, 1999; No-Till Farmer, 2017; Soil and Water Conservation Society, 2017).

4 Case studies in soil health: strengths and limitations

Case studies are frequently employed by government agencies and extension agents to illustrate the costs and benefits of adopting a particular practice. Table 3 presents a number of soil health case studies that are publicly available to farmers and researchers. Another publicly available source of case studies is the 'Profiles in Soil Health' published by USDA Natural Resources Conservation Service (NRCS) (Natural Resources Conservation Service, 2016). These compelling discussions with farmers highlight the cases where managing for soil health resulted in large soil health gains and was profitable for the operations.

Case studies are important in disseminating real-world experiences with soil health practices or combinations of practices. They are also written so as to be accessible to farmers and practitioners, and to focus on things that farmers care about (rather than academics). Another strength of case studies is that they are often inexpensive to research and write, and can provide early intuition about the economic variables that affect farmer decision-making. But they have the disadvantage of being a single data point (or few data points, if farmers engage in small on-farm trials), which makes it difficult

Table 3 Selected economic case studies in soil health

Case study author(s)*	Farmer/Farm and location	Practices adopted	Challenges/Costs	Benefits and economic data points
Susan Tallman, National Center for Appropriate Technology	Gabe Brown (North Dakota)	• No-till • Diverse cover crops • Intensive rotational grazing		• SOM increased from 2 to 4% • Improved water infiltration • Reduced fertilizer and herbicide use • Average yield much higher than county average
Madalene Ransom, USDA NRCS	Russell Hedrick (North Carolina)	Cover crops added to no-till corn and soybeans	• Learning costs in year 1 • High costs for seed and herbicides	• 927 kg/ha higher yield for corn (9%) • 1,201 kg/ha higher yield for soybeans (35%)
Susan Tallman, National Center for Appropriate Technology	Bauer Farm (North Dakota)	Cover crops and grazing (for corn) added to no-till wheat and corn	• Adapting rotations to no-till	• No difference in wheat yields between CC field and control • Corn yields for CC + grazing were 6,178 kg/ha (20% higher) compared to 5,127 kg/ha in control
Susan Tallman, National Center for Appropriate Technology	Richter Farm (North Dakota)	Grazing + cover crops in forage based system, followed by corn		• Corn with cover crop had a net return of 153.85 USD/ha vs. no cover crop of 34.03 USD/ha (452% higher net return)
Natural Resources Conservation Service, New York	Dave Magos (New York)	Cover crops and no-till in corn silage and alfalfa rotation		• Reduced soil erosion, and pesticide and fertilizer use • 5% increase in corn silage yields • Net return on cover crops is 153 USD/ha
Conservation Cropping Systems Initiative	Rulon Enterprises (Indiana)	Cover crops added to no-till corn and soybeans	Working with Purdue and NRCS in initial trials may have helped with learning costs	• Yield over county yield higher during drought years • 439 kg/ha yield boost for corn and 133 kg/ha for soybeans over three years • Net benefit of 200 USD/ha for cover crops inc. CSP payment, expected yield boost in drought year, value on reduced soil erosion
Lauren Cartwright, MO State Economist, NRCS	(Anonymous)(Missouri)	Cover crops added to no-till corn and soybeans		• Yield increase for corn but not soybeans • Short term annual net cost • Herbicide reduction for soybeans • N credit for corn

* Full case study reference and link to online version (if available) provided in References.

to draw statistical conclusions or generalize the results to other farmers, regions or policy situations. Methods are also not commonly standardized across case studies, which means that the quantification of costs and benefits and any estimates of net costs and benefits are highly dependent on the assumptions used and information collected as part of the case. Developing standards and guidelines for future case studies will be essential for comparability across case studies in soil health, though a meta-analysis may also prove useful. See Knight (2016) for an example of guidelines for developing case studies.

Case studies are also prone to issues of bias. In particular, farmers that manage for soil health and adopt soil health practices (as well as farmers that are willing to participate in case studies) are often not representative of other farmers. They may be more 'progressive' in their farming practices, be more accepting of risk or uncertainty, take more pride in environmental stewardship or be better or more holistic land managers than their peers (Dumanski et al., 2014). Early adopters are also typically those who benefit the most from adopting soil health practices (see Wu and Babcock (1998)). As a result, economic estimates of the costs and benefits of soil health practices based on these farmers are likely to be biased. In addition to the selection bias associated with the adoption of practices, case studies may also suffer from a form of publication bias. For example, because the purpose of a case study might be to provide an example of a success with a practice so that other farmers can overcome their misgivings about adopting a new technology or practice, it may be more difficult to find published examples of case studies where the outcome was not as successful.

Much of this research is conducted with an end goal of providing useful information to producers about the costs, benefits and profitability of soil health practices. To this end, a number of organizations and research groups have developed online decision tools or spreadsheet models that allow producers to input parameters that describe their system and receive outputs that reflect the impacts of different choices or practices on costs, yields or profitability. Box 2 'Economic decision tools for soil health practices', contains a short list of some publicly available decision tools related to soil health economics and management.

Box 2 Economic decision tools for soil health practices in the United States

Cornell: Cover Crops for Vegetable Growers
http://covercrops.cals.cornell.edu/decision-tool.php
NRCS: Cover Crops Economics Tool
https://www.nrcs.usda.gov/wps/portal/nrcs/detailfull/il/soils/health/
Midwest Cover Crops Council: Selector Tool for Cover Crops
http://mccc.msu.edu/selector-tool/
Ohio State: Soil Organic Matter Calculator
https://southcenters.osu.edu/soil-water-bioenergy/extention/som-soil-organic-matter-calculator
Kansas State: Cover Crop Cost-Return Budget
http://www.agmanager.info/production-economics/farm-management -guides /livestock-budgets/ksu-cover-crop-cost-return-budget
Useful to Useable: Corn Split-N tool
http://mrcc.isws.illinois.edu/U2U/splitn/

---

5 Public benefits of soil health and soil health management practices

In addition to the costs and benefits of changes in soil health and soil health practices that are relevant to the farmer, soil health can have important public benefits. These can include reduced greenhouse gas emissions from agriculture, increased carbon storage or sequestration, improved water quality through reduced nutrient and sediment pollution and impacts on water quantity/availability (Dale and Polaskly, 2007; Hansen and Ribaudo, 2008; Hatfield et al., 2001; Lal, 2004; Ribaudo et al., 2010; Swinton et al., 2007). Quantifying and valuing these public benefits is difficult, because they accrue at different scales and are difficult to measure. Nonetheless, there is theory and evidence to support the fact that these benefits could be significant, and valuing them is important when considering the costs and benefits of public policy (see Robinson et al. (2012) for a framework for valuing the natural capital of soil). In particular, the relationship between soil health and water quantity/quality has the potential to generate significant public benefits due to improved infiltration rates and increased soil water storage capacity (Hudson, 1994; Rawls et al., 2003: Kemper et al., 2011). This may contribute to the following:

- Reduced severity of flooding during extreme precipitation events.
- Reduced sedimentation in waterways and reduced costs associated with dredging public reservoirs.
- Reduced demand for irrigation water, which could have a positive impact on groundwater and surface water supply, as well as demand for water deliveries from federal or state programmes.
- Improved resilience to drought, which could reduce federal indemnity payments for yield losses incurred during drought years.
- Reduced nutrient pollution at local, regional and watershed scales due to reduced runoff from farm fields.

In the early 2000s, USDA began the Conservation Effects Assessment Project (CEAP) with the aim of understanding the environmental benefits of USDA conservation programmes (Mausbach and Dedrick, 2004). As part of this research, ARS completed several 'benchmark watershed assessments', which were designed specifically to look at the impact of agricultural conservation practices on water quality at the watershed scale. Tomer and Locke (2011) summarize the results of the assessments in the CEAP benchmark watersheds. Models using results from the project predict that a number of conservation practices that are also soil health practices – such as conservation cover, cover crops, nutrient management, no-till and pasture rotation – will contribute to benefits such as reductions in sediment concentration and nitrate and phosphorus leaching and loading at multiple scales. However, they also discuss the limitations of the project in terms of available data and ability to link changes in practices to water quality outcomes. In particular, they note that monitoring water quality alone may not allow researchers to detect the specific effects of practices given climatic variation, monitoring issues associated with non-point sources such as agricultural fields and farms, and time lags from practice implementation to water quality effect detection.

In a 2016 report, USDA NRCS estimated that all voluntary conservation practices (including structural, nutrient management, cover crops and others) in the Western Lake

Erie Basin (as of 2012) reduce annual sediment losses by 81%, nitrogen by 36% and phosphorus by 75% relative to a baseline of no conservation practices (U.S. Department of Agriculture, 2016). Though not all these practices were soil-health-related, these results illustrate the magnitude of the potential public benefits that can be achieved through conservation practice adoption.

In 2012, the Australian government sponsored a report summarizing the relationship between soil health management practices and ecosystem service provision (Cork et al., 2012). Some key findings of this report were that soil health management practices have slowed the loss of soil carbon and that this has contributed to public and private benefits; that maintaining high levels of residue cover (50–70%) has reduced wind and water erosion; and that improved management has reduced dust storms, which have significant private and public costs. See Robinson et al. (2012), Bennett et al. (2010) and Huguenin et al. (2006) for additional resources that discuss on valuing the public benefits (ecosystem services) associated with soil health and soil health management practices in different regions.

6 Barriers to adoption of soil health practices

Although there are public and private benefits to the adoption of soil health practices and management systems, they can also be costly to implement. A straightforward reason that farmers might not adopt soil health practices is that they may not be profitable on a particular farm at a particular point in time, or the farmers do not anticipate that they will be profitable in the long run. This could be due to a number of factors already mentioned in this chapter, including fixed or variable costs associated with soil health practices. Important among these may be investments in capital, such as new planters, or investments in human capital required for a farmer to learn how to implement a new practice (Arbuckle and Ferrell, 2012; Carlisle, 2016). Risk or perceived risk of adopting soil health practices also influences farmer decision making (see the Box 1 'Risk and farmers' decision making'). Arbuckle and Roesch-McNally (2015) found that farmers who do not adopt cover crops often perceive higher levels of risk associated with the practice. The authors suggest that this highlights a need for improved communication with farmers about both the potential benefits and the risks of the practice, as well as a clear need for technical assistance and education.

There are also behavioural or informational barriers to the adoption of soil health practices and systems. These include social barriers to adoption, such as when practices are not viewed as compatible with current farming practices, or when low adoption rates influence attitudes towards a practice within a social network (Rogers, 2003; Reimer et al., 2012). Lack of information about the costs and benefits of soil health practices seems to be a significant barrier. Dunn et al. (2016) use a national survey of cover crop users to look at patterns in self-funding vs. programme participation, as well as perceived barriers to adoption and factors affecting continuous use of cover crops. The authors find that many see the high cost of cover crops, combined with uncertain benefits, to be a major barrier – and that those who use cover crops continuously are comfortable with engaging in a long learning process by trial and error about what works best on their farm. While trial and error may be an important tool for farmers, it implies that farmers have uncertainty and limited information about how to implement cover crops profitably on their operation.

Another important barrier in this category relates to the time lags in benefits or profitability associated with practices. Though the costs of implementing soil health practices (such as investments in human or physical capital and cover crop seed) are borne in the present, the farmer may not immediately reap the benefits of changing their practices. Finally, institutional and structural barriers can also impact adoption, such as transaction costs associated with participating in government programmes and land tenure issues that affect short- and long-term decision-making of land owners and operators. Carlisle (2016) provides an excellent narrative review of the literature on the barriers to adoption of soil health practices.

7 Evaluating the role of federal and state regulations, policies and incentive programmes

In part because of the public benefits associated with soil health and soil conservation, it may make sense to incentivize farmers to adopt soil health practices when they might not adopt them on their own. There are many United States policies that target farm management practices related to soil health. Several types of regulations directly affect farmer decision-making about soil health practices. Perhaps the best example of a United States regulation that affects soil health is 'conservation compliance', or provisions that require farmers to implement soil conservation practices on highly erodible cropland in order to retain eligibility for federal farm payment programmes such as conservation payments, crop insurance, disaster payments and loan programmes. Conservation compliance provisions were implemented as part of the 1985 Farm Act. During the years following the implementation of conservation compliance, erosion declined sharply on highly erodible cropland with about 25% of the total erosion reductions occurring on highly erodible cropland (Claassen et al., 2004). Because erosion also declined on the land that was not highly erodible or subject to compliance, it is difficult to evaluate the role of other factors affecting soil erosion during the same time period (see Box 3 'The challenge of additionality'). However, Claassen et al. (2017) showed that conservation compliance contributed to decreasing water erosion on cropland subject to compliance.

Voluntary national programmes can also induce farmers to engage in soil health management practices, for example, by providing a cost-share or financial incentive to a farmer to adopt a specific practice for a defined length of time. In the United States, two programmes that provide financial incentives to farmers to engage in soil health practices or 'enhancements' (as described because they take conservation practices to a higher level of stewardship) are the Environmental Quality Incentives Program (EQIP)

Box 3 The challenge of additionality

Economists and social scientists face a research challenge: many farmers implement soil health or soil conservation practices even in the absence of regulations or incentive programmes, which makes it difficult to quantify how the farmer behaviour changed in response to the regulation or programme. The additional effect of a policy, above and beyond baseline adoption, is known as 'additionality'. Research on additionality in conservation programmes is an important tool for evaluating how well they work, and whether they are cost-effective (see Claassen et al. (2013)).

and the Conservation Stewardship Program (CSP). These programmes are managed by the NRCS of the USDA. For both programmes, farmers engage in a contract with NRCS for specific practices/enhancements on specific acres over a defined time period (e.g. many EQIP no-till contracts are three-year contracts). The most popular EQIP practices for soil health include nutrient management, cover cropping, no-till, reduced tillage, conservation crop rotations, mulching and seasonal residue management. Between 2009 and 2013, NRCS increased the land area enrolled in EQIP for cover crops, as well as the total expenditures for the practice. At the same time, there was a decrease in the amount of land receiving payments and in total no-till payments. This may be because no-till has become more widely adopted, which has presented opportunities for the programme to focus on other practices that promote soil health but may have a higher cost to the farmer, such as cover crops (Bowman et al., 2016). Table 4 lists some enhancements that are funded through CSP. The number of hectares receiving CSP payments for at least one soil health practice or enhancement grew from under 3 million ha to more than 12 million ha between 2010 and 2014 (Bowman et al., 2016). Zimmerman and Britz (2016) evaluate the participation of farms in the EU in agri-environmental measures or voluntary conservation programmes that are implemented as part of the EU's Common Agricultural Policy.

In the United States, state-level incentives are another important source of financial assistance for soil health practices. For example, the state of Maryland's Agricultural Water Quality Cost-Share Program provides cost-share payments that cover up to 87.5% of the costs of installing a number of conservation practices, including cover crops (Maryland Agricultural Water Quality Cost-Share Program, 2016). In 2004, the programme paid farmers a minimum of 49.40 USD/ha to plant a cover crop. In that year, Maryland enrolled only 46,750 ha – but after raising the minimum payment to 111.20 USD/ha, more than 192,452 ha were enrolled in 2015 (Bowman et al., 2016). In 2016,

Table 4 Examples of soil-health-related Conservation Stewardship Program (CSP) enhancements for United States cropland and rangeland

Cropland	Rangeland
• ENR10: Using Nitrogen (N) provided by legumes, manure or compost • ENR12: Use of legume cover crops as N source • PLT20: High residue cover crops for weed suppression and soil health • SOE05: Intensive no-till • SQL01: Controlled traffic system • SQL04: Use of cover crop mixes • SQL08: Intercropping to improve soil quality and increase biodiversity • SQL12: Intensive cover cropping in annual crops • SQL18: Soil health crop rotation • WQL10: Plant a cover crop that will scavenge residual N • CCR98: Improved Resource Conserving Crop Rotation • CCR99: Resource Conserving Crop Rotation	• ANM03: Incorporate native grasses and/or legumes into 15% or more of herbage dry matter productivity • ANM21: Prairie restoration for grazing and wildlife habitat • ANM37: Prescriptive grazing management • PLT02: Monitoring key grazing areas to improve grazing management • PLT16: Intensive rotational grazing • PLT30: Monitor pasture health using pasture condition scores • SQL16: High species diversity grazing lands • SQL17: Placement of hay feeding areas on low fertility soils • SQL19: Management for rangeland soil health

farmers received up to 222.39 USD/ha to cover crop costs for traditional cover crops, and up to 86.49 USD/ha to grow a rye cover crop (Maryland Agricultural Water Quality Cost-Share Program, 2016).

In addition to federal and state policies and programmes, the potential for private markets or market-driven initiatives can also provide incentives to farmers and ranchers to build soil health and adopt soil health practices. For example in 2009–10, approximately 1,000 ranchers enrolled more than two million hectares of rangeland in voluntary carbon credit trading through the Chicago Climate Exchange (CCX) (Gosnell et al., 2011). The sale of carbon credits from rangeland was developed through the CCX's Rangeland Soil Carbon Offset Protocol. Though the CCX ultimately collapsed, it set a precedent for 'carbon ranching', and a sense of promise that ranchers might be rewarded for the greenhouse gas implications of good stewardship via the private market. Nonetheless, researchers caution that there are a number of challenges associated with such initiatives, including the difficulties of measuring and monitoring soil carbon sequestration, concerns about additionality and permanence in dynamic systems and the fact that net emission reductions are not guaranteed (Booker et al., 2013).

Major buyers of agricultural products, such as grocery stores, restaurant chains and food service companies, can also influence farmer decision-making about soil health practices. The media have focused more heavily on major buyers' influence on changing animal-raising practices, such as the use of gestation grates for sows, the use of cages for egg-laying hens and the use of antibiotics for growth promotion and disease prevention in livestock. Though reducing emissions and fertilizer applications are not inherently synonymous with building soil health, initiatives promoting more sustainable crop production practices which contribute to soil health are also a target of major suppliers and retailers. For example, Walmart is working with suppliers and non-profit organizations such as the Environmental Defense Fund to encourage the adoption of nitrogen management tools and BMPs in crop production, and the Environmental Defense Fund is working with rice, corn and almond growers in California to help them gain access to greenhouse gas emission offset programmes (Pollock, 2015). United Suppliers, a wholesaler and retailer of agricultural inputs in the United States, developed the 'Sustain' platform/programme that requires that farmers adopt some combination of BMPs, such as no-till or cover crops in addition to optimizing and reducing fertilizer application (Betts, 2016). Field to Market, a non-profit organization based in Washington D.C., has also developed a sustainability tool called the Fieldprint® Calculator that helps growers measure and market the sustainability of their operation based upon a number of metrics (Miller, 2014).

7.1 Designing cost-effective policies for conservation practices in the USA

When a farmer participates in a federal or state incentive or cost-share programme, economists know that payment to the farmer is at least as large as the farmers' 'willingness to accept' (WTA, or the minimum amount they would need to be paid to be compensated to engage in a practice or change their behaviour). This means that the value of the benefit the farmer receives for implementing the practice less the costs of implementing the practice plus the value of the payment from the programme is greater than zero. However, we do not usually directly observe the costs or benefits of the practice to the farmer, which means that we cannot observe a farmer's WTA for implementing a practice. One potential solution to this issue is to design conservation payment programmes as reverse auctions, where farmers make offers to sell an environmental service or implement a practice or combination of practices in exchange for a value that they bid. Because they are competing against other farmers who may have lower offers, there is an incentive to offer a value close to the farmer's true WTA. However, the way in which reverse auctions are implemented can impact the cost-effectiveness and total environmental benefit of the programme (Claassen et al., 2008; Hellerstein et al., 2015). The U.S. Conservation Reserve Program is an example of a programme that uses a reverse auction design in its general signup, though economists continue to research the economic implications of CRP design and implementation (Hellerstein et al., 2015). This type of research is an example of how behavioural and experimental economics can be used to assess the impact of behavioural nudges on the practice and programme participation, including farmer participation in the conservation practices and programmes.

8 Future trends and conclusion

8.1 Researching the economics of soil health management systems

This chapter began by discussing some approaches economists use to empirically model farmer decision-making. In many cases, these approaches focus on modelling the factors that affect a farmer's decision to adopt a single practice – for example, no-till or reduced tillage. By controlling for farmer characteristics, input and output prices, crop choices and other variables, economists take into account the fact that these variables are related in an economic model. However, they are not necessarily and explicitly modelling the fact that the farmer may be making decisions about multiple soil health management practices at once, or how the costs or benefits of implementing one soil health practice may be affected by adopting other practices at the same time. For example, if the farmers are implementing several practices simultaneously as part of a soil conservation plan, the cost and benefit to the farmer of practice implementation depend upon whether the practices being implemented are substitutes or complements. This is also important from a public policy perspective, because it implies that paying farmers to implement some practices may also induce the adoption of other complementary practices, and that for practices that are substitutes, it may be more cost-effective to pay for a practice that is cheaper to implement but provides similar public benefits (Lichtenberg, 2004). Many existing cost-share and incentive programmes that pay farmers to engage in soil health practices pay fixed amounts for the adoption of individual practices, but might be more cost-effective if they considered the existence of complementarity or synergies between practices.

Several previous economic studies look at the farmer's decision to adopt more than one soil-health-related practice (joint or sequential adoption of practices). For example, Khanna (2001) considers the adoption of soil testing and variable rate technology, and finds that farmers with below-average soil quality realize the greatest gains in nitrogen productivity. In addition, her results suggest that for farmers that adopt the technologies, productivity is higher for full-time farm operators. This implies that for these farmers, the adoption of the two technologies might help the farmer to overcome risk aversion and uncertainty about soil quality, thereby reducing over-application of nitrogen. Wu and Babcock (1998) look at farmers' decision-making about tillage, crop rotations and soil testing. The authors find that soil testing alone does not reduce nitrogen application rates, but that a combination of soil testing with crop rotation or conservation tillage does. Future research is needed that looks at the economic and ecological implications of the adoption of multiple practices. For example, how does the cover crop termination method affect the marginal costs of planting a cover crop, and the profitability of no-till production following a cover crop? Does the joint adoption of cover crops and no-till increase herbicide use, and if so, does this have a trade-off in terms of costs of herbicide resistance, or a negative impact on soil biota (and, in turn, soil health)?

8.2 Linking management practices to changes in soil health and economic outcomes

Much previous agronomy and soil science research has focused on linking changes in management practices to changes in yield or profit. In addition, many of these studies are implemented as controlled trials, which makes it difficult to evaluate whether the effects can be generalized to diverse farms and growing conditions. At the same time, economic studies that look at the impact of practice adoption on farms have to control for selection in their studies, or the fact that some farmers are more likely to adopt practices due to variables the researcher cannot observe, such as soil quality or other edaphic variables.

Researchers can develop an understanding of how management practices affect soil health and economic outcomes for farmers by collecting and analysing data over time on management practices, soil health indicators, and economic variables. Such variables include input use and expenditures (e.g. labour, equipment, chemical and water use), crop choice, yields and revenues, and other farm business activities that impact the farm operation's bottom line. It is also important to understand how farmers and land managers make decisions about when to diagnose and treat soil health concerns, and whether soil health monitoring and testing are important tools in this process. Such studies are data-intensive, require data collection over time and necessitate interdisciplinary collaboration. This makes them difficult and costly to conduct. But, without these types of studies being conducted in different locations across the United States and the world, it is difficult to assess the private costs and benefits of implementing soil health practices and how to develop and target public policy incentives that would be appropriate to induce the adoption of practices with public benefits.

8.3 Valuing the public benefits of changes in soil health at multiple scales

Linking farm practices and changes in soil health to the provision of public benefits and valuing these benefits require more research. Does a change in a practice on one farm affect the quality of a neighbour's well water, or does it require all farmers within a 10-mile radius to adopt several management practices? If all farmers in the state of Maryland plant a winter cover crop or stop applying manure to fields in the fall, does this have a measurable impact on the water quality of the Chesapeake Bay, and what would be the costs of such a policy? Would the payments needed to induce universal adoption make such a programme cost-effective relative to the expected benefit to water quality?

Published by Burleigh Dodds Science Publishing Limited, 2018.

9 Where to look for further information

U.S. National Agricultural Statistics Service 'Quick Stats' Interface
(www.quickstats.nass.usda.gov)

USDA's National Agricultural Statistics Service provides summary information for the U.S. Agricultural Census and several other surveys via the Quick Stats interface. These include statistics on national- and state-level adoption of several soil health practices, such as cover crops.

Eurostat Agri-Environmental Indicator – Commitments
(http://ec.europa.eu/eurostat/statistics-explained/index.php/Agri-environmental_indicator_-_commitments)
Statistics summarizing coverage of agri-environmental measures in the EU, including land area, expenditures and other measures.

Conservation Technology Information Center
(www.ctic.org)

The Conservation Technology Information Center (CTIC) is a non-profit, public–private partnership that uses public funding to 'champion, promote and provide information on technologies and sustainable agricultural systems that conserve and enhance soil, water, air and wildlife resources and are productive and profitable'. Among other activities, CTIC provides training for industry consultants and professionals, and collects data and conducts surveys about the use of conservation practices, such as no-till agriculture and cover crops.

Sustainable Agriculture Research and Education
(www.sare.org)

Sustainable Agriculture Research and Education (SARE) partners with the University of Maryland and receives funding from the USDA's National Institute for Food and Agriculture to provide information and financial and technical assistance to farmers and ranchers to support sustainable agricultural practices. SARE has published several books on soil health and soil health practices, and has diverse outreach and extension resources available on its website. Since 2012, SARE and CTIC and partners have been conducting an annual survey that asks farmers about their experiences with cover crops.

Center for Behavioural and Experimental Agricultural Research
(www.centerbear.org)
Center for Behavioural and Experimental Agricultural Research is a collaboration among economists at the University of Delaware, Johns Hopkins University and Cornell University to conduct and fund behavioural and experimental research that relates to agricultural and environmental issues. This includes research on USDA conservation programme performance, as well as farmer adoption of conservation and soil health practices.

Soil Health Institute's Research Landscape Tool
(www.soilhealthinstituteresearch.org)

The Soil Health Institute is an independent non-profit organization dedicated to supporting soil health and soil health research. In addition to seeking out and distributing funds for soil health research, the institute has developed a number of resources on its website. One of these resources is a virtual repository of soil health research article abstracts (and links to full publications, though some are only available through subscription or paid services), including research on the economics of soil health.

National Sustainable Agriculture Information Service Case Studies
(www.attra.ncat.org)

The National Sustainable Agriculture Information Service (ATTRA) has published a number of case studies and other resources that address questions about the economics of soil health. In order to find these case studies on their website, go to 'Publications' and click on 'Soils and Compost'.

Landcare Australia
(https://landcareaustralia.org.au/)

Landcare was established as a national programme in Australia in 1989. Its cornerstones are 'being community owned and driven, being bi-partisan in nature and encouraging integrated management of environmental assets, including productive farmland and a sustainable approach to private land management'. The movement has spread to more than 20 other countries globally.

10 References

Al-Kaisi, M., Archontoulis, S. and Kwaw-Mensah, D. 2016. Soybean spatiotemporal yield and economic variability as affected by tillage and crop rotation. *Agronomy Journal* 108: 1267–80.
Arbuckle, J. G. and Ferrell, J. 2012. Attitudes Toward Cover Crops in Iowa: Benefits and Barriers. Extension and Outreach Publications. 22. http://lib.dr.iastate.edu/extension_pubs/22.
Arbuckle, J. G. and Roesch-McNally, G. 2015. Cover crop adoption in Iowa: The role of perceived practice characteristics. *Journal of Soil and Water Conservation* 70(6): 418–29.
Baumgart-Getz, A., Prokopy, L.S. and Floress, K. 2012. Why farmers adopt best management practice in the United States: A meta-analysis of the adoption literature. *Journal of Environmental Management* 96(1): 17–25.
Bennett, L. T., Mele, P. M., Annett, S. and Kasel, S. 2010. Examining links between soil management, soil health, and public benefits in agricultural landscapes: An Australian perspective. *Agriculture, Ecosystems & Environment* 139(1): 1–12.
Betts, L. 2016. Corn and soybean producers find retailer who supports sustainable farming. *Corn and Soybean Digest*. Viewed 5 April 2017. http://www.cornandsoybeandigest.com/corn/corn-and-soybean-producers-find-retailer-who-supports-sustainable-farming
Booker, K., Huntsinger, L., Bartolome, J. W., Sayre, N. F. and Stewart, W. 2013. What can ecological science tell us about opportunities for carbon sequestration on arid rangelands in the United States? *Global Environmental Change* 23(1): 240–51.

Bowman, M., Wallander, S. and Lynch, L. 2016. An economic perspective on soil health. *Amber Waves*. September 2016. https://www.ers.usda.gov/amber-waves/2016/september/an-economic-perspective-on-soil-health/
Carlisle, L. 2016. Factors influencing farmer adoption of soil health practices in the United States: A narrative review. *Agroecology and Sustainable Food Systems* 40(6): 583–613.
Cartwright, L. 2015. Adding cover crops to a corn-soybean rotation. Missouri Cover Crop Economics Case Study 1. Viewed 14 April 2017. https://www.nrcs.usda.gov/wps/PA_NRCSConsumption/download?cid=nrcseprd352827&ext=pdf
Chavas, J. P., Chambers, R. G. and Pope, R. D. 2010. Production economics and farm management: A century of contributions. *American Journal of Agricultural Economics* 92: 356–75.
Claassen, R., Duquette, E. and Horowitz, J. 2013. Additionality in agricultural conservation payment programs. *Journal of Soil and Water Conservation* 68(3):74–8.
Claassen R., Breneman, V., Bucholtz, S., Cattaneo, A., Johansson, R. and Morehart, M. 2004. *Environmental Compliance in Agricultural Policy: Past Performance and Future Potential.* Agricultural Economic Report AER-832, U.S. Department of Agriculture, Economic Research Service, June, Washington, DC. .
Claassen, R., Cattaneo, A. and Johansson, R. 2008. Cost-effective design of agri-environmental payment programs: U.S. experience in theory and practice. *Ecological Economics* 4(1): 737–52.
Claassen, R., Bowman, M., Breneman, V., Wade, T., Williams, R., Fooks, J., Hansen, L., Lovanna, R. and Loesch, C. 2017. Conservation Compliance: How Farmer Incentives are Changing in the Crop Insurance Era. Economic Research Report ERR-234, U.S. Department of Agriculture, Economic Research Service, July.
Conservation Cropping Systems Initiative. 2016. A 321% Return on Investment. *Economics of Conservation Series.* Viewed 14 April 2017. http://ccsin.iaswcd.org/wp-content/uploads/2016/07/CaseStudy-Rulon4.pdf
Cook, R. L. and Trlica, A. 2016. Tillage and fertilizer effects on crop yield and soil properties over 45 years in Southern Illinois. *Agronomy Journal* 108: 415–26.
Cork, S. T., Eadie, L. A., Mele, P. A., Price, R. I. and Yule, D. 2012. The relationships between land management practices and soil condition and the quality of ecosystem services delivered from agricultural land in Australia. Kiri-ganai Research Pty Ltd. Camberra, Australia. 2012 September.
Dale, V. H. and Polasky, S. 2007. Measures of the effects of agricultural practices on ecosystem services. *Ecological Economics* 64(2): 286–96.
Dumanski, J., Reicosky, D. C. and Peiretti, R. A. 2014. Global pioneers in soil conservation: Common elements and lessons learned. *International Soil and Water Conservation Research* 2(1): 1–4.
Dunn, M., Ulrich-Schad, J. D., Prokopy, L. S., Myers, R. L., Watts, C. R. and Scanlon, K. 2016. Perceptions and use of cover crops among early adopters: Findings from a national survey. *Journal of Soil and Water Conservation* 71(1): 29–40.
Foster, A. D. and Rosenzweig, M. R. 1995. Learning by doing and learning from others: Human capital and technical change in agriculture. *Journal of Political Economy* 103(6):1176–209.
Gosnell, H., Robinson-Maness, N. and Charnley, S. 2011. Engaging ranchers in market-based approaches to climate change mitigation: Opportunities, challenges, and policy implications. *Rangelands* 33(5): 20–4.
Hansen, L. and Ribaudo, M. 2008. Economic measures of soil conservation benefits: Regional values for policy assessment. USDA Technical Bulletin No. 1922, 32p, Washington DC .
Hanson, J. C., Lichtenberg, E., Decker, M. and Clark, A. 1993. The profitability of no-tillage corn following a hairy vetch cover crop in the mid-Atlantic region. *Journal of Production Agriculture* 6: 333–437.
Hatfield, J. L., Sauer, T. J. and Prueger, J. H. 2001. Managing soils to achieve greater water use efficiency. *Agronomy Journal* 93(2): 271–80.
Hellerstein, D., Higgins, N. and Roberts, M. 2015. Options for Improving Conservation Programs: Insights from Auction Theory and Economic Experiments. Economic Research Report No. (ERR-181), 53p, Washington DC.

Hudson, B. 1994. Soil water and available water capacity. *Journal of Soil and Water Conservation* 49(2): 189–94.
Huguenin, M. T., Leggett, C. G. and Paterson, R. W. 2006. Economic valuation of soil fauna. *European Journal of Soil Biology* 42: S16–S22.
Karlen, D. L., Cambardella, C. A., Kovar, J. L. and Colvin, T. S. 2013a. Soil quality response to long-term tillage and crop rotation practices. *Soil and Tillage Research* 133: 54–64.
Karlen, D. L., Kovar, J. L., Cambardella, C. A. and Colvin, T. S. 2013b. Thirty-year tillage effects on crop yield and soil fertility indicators. *Soil and Tillage Research* 130: 24–41.
Kemper, W. D., Schneider, N. N. and Sinclair, T. R. 2011. No-till can increase earthworm populations and rooting depths. *Journal of Soil and Water Conservation* 66: 13A–17A.
Khanna, M. 2001. Sequential adoption of site-specific technologies and its implications for nitrogen productivity: A double selectivity model. *American Journal of Agricultural Economics* 83(1): 35–51.
Knight, L. 2016. Developing Conservation Case Studies for Decision-making. Economics Technical Note 200-ECN-4. USDA Natural Resources Conservation Service. Viewed 10 July 2017. https://www.nrcs.usda.gov/wps/PA_NRCSConsumption/download?cid=nrcseprd1298424&ext=docx
Lal, R. 2004. Soil carbon sequestration impacts on global climate change and food security. *Science* 304(5677): 1623–7.
Lichtenberg, E. 2004. Cost-responsiveness of conservation practice adoption: A revealed preference approach. *Journal of Agricultural and Resource Economics* 29(3): 420–35.
Lichtenberg, E., Hanson, J. C., Decker, M. and Clark, A. 1994. The profitability of legume cover crops in the mid-Atlantic region. *Journal of Soil and Water Conservation* 49: 582–5.
Maryland Agricultural Water Quality Cost-Share Program, 2016. Opening doors for Maryland's Conservation Farmers: MACS Annual Report. http://mda.maryland.gov/resource_conservation/counties/MACS_FINAL.pdf.
Mausbach, M. J. and Dedrick, A. R. 2004. The length we go. *Journal of Soil and Water Conservation* 59(5): 96–103.
Meitner, K. 1999. Making the most of it! *No-Till Farmer*. Viewed 11 July 2017. https://www.no-tillfarmer.com/articles/3739-making-the-most-of-it?v=preview
Miller, S. 2014. Fieldprint calculator uses a USDA tool to help farmers track water quality improvement. *USDA Blog*. Viewed 6 April 2017. https://www.usda.gov/media/blog/2014/04/16/fieldprint-calculator-uses-usda-tool-help-farmers-track-water-quality
No-Till Farmers. 2017. National No-Tillage Conference. Viewed 11 July 2017. https://www.no-tillfarmer.com/nntc
Pollock, D. 2015. Pilot project to generate greenhouse gas credits from almonds, corn. *Western Farm Press*. Viewed 8 April 2017. http://www.westernfarmpress.com/tree-nuts/pilot-project-generate-greenhouse-gas-credits-almonds-corn
Prokopy, L. S., Floress, K., Klotthor-Weinkauf, D. and Baumgart-Getz, A. 2008. Determinants of agricultural best management practice adoption: Evidence from the literature. *Journal of Soil and Water Conservation* 63(5): 300–11.
Ransom, M. 2014. Case Study: Transition to Cover Crop for a Beginning Farmer. Viewed 10 April 2017. https://www.nrcs.usda.gov/wps/PA_NRCSConsumption/download?cid=stelprdb1256123&ext=pdf
Rawls, W. J., Pachepsky, Y. A., Ritchie, J. C., Sobecki, T. M. and Bloodworth, H. 2003. Effect of soil organic carbon on soil water retention. *Geoderma* 116(1): 61–76.
Reimer, A. P., Weinkauf, D. K. and Prokopy, L. S. 2012. The influence of perceptions of practice characteristics: An examination of agricultural best management practice adoption in two Indiana watersheds. *Journal of Rural Studies* 28(1): 118–28.
Ribaudo, M., Greene, C., Hansen, L. and Hellerstein, D. 2010. Ecosystem services from agriculture: steps for expanding markets. *Ecological Economics* 69(11): 2085–92.
Robinson, D. A., Hockley, N., Dominati, E., Lebron, I., Scow, K. M., Reynolds, B., Emmett, B. A., Keith, A. M., de Jonge, L. W., Schjonning, P., Moldrup, P., Jones, S. B. and Tuller, M. 2012. Natural capital, ecosystem services, and soil change: why soil science must embrace an ecosystems approach. *Vadose Zone Journal* 11(1).

Rogers, E. M. 2003. *Diffusion of Innovations*, 5th Edition. Free Press, New York, NY.

Schoengold, Karina, Ding, Y. and Headlee, R. 2015. The impact of AD HOC disaster and crop insurance programs on the use of risk-reducing conservation tillage practices. *American Journal of Agricultural Economics* 97(3): 897–919.

Snapp, S. S., Swinton, S. M., Labarta, R., Mutch, D., Black, J. R., Leep, R., Nyiraneza, J. and O'Neil, K. 2005. Evaluating cover crops for benefits, costs and performance within cropping system niches. *Agronomy Journal* 97(1): 322–32.

Soil and Water Conservation Society. 2017. 2017 National Conference on Cover Crops. Viewed 10 July 2017. http://www.swcs.org/en/conferences/2017_national_conference_on_cover_crops/

Sunding, D. and Zilberman, D. 2001. Chapter 4: The agricultural innovation process: Research and technology adoption in a changing agricultural sector. . In: *Handbook of Agricultural Economics*, Volume 1 (Part A), North Holland, pp. 207–61.

Swinton, S., Frank Lupi, Philip Robertson, G., Hamilton, S. K. 2007. Ecosystem services and agriculture: Cultivating agricultural ecosystems for diverse benefits. *Ecological Economics* 64(2): 245–52

Tallman, S. 2012a. No-till case study, Brown's Ranch: Improving soil health improves the bottom line. National Sustainable Agriculture Information Service, National Center for Appropriate Technology. Viewed 6 April 2017. https://attra.ncat.org/attra-pub/download.php?id=414

Tallman, S. 2012b. No-till case study, Bauer Farm: Cover crop cocktails on former CRP land. National Sustainable Agriculture Information Service, National Center for Appropriate Technology. Viewed 6 April 2017. https://attra.ncat.org/attra-pub/download.php?id=420

Tallman, S. 2012c. No-Till Case Study, Richter Farm: Cover Crop Cocktails in a Forage-Based System . National Sustainable Agriculture Information Service, National Center for Appropriate Technology. Viewed 14 July 2017. https://attra.ncat.org/attra-pub/download.php?id=417

Tomer, M. D. and Locke, M. A. 2011. The challenge of documenting water quality benefits of conservation practices: A review of USDA-ARS's conservation effects assessment project watershed studies. *Water Science and Technology* 64(1): 300–10.

USDA Natural Resources Conservation Service. 2016. Profiles in Soil Health. Viewed 14 April 2017. https://www.nrcs.usda.gov/wps/portal/nrcs/detail/national/soils/health/?cid=stelprdb1083183

USDA Natural Resources Conservation Service, New York. 2016. Economic Case Study – Farmer Profile, Dave Magos. Viewed 3 April 2017. https://www.nrcs.usda.gov/wps/portal/nrcs/detail/ny/soils/health/?cid=nrcseprd1298872

USDA Natural Resources Conservation Service. 2017. Healthy, productive soils checklist for growers. Fact Sheet. Viewed 10 July 2017. https://www.nrcs.usda.gov/wps/PA_NRCSConsumption/download?cid=nrcseprd1318014&ext=pdf

U.S. Department of Agriculture, Natural Resources Conservation Service. 2016. Effects of Conservation Practice Adoption on Cultivated Cropland Acres in Western Lake Erie Basin, 2003–06 and 2012, 120p https://www.nrcs.usda.gov/Internet/FSE_DOCUMENTS/nrcseprd889806.pdf.

Wade, T., Claassen, R. and Wallander, S. 2015. Conservation-Practice Adoption Rates Vary Widely by Crop and Region. Economic Research Service EIB 147, Washington DC., 40p.

Wallander, S., Aillery, M., Hellerstein, D. and Hand, M. 2013. The Role of Conservation Programs in Drought Risk Adaptation. Economic Research Service ERR 148, Washington DC.

Watkins, B., Anders, M. M. and Windham, T. E. 2004. An Economic Comparison of Alternative Rice Production Systems in Arkansas. *Journal of Sustainable Agriculture* 24(4): 57-78.

Wu, J. J. and Babcock, B. A. 1998. The choice of tillage, rotation, and soil testing practices: Economic and environmental implications. *American Journal of Agricultural Economics* 80 (3): 494–511.

Zimmermann, A. and Britz, W. 2016. European farms' participation in agri-environmental measures. *Land Use Policy* 50: 214–28.

Part 2

Soil structure and composition

Chapter 6

Soil texture and structure: role in soil health

Rainer Horn, Heiner Fleige and Iris Zimmermann, Institute for Plant Nutrition and Soil Science, Christian-Albrechts-University of Kiel, Germany

1 Introduction

Soils are complex bodies formed on the land surface consisting of liquid, gaseous and solid components. They contain inorganic and organic materials, including a large number and variety of living organisms. They must be used carefully in accordance with their resilience and elasticity for the long-term maintenance of key properties and processes of the soil, if they are to meet the demands of a growing world population (Lal and Shukla 2004).

Soils are the most complex biomaterials on the planet, and soil health is governed by the self-regulating biological processes within the soil system. Poor land use and soil management will result in degradation and reduced soil health which cannot be quickly regained. Because soil amelioration approaches are seldom effective, or effective only over very long periods, the risk of losing the basis for sustaining our future life is considerable. Taking into account the expected growth in world population, a diminishing area of soils will need to deliver at least 70% more food than is produced today. Such requirements can only be met by a steep increase in soil productivity, properties and functions within their resilience limits. Consequently, more understanding of the available soil parameters and their relevance for soil health is necessary in order to understand soil processes and to forecast functions such as filter and buffer ability, soil strength and pore rigidity.

2 Particle size and soil health

Soils are natural bodies characterized by properties such as texture. With regard to soil texture or grain size distribution, grain sizes are typically divided into two main subunits: fine earth particles (diameter smaller than 2 mm) and skeleton particles (diameter exceeding 2 mm). Within these units, a further subdivision differentiates between clay, silt, sand, gravel, stone, boulders and coarser particles (Fig. 1).

http://dx.doi.org/10.19103/AS.2017.0033.07

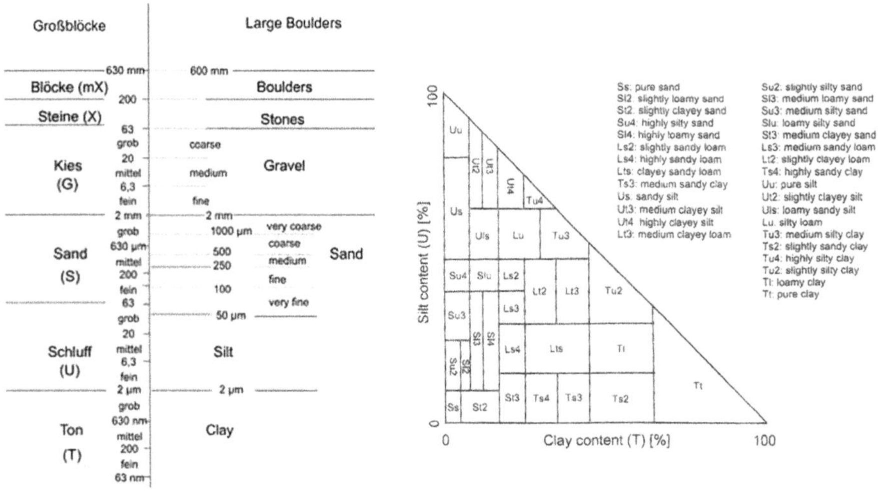

Figure 1 Classification of particle size fractions. German (left) and American nomenclature (right) (Scheffer and Schachtschabel, 2016), and German soil texture classification; T = clay, S = sand, U = silt, L = loam; 2 = slight, 3 = medium, 4 = high (sources: Ad-hoc-AG Boden, 2005; Bormann, 2007).

A mixture of particle sizes rather than a single grain size is normally found in soils. These mixtures cause different particle arrangements within a given soil volume and, as a result, the pores in the matrix will vary in size, arrangement and continuity. In addition to these primary effects of particle formation and arrangement within a given soil volume, secondary transport and rearrangement within the soil causes alteration in ecological properties and functions. Due to slaking, displacement and precipitation, a coarser grain size distribution in the topsoil and a finer clay-enriched subsoil result in microscopic layering close to the pore walls (For more information see Scheffer and Schachtschabel, 2016; Hartge and Horn, 2016; Jury and Horton, 2004). The following sections discuss the texture-dependent chemical and physical soil properties. These properties are the basis for understanding the interactions between texture, structure, swelling/shrinkage and biological activity of soils which are dealt with thereafter.

2.1 Particle size and cation exchange capacity

The specific particle surface S ($m^2\ g^{-1}$) increases with decreasing particle diameter, with a further differentiation at a given diameter, depending on the clay mineralogy (Table 1). As a result, the cation exchange capacity (CEC – the capacity of the soil to hold exchangeable cations, the form many nutrients take) is higher when the clay content is greater, and when the clay mineral layers are widened. This results in a higher nutrient storage capacity. The soil buffering capacity (the ability to absorb nutrients) is also higher when particles are smaller. Based on several thousand measurements, Table 2 shows the CEC of various soil texture classes (as shown in Fig. 1), with the CEC of clay soils being the highest.

Table 1 Specific surface area of ideally shaped particles (top) and specific surface area as well as cation exchange capacity of various clay minerals (below) (taken from Yong and Warkentin (1966))

Specific surface area S of idealized shapes representing prototype				
Particles found in soil				
Particle	d_{eff} (cm)	Mass (g)	Area (m^2)	S (m^2 g^{-1})
Gravel	2×10^{-1}	1.13×10^{-2}	1.26×10^{-5}	0.011
Sand	5×10^{-3}	1.77×10^{-7}	7.85×10^{-9}	0.044
Silt	2×10^{-4}	1.13×10^{-11}	1.26×10^{-11}	1.11
Clay	2×10^{-4}	8.48×10^{-15}	6.28×10^{-12}	740
Specific surface area S, cation exchange capacity CEC of various clay minerals				
Clay mineral	S (m^2 g^{-1})	CEC ($cmol_c$ kg^{-1})		
Kaolinites	5–20	3–15		
Micas (illites)	100–200	10–40		
Vermiculites	300–500	100–150		
Smectites	700–800	80–150		

Table 2 Potential cation exchange capacity (CEC_{pot}, $cmol_c$ kg^{-1} soil) for various soil texture classes (German classification) (Ad-hoc-AG Boden 2005)

Texture	CEC_{pot} ($cmol_c$ kg^{-1})	Texture	CEC_{pot} ($cmol_c$ kg^{-1})
Ss, Su2	2	Lu, Ts4	15
Su3, Su4, Sl2	4	Lt2, Tu4	17
Us	5	Lts	19
St2, Sl3, Uu	6	Ts3	20
Slu, Sl4, Ut2, Uls	9	Tu3	21
Ut3, St3	11	Lt3	22
Ls3, Ls4	12	Ts2, Tu2	28
Ls2	13	Tl	29
Ut4	14	Tt	38

2.2 Particle size and hydraulic conductivity

A direct link between grain size and pore diameters results in varying pore size distribution and hydraulic properties in differently textured soils. The plant available water capacity is smaller where the grains are coarser (Table 3). Hydraulic conductivity is a measure of the ability of a soil to allow water movement. With decreasing particle diameter, the air capacity declines, while silty textured soils store the highest amount of plant available water. The saturated hydraulic conductivity of homogeneous silt and clay soils is lower

Table 3 Air capacity and plant available water capacity as a function of bulk density and texture (Ad-hoc-AG Boden 2005)

Texture	Air capacity Pore ø > 50 μm (pF < 1.8)			Plant available, water capacity Pore ø 0.2–50 μm (pF 4.2–1.8)		
Bulk density (g cm^{-3})	<1.4	1.4–1.6	>1.6	<1.4	1.4–1.6	>1.6
Ss	36	32	27	9	7	7
Sl2	23	18	13	20	18	17
Sl3	18	15	10	22	18	17
Sl4	18	12	8	22	18	15
Slu	14	10	7	23	21	19
St2	24	20	15	18	16	13
St3	18	14	9	18	15	12
Su2	24	21	15	20	18	17
Su3	17	14	10	25	21	20
Su4	14	11	8	27	23	21
Ls2	13	9	6	21	16	14
Ls3	15	9	6	21	16	14
Ls4	15	11	7	20	16	13
Lt2	11	7	5	18	14	11
Lt3	8	5	3	17	12	10
Lts	10	6	5	17	14	11
Lu	12	7	4	21	17	15
Uu	10	7	3	30	26	23
Uls	13	8	5	24	22	21
Us	11	9	4	28	25	22
Ut2	10	6	3	28	26	23
Ut3	11	6	3	26	25	23
Ut4	12	7	3	23	21	19
Tt	4	3	2	15	13	12
Tl	5	4	3	15	13	11
Tu2	5	4	3	16	12	10
Tu3	8	6	3	17	13	10
Tu4	10	6	3	19	17	16
Ts2	5	4	3	16	13	12
Ts3	7	6	5	16	13	11
Ts4	13	10	6	17	14	11
Sand						
fS, fSms, fSgs	34	31	23	10	9	8
mS, mSfs, mSgs	36	32	26	9	6	5
gS	38	33	29	8	5	4

than that of sandy soils (Table 4). In sandy soils the decline in the unsaturated hydraulic conductivity is much greater than in the silt or clay soils. The capillary rise is therefore smaller in sandy soils than in silt, while clay soils again only allow a smaller capillary rise. Table 5 shows the texture- and bulk density-dependent capillary rise (mm d^{-1}) as a function of the distance between the groundwater surface and the effective rooting depth (dm). This shows that silty soils allow water flow from deeper soil levels, while both sandy and clay soils provide less water for plant uptake, and only from a shallower depth. These differences are explained by the unsaturated hydraulic conductivity curve pattern, which remains nearly constant in silty soils until the drying is well advanced, whereas the greater proportion of air-filled pores in coarser soils results in a more rapid decline.

2.3 Particle size and gas transport function

The coarser the particles, the better aerated the soil (i.e. the air capacity is greater). Studies on the effects of soil structure on gas transport reveal a strong correlation between the pore size distribution and the air entry value. The air entry value is the soil water potential at which gas diffusion to the aggregate centre increases because water menisci are removed from continuous pores. In prisms of sandy-loamy texture, O_2 partial pressure increased at a soil water potential of about −15 kPa, while polyhedrons with loamy-clay texture became aerated at more negative water potentials of < −60 kPa (Zausig et al. 1993). According to Fick's law, gas flow depends on the diffusion coefficient and concentration gradient. Assuming oxygen content of the air surrounding the aggregate to be 21%, oxygen content in aggregate centre will be lower as the transport within the aggregate would require additional time. If chemical and microbial oxygen demand inside the aggregate exceeds the oxygen supply, zones of low oxygen partial pressure or even anoxic microsites (i.e. devoid of oxygen) may develop.

The intensity of anoxia and the diameter of anoxic centres can be controlled not only by microbial and chemical oxygen demand, but also by parameters such as aggregate hydraulic conductivity and pore size distribution, that is, by the soil texture. Haas et al. (2016) detected a strong correlation between aeration and mechanical stresses applied during soil management. Depending on the internal soil strength, even methane emissions occurred with stresses exceeding the internal soil strength. Thus, sink–source effects of soils caused by land management alter their health status and must be considered in more detail in order to avoid further soil degradation.

2.4 Particle size and mechanical strength

An approach often used to quantify the mechanical strength of homogenized soils is based on the grain size distribution pattern where the degree of unevenness U = D60/D10. D60 is the diameter at 60% of the grain size distribution curve, while D10 is the corresponding smaller diameter. The greater the value (U > 15), the more compressible the samples are while for values < 5, the samples are mostly incompressible. Thus, distributions which are almost mono-grain sized are defined as rigid, while loamy material is highly compactable. However, this method is only valid for coarse-grained substrates. If soils contain more than 10% clay, the corresponding lower diameter cannot be determined using the normal measurement technique applied, because the texture analysis is limited to 2 µm.

The physical correlation between mechanical strength and soil texture is based on multi-regression analyses, which give reliable data for further extrapolations under *in situ* conditions (Horn and Fleige 2003). The outcome can be used to quantify anthropo-,

Table 4 Saturated hydraulic conductivity as a function of texture and bulk density (Ad-hoc-AG Boden 2005)

Saturated hydraulic conductivity (cm d^{-1})			
Bulk density (g cm^{-3})	<1.4	1.4–1.6	>1.6
Ss	375	340	230
Sl2	161	98	52
Sl3	98	65	29
Sl4	106	42	21
Slu	60	28	13
St2	179	118	68
St3	114	42	24
Su2	174	127	66
Su3	88	59	31
Su4	58	38	17
Ls2	53	23	10
Ls3	74	23	11
Ls4	68	36	11
Lt2	33	13	6
Lt3	20	7	3
Lts	31	10	7
Lu	45	16	6
Uu	32	13	2
Uls	49	20	7
Us	37	22	5
Ut2	32	12	2
Ut3	41	12	3
Ut4	45	13	3
Tt	4	3	2
Tl	8	6	2
Tu2	8	3	2
Tu3	18	9	3
Tu4	33	12	3
Ts2	8	5	3
Ts3	15	11	8
Ts4	51	38	8
Sand			
fS, fSms, fSgs	410	300	200
mS, mSfs, mSgs	675	490	250
gS	2100	835	335

Table 5 Capillary rise (mm d^{-1}) as a function of the distance between the groundwater surface and the end of the effective rooting depth (dm) (Ad-hoc-AG Boden 2005)

	Capillary rise (mm d^{-1})									
	Distance between the groundwater surface and the effective rooting depth (dm)									
Texture	1	3	4	6	8	10	12	14	20	25
Ss	>5	5.0	3.5	1.1	0.3	0.1	–	–		
Sl2	>5	>5	3.1	1.0	0.4	0.2	–	–	–	–
Sl3	>5	5.0	2.7	0.9	0.4	0.2	0.1	–	–	–
Sl4	>5	4.5	3.1	0.9	0.4	0.2	0.1	–	–	–
Slu	>5	4.3	2.4	1.0	0.4	0.2	0.1			
St2	>5	4.8	2.2	0.6	0.2	0.1	–	–	–	–
St3	>5	5.0	2.7	0.3	0.1	–	–	–	–	–
Su2	>5	>5	4.0	1.3	0.5	0.2	–	–	–	–
Su3	>5	5.0	2.6	0.9	0.4	0.2	0.1			
Su4	>5	>5	4.2	1.6	0.8	0.4	0.2	0.1	–	–
Ls2	>5	>5	2.6	1.3	0.7	0.2	–	–	–	–
Ls3	>5	>5	2.1	1.1	0.5	0.2	–	–	–	–
Ls4	>5	>5	2.0	0.9	0.5	0.1	–	–	–	–
Lt2	>5	2.3	1.2	0.4	0.2	0.1	–	–	–	–
Lt3	>5	2.9	1.6	0.6	0.3	0.1	–	–	–	–
Lts	>5	2.4	1.3	0.5	0.2	0.1	–	–	–	–
Lu	>5	>5	>5	3.1	1.5	0.8	0.4	0.2	–	–
Uu	>5	>5	>5	>5	>5	5.0	3.4	2.4	0.9	0.4
Uls	>5	>5	>5	3.2	1.7	1.0	0.6	0.4	0.1	–
Us	>5	>5	>5	>5	4.4	2.7	1.7	1.1	0.3	0.1
Ut2	>5	>5	>5	>5	>5	4.4	3.0	2.1	0.8	0.3
Ut3	>5	>5	>5	>5	>5	4.2	2.9	2.0	0.8	0.3
Ut4	>5	>5	>5	>5	3.3	2.1	1.4	1.0	0.4	0.1
Tt	4.9	1.2	0.8	0.1	0.2	0.1	–	–	–	–
Tl	5.0	1.3	0.7	0.3	0.1	–	–	–	–	–
Tu2	5.0	1.3	0.7	0.3	0.2	0.1	–	–	–	–
Tu3	>5	3.1	2.4	0.7	0.3	0.2	–	–	–	–
Tu4	>5	>5	5.0	2.7	0.9	0.4	0.2	0.1	–	–
Ts2	5.0	1.5	1.0	0.3	0.1	–	–	–	–	–
Ts3	>5	2.5	1.5	0.3	0.1	–	–	–	–	–
Ts4	>5	3.5	2.0	0.3	0.1	–	–	–	–	–
Sand										
fS, fSms, fSgs	>5	>5	>5	1.9	0.4	0.2	–	–	–	–
mS, mSfs, mSgs	>5	>5	3.0	0.5	0.2	–	–	–	–	–
gS	>5	1.5	0.9	0.2	–	–	–	–	–	–

pedo- and geogenic impacts on soil strength and pore rigidity. They also define the limits for sustainable soil management and soil health. The regression equations are valid for soil texture classes. General soil properties, such as organic carbon, bulk density, inorganic carbon and most importantly, cohesion and angle of internal friction determined by the Mohr–Coulomb equation are applied as independent parameters. If the parameters of cohesion and angle of internal friction for various texture classes are the prime considerations, it is clear that dry sandy soils are almost without cohesion and have a medium angle of internal friction of approx. 25–30°, which resembles the slope of dry sand dunes.

With decreasing particle diameter (from sand to silt and clay), the shape of the particles as well as their arrangement becomes more important, especially the angle of internal friction. Irrespective of the arrangement, the cohesion value under saturated conditions remains close to 0 kPa, while the angle of internal friction decreases for clayey material to approx. 15–20°. If, however, the water menisci in textured soil samples increase due to more negative matric potentials (pF 1.8 versus pF 2.5), the cohesion and angle of internal friction values also increase (see Table 6). These data prove that soil structure has an important effect on internal soil strength as will be discussed in the following section.

2.5 Optimum particle size

The preceding sections show that silty soils can be defined as optimal when considering the definition of healthy soils with respect to plant growth, filter and buffer processes. They guarantee sufficient aeration, high levels of plant available water as well as perfect saturated and especially unsaturated hydraulic conductivity. They also provide very good cation exchange and ion buffering capacity. This evaluation will, however, be modified if *in situ* conditions are included. Climatic conditions, wetting/drying-induced swell/shrink processes as well as biological interactions in soils alter the mechanical strength, especially the accessibility and availability of particle or pore wall surfaces (Hillel 1998).

3 Soil aggregate formation and soil health

Soils containing clay (particle size < 2 μm), or even pure sand soils with a certain salt content, tend to form aggregates. Usually, the process occurs when soils shrink and swell and is further enhanced by biological activity. Aggregates may show great variation in size from crumbs (diameter < 2 mm) to angular (polyhedral) or subangular blocks of 0.005–0.02 m, or even, to prisms or columns of more than 0.1 m (Fig. 2).

During the initial shrinkage, capillary forces pull mineral particles together, increase the number of contact points and result in a higher bulk density. The initial aggregates always have rectangular-shaped edges because under these conditions, stress release occurs perpendicular to the initial crack and stress remains parallel to the crack (strain-induced fracturing). However, due to the increased mechanical strength, particle mobility declines and causes the formation of non-rectangular shear plains in the following crack generation. These are created after repeated swelling and shrinking processes and result in fractures in which the value of the angle of internal friction determines the deviation from 90° (Or and Ghezzehei 2002).

Table 6 Effect of soil texture and structure on cohesion (c) and angle of internal friction (φ) as a function of pre-drying (pF value) (Horn and Fleige 2003)

		pF 1.8		pF 2.5	
Soil texture classes	Soil structure	c (kPa)	φ (′)	c (kPa)	φ (′)
X	sin	0	25	0	26
Ss, Su2-4	sin	8	26	12	28
Sl2-4	sin	8	30	10	32
Slu	coh/pri	12	34	14	37
	pol	15	39	18	41
St2-3	coh/pri	13	38	15	40
	pol	20	41	23	43
Uu, Us	sin	2	29	4	30
Us	coh/pri	10	35	15	39
	cru	8	37	12	40
Ut2-4	coh/pri	12	35	26	37
Uls	pol	18	35	20	38
	sub/cru	44/40	40/43	50/45	42/45
Lu	coh/pri	10/12	30/35	15	32/36
	pol	16	35	18	37
	sub/cru	34	38/39	44	40/46
Ls2-4	coh/pri	10	22/25	14	31/33
Lts	pol	19	30	26	35
	sub/cru	26/22	36/38	38/33	39/42
Lt2-3	sin	1	19	2	23
	coh/pri	15	28/32	26/34	36/38
	pol	30	36	41	40
	sub	46	39	66	43
Tu	coh/pri	32	22/26	45	30/32
	pol	40	30	70	34
	sub	45	36	40	42
Tl	sin	0	16	0	20
Tl/Tt	coh/pri	30/40	24/32	34/45	38/42
Ts2-4	pol	50	44	60	48
	sub	50	48	70	56

X = coarse fragment (> 80 vol.-%), sin = single grain, coh = coherent, pri = prismatic. pol = blocky, sub = subangular, cru = crumb.

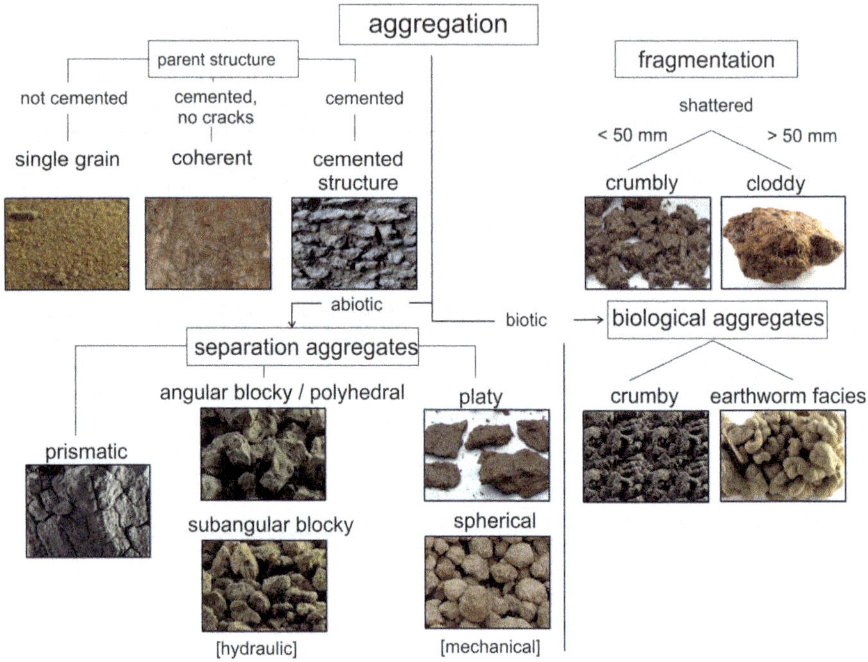

Figure 2 Schematic diagram of various aggregate types.

In newly formed aggregates, the number of contact points depends on the range of water potential and on the distribution of particle sizes as well as on their mobility (i.e. state of dispersion, flocculation and cementation). Soil shrinkage, including crack formation, increases the bulk density of aggregates. The increase in bulk density with the initial wetting and drying of the soil permits the aggregates to withstand structural collapse. The determination of soil and/or aggregate strength always has to be subdivided into (1) mechanically, hydraulically or chemically prestressed and (2) virgin conditions, which also ultimately affects the predictability of physical properties. These general ideas have been described by Hartge and Horn (2016), Baumgartl and Horn (1999), Groenevelt and Grant (2002) and Peng and Horn (2005, 2007).

3.1 Mechanical properties and soil health

The mechanical strength of soils increases with swelling and shrinkage and is enhanced by biological 'clogging' processes. This results in a very rigid soil structure and good predictability of ecological functions (Horn 1994). The data for cohesion as well as the angle of internal friction in Table 6 underline that the more pronounced the aggregation, the higher the resistance against any kind of deformation. It also verifies the often-mentioned higher strength and enhanced hydraulic, gaseous or chemical fluxes under conservation tillage conditions, which is even more pronounced under zero tillage when compared to conventional soil management.

3.2 Hydraulic properties and soil health

Aggregate formation always results in the rearrangement of particles and the formation of inter-aggregate cracks and smaller intra-aggregate pores with varying pore continuity. The general variations of water retention curve shapes in homogenized or compacted soils (similar to the formation of a platy structure) in comparison with natural, well-structured soils are shown in Fig. 3.

If the general progression of both the water retention and the hydraulic conductivity/matric potential curves for soils with varying aggregate formation is compared, a direct link between the curve patterns and the soil structure effects is observed (Fig. 4). The formation of aggregates also results in modified tensorial properties, also defined as isotropy, leading to vertical or horizontal anisotropic fluxes (Fig. 5). These factors affect hydraulic conductivity.

3.3 Cation exchange intensity

Aggregate formation alters the accessibility and availability of particle surfaces, which results in altered cation exchange options (Fig. 6). During aggregate formation, the particle contraction initially diminishes the cation accessibility accordingly. Only well-developed aggregates result in macroscopically almost homogeneous and rigid pore systems with evenly accessible sorption places. The latter are subangular blocks as well as crumbs and are mostly available in topsoil under conservation tillage or long-term grassland management.

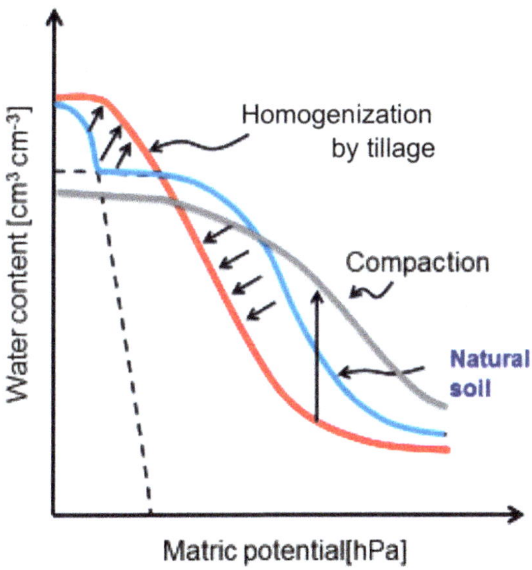

Figure 3 Water retention curve pattern of natural soil (blue) in comparison with the tillage effects (red) and those of the compaction (grey) due to the formation of dense platy structure (schematic patterns).

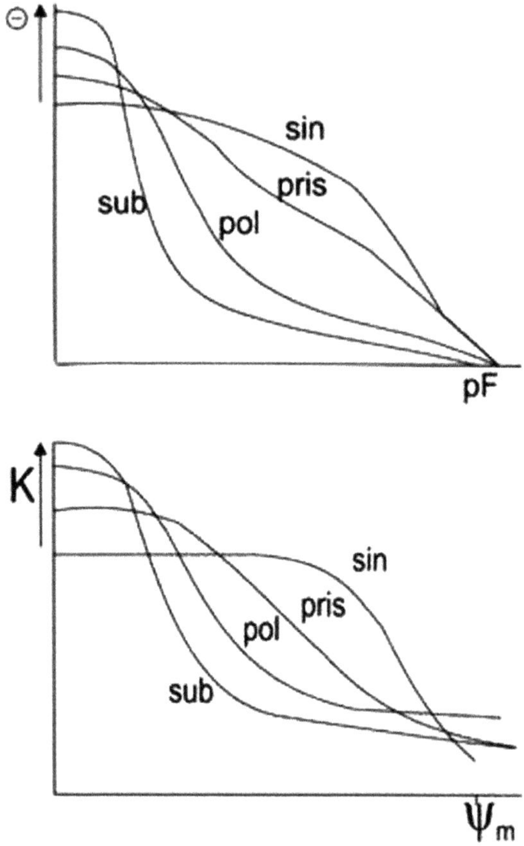

Figure 4 Water retention curve (above) and hydraulic conductivity (K)/matric potential (ψ_m) patterns (below) as a function of aggregation (single grain (sin), prismatic (pris), polyhedral/angular blocky (pol), subangular blocky (sub) schematic pattern).

3.4 Optimal aggregate formation

The properties and processes described earlier clearly illustrate the interactions between particle properties, changes due to aggregate formation and interactions with different types of soil management. Aggregate formation alters the accessibility and availability of exchange surfaces and the inter- or intra-aggregate pore system as well as mechanical or hydraulic strength. Accessibility and availability improve with advanced aggregate formation and minimized disturbance of the pore system during tillage. Subangular blocky or crumbly structured soil horizons are therefore optimal for long-term maintenance of chemical, physical and biological properties and functions and may be defined as synonyms for healthy soil properties. Dynamic stress input such as conventional tillage results in soil weakening and a reduction in pore continuity. This particularly affects gas exchange, water flux, associated physico-chemical processes and biological activity.

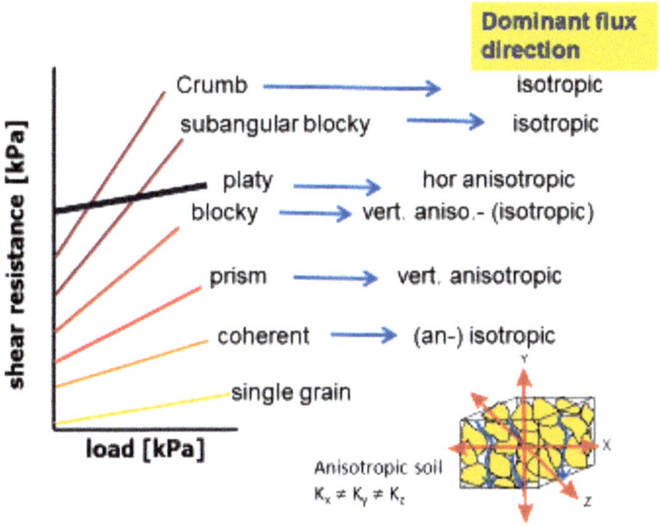

Figure 5 Aggregate-dependent shear strength and dominant flow directions.

Obviously, these positive soil properties and functions are restricted to certain boundary conditions. Exceeding these boundaries causes deterioration or alteration in aggregates, their rigidity and inter- as well as intra-aggregate properties and functions. The definition of healthy soil properties is therefore only valid for these ranges. Soil level-specific properties and the limitations applied to corresponding prediction models are listed in Table 7.

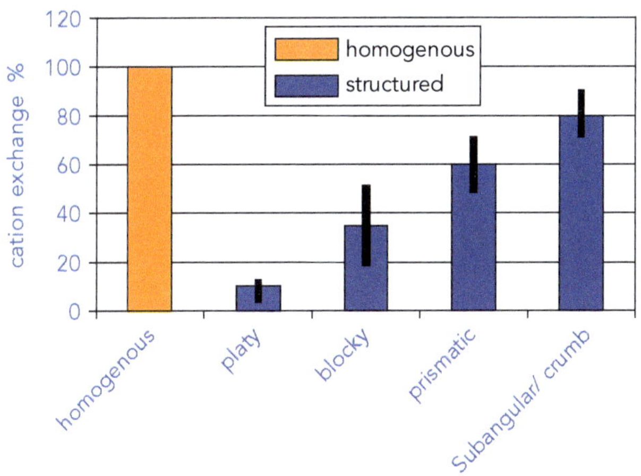

Figure 6 Ecological consequences of soil management for the accessibility of chemical exchange places.

Table 7 Effect of soil structure in various soil horizons on properties and options to model physical processes

	Soil characteristic			Time dependence of functions		Time dependence of process	Modelling
	Typical horizons	Typical structure	Rigidity of pore system	Pore system	Hydraulic properties	Pore water pressure (due to compaction)	Application of (existing) models
Ploughed horizon	Ap	Coherent	Small	Very high	Extreme	Extreme	Impossible
Plough pan	App	Platy	Extremely high	Small	Small	High	++
Subsoil	E (clay depletion)	Coherent	Small	High	High	Medium	–
	E (eluviation)	Singular	Small	Medium	High		–
	Bt (clay accumulation)	Subangular blocky – prism	Very high – medium	Small – medium	Small – medium		++
	Bhs (illuviation of organic compounds, sesquioxides)	Coherent, cementation	Small	Small	Medium		+
	Bw (weathering, oxidation)	Prismatic – polyhedral	Medium – high	Medium	High		–
	C parent material	Coherent	Small – high (glacial till)	High Small	Very High medium	Small	– +
		Solid (rock)	Extremely high	Extremely small	Very small		(++)

4 Conclusions

The preceding discussion suggests the following general conclusions:

- The prediction of physical properties of soils requires information about texture and structure processes.
- Soil texture is a major parameter in quantifying basic soil properties under the assumption of complete surface accessibility, and as an expression of the soil matrix and the inter-particle pore system.
- The process of aggregation changes mechanical, hydraulic, gaseous and chemical soil functions. Accessibility, pore continuity and pore functions depend directly on the aggregate properties, and this results in very distinct differences between texture- and structure-dependent soil properties and functions.
- Soil structure formation affects the rigidity of pore functionality and restricts the possibility of modelling soil properties and processes.
- Healthy soils are well aggregated and provide a continuous pore system with high surface accessibility. They are well aerated, contain sufficient plant available water and provide a reliable filter and buffer properties. Because healthy soils possess a high mechanical rigidity, they warrant intensified sustainability and the best possible protection against further soil degradation.
- Structure-dependent soil properties and functions are best maintained with minimal soil disturbance.
- Healthy soils must be defined site-specifically, including the history/memory effects, as their characteristics are lost as soon as external or internal forces exceed the internal soil strength or virgin compression processes occur.

Agricultural soil management is critical to maintain our food security in view of the expanding global population and finite amount of natural resources available for food production. The detailed physical principles and concepts discussed here suggest better soil management to enhance soil biology and its impact on soil health. With good soil management and healthy soils, future generations will continue to enjoy food security.

5 Future trends

Future research for a better understanding of soil health topics needs to include the quantification of the availability and accessibility of soil pores, particle or aggregate surfaces in order to optimize fertilizer application and its timing, and biodiversity issues in order to get a better understanding of sustainability limits for soil management under various climatic conditions. The actual daily loss of approx. 300 km^2 of land worldwide together with the growing population (2050: 9 Bill. population) require a careful sustainable land management worldwide.

Most of the capacity parameters documented thus far need to be modified in order to quantify actual intensity parameters and functions which finally provide a reliable basis for soil protection laws and/or recommendations for land use and management. Thus, there is an urgent need to include this information as a prerequisite for legislative activities both in ongoing scientific research and extension work analyses.

6 References

Ad-hoc-AG Boden, 2005. *Bodenkundliche Kartieranleitung. Bundesanstalt für Geowissenschaften und Rohstoffe in Zusammenarbeit mit den Staatlichen Geologischen Diensten*. Hannover. ISBN: 3-510-95920-5.

Baumgartl, T. and Horn, R., 1999. Influence of mechanical and hydraulic stresses on hydraulic properties of swelling soils. In: van Genuchten, M., Leij, F. J. and Wu, L. (Eds) *Characterization and Measurement of the Hydraulic Properties of Unsaturated Porous Media*. University of California, Riverside Publ., 1st edition, Riverside, USA. pp. 449–458.

Bormann, H., 2007. Analysis of the suitability of the German soil texture classification for the regional scale application of a physical-based hydrological model. *Adv. Geosci.*, 11, 7–13.

Groenevelt, P. H. and Grant, C. D., 2002. Curvature of shrinkage lines in relation to the consistency and structure of a Norwegian clay soil. *Geoderma*, 106, 235–47.

Haas, C., Holthusen, D., Mordhorst, A., Lipiec, J. and Horn, R., 2016. Elastic and plastic soil deformation and its influence on emission of greenhouse gases. *Int.* Agrophys., 30, 173–84.

Hartge, K. H. and Horn, R., 2016. *Essential Soil Physics*. Schweizerbart, Stuttgart.

Hillel, D., 1998. *Environmental Soil Physics*. Elsevier, Amsterdam.

Horn, R., 1994. The effect of aggregation of soils on water, gas and heat transport. In: Schulze, E. D. (Ed.) *Flux Control in Biological Systems*. Academic Press, 10, pp. 335–61.

Horn, R. and Fleige, H., 2003. A method for assessing the impact of load on mechanical stability and on physical properties of soils. *Soil Till. Res.*, , 1st edition, San Diego, USA. 73, 89–99.

Jury, W. and Horton, R., 2004. *Soil Physics*. 6th Ed. John Wiley & Sons, New Jersey, USA, 370p.

Lal, R. and Shukla, M. K., 2004. *Principles of Soil Physics*. Marcel Dekker, New York.

Or, D. and Ghezzehei, T. A., 2002. Modeling post-tillage soil structural dynamics: A review. *Soil Till. Res.*, 64, 41–59

Peng, X. and Horn, R., 2005. A new model for soil shrinkage curve at a wide range of soil types. *Soil Sci. Soc. Am. J.* 69, 584–92.

Peng, X. and Horn, R., 2007. Anisotropic shrinkage and swelling of some organic and inorganic soils. *Eur. J. Soil Sci.*, 58, 98–107.

Scheffer, F. and Schachtschabel, P., 2016. *Soil Sciences*. Springer Publ, Berlin Heidelberg.

Yong, R. and Warkentin, B., 1966. *Introduction to Soil Behavior*. Macmillan Series in Civil Engineering. Macmillan Publ., New York.

Zausig, J., Stepniewski, W. and Horn, R., 1993. Oxygen concentration and redox potential gradients in different model soil aggregates at a range of low moisture tensions. *Soil Sci. Soc. Am. J.*, 57, 908–16.

Chapter 7

Chemical composition of soils: role in soil health

Samira Daroub and Claire Friedrichsen, University of Florida, USA

1 Introduction: the chemical composition of soils and major issues impacting soil health

Chemical, physical and biological soil properties are the building blocks for promoting and maintaining soil health (USDA, 1996). The United States Department of Agriculture (USDA) defines soil health or soil quality as the ability of the soil to perform the function necessary for its intended use (USDA, 2015). Soil health is best expressed through the perspective of soil as a living biological system that requires holistic management similar to that of human health (Reicosky, 2015). In agriculture, the intended use of soil is to support sustainable, high production yields with limited environmental consequence. Soil chemical components are important for soil health because they provide essential plant nutrients and support soil biological growth. In addition, healthy soil provides the appropriate chemical environment for the crop and supports microbial communities to thrive by maintaining optimal soil pH and low salt-affected conditions (Lehman et al., 2015). Chemical indicators of soil health include pH, salinity (electrical conductivity),

http://dx.doi.org/10.19103/AS.2017.0033.08

organic matter (reactive carbon), phosphorus concentrations, cation exchange capacity (CEC), nutrient cycling (soil nitrate) and concentrations of elements that may be potential contaminants (heavy metals, radioactive compounds, etc.) or those that are needed for plant growth and development (USDA, 1996).

Plant nutrient levels and cycles are critical to soil health (Kibblewhite et al., 2008). The 17 essential nutrients required for plant growth are divided into macro- and micronutrients according to their chemical concentrations in plants. Concentration of macronutrients in plant tissue is higher than 0.1%, and these nutrients include nitrogen, phosphorus, potassium, calcium, magnesium and sulphur. Micronutrients have a concentration of less than 0.1% in the plant and include iron, zinc, copper, manganese, boron, chlorine and molybdenum. In addition, cobalt is essential to legumes, and silicon is a beneficial mineral element. A healthy soil provides balanced nutrients to plants without toxic concentrations of any element.

Deficiencies and excesses of nitrogen (N) have major impacts on the health and productivity of the world's ecosystems (Weil and Brady, 2017). Management of organic sources of N in soils, including plant residues, manure and other sources, is crucial to reduce the reliance on inorganic fertilizers that may not be readily available or economical in certain countries. There are many reactions in the soil N cycle, including nitrate leaching, ammonia volatilization and nitrous oxide gas release to the atmosphere, leading to the loss of nitrogen and thus making soil processes central to the global N cycle (Mullen, 2011; Weil and Brady, 2017). Proper management of both inorganic and organic sources of N is not only necessary for healthy crops, but equally important to reduce losses and contamination of air and water for healthy soils and ecosystems. One of the vital reactions in the N cycle is that of biological dinitrogen (N_2) fixation that enriches soil with N without the use of external fertilizers. Soil nitrate is the chemical indicator with a primary function to show the availability of N for plants and soil microorganisms.

Phosphorus (P) is second only to N in its impact on the productivity and health of terrestrial and aquatic ecosystems (Weil and Brady, 2017). Phosphorus is a complex element in soil. Depending on the soil and ecosystem, the management of phosphorous may need to focus on improving low availability or curbing excessive application of P-containing fertilizers and organic waste, which can cause environmental degradation. In sub-Saharan Africa, P-deficient soils and inadequate supplies of P fertilizers have contributed to this area being the main region of the world where food supplies remain insecure (Weil and Brady, 2017).

This chapter gives a brief overview of the chemical components in soils or indicators related to soil health. The chapter covers the macronutrients with a focus on role of N and P in soil health, plant growth, soil chemistry, loss mechanisms and factors affecting availability of these nutrients. The chapter discusses key issues, challenges and solutions for both N and P management in soils. The case study will expand the discussion of soil health and chemical components by discussing the impact of micronutrients and sodicity on crop yields and soil health from research done in India (Sawargaonkar et al., 2016; Wani, 2016).

2 Key chemical indicators of soil health

Indices for soil health will be most readily adopted if they are easily measured, inexpensive, comprehensible and useful to land managers, responsive to management-induced changes and relevant across sites and time (Doran and Zeiss, 2000; Schoenholtz et al., 2000). Soil chemical properties or indicators as defined by USDA include pH, P, reactive carbon (C), soil electrical conductivity (EC) and soil nitrate (USDA, 2015). All of these are measured in the field except for reactive C that is measured in the laboratory. Patra et al. (2015) proposed a set of soil properties for minimum data set in soil quality assessment for Indian soils that includes biological, chemical and physical properties. The chemical properties include soil organic matter (SOM), pH, electrical conductivity and extractable N, P and K (Patra et al., 2015).

Soil pH is a master variable that affects the soil's chemical environment to support plant growth. Soil pH regulates the availability of many nutrients and other elements that can be toxic to plants. For example, P solubility and concentration in soil solution is impacted by high- and low-pH soils. Micronutrients including Fe, Mn and Zn become increasingly soluble with pH decrease, and Al can be highly soluble and toxic in low-pH soils (Weil and Brady, 2017). Extremes in soil pH (<4.5 and >8.0) indicate unhealthy soils for plant production and biological activity.

Another variable that affects the soil's chemical capacity to function by supporting thriving crop production and soil organisms is salinization. It is estimated that 1 billion acres globally is degraded due to salinization (Shrivastava and Kumar, 2015). EC is a measure of soluble salts in the soil solution. Salt-affected soils reduce soil microbial activity (Smith and Doran, 1996) and availability of plant nutrients (phosphorus solubility, etc.), thus increasing ion toxicity and osmotic stress (Shrivastava and Kumar, 2015). Soils are classified as saline, saline sodic or sodic soils depending on their EC, pH and Na concentration relative to Ca and Mg. Saline soils have EC values greater than 4 dS m^{-1} and growth of crops sensitive to salinity is restricted on these soils. Sodium Adsorption Ratio (SAR) is a measure of soluble Na relative to Ca and Mg in soil solution. An SAR greater than 13 and pH > 8.5 indicate a sodic soil (USDA, 1954). Exchangeable sodium percentage (ESP) is the percentage of exchangeable Na occupying the CEC of the soil. An ESP greater than 15 indicates a sodic soil (Weil and Brady, 2017). Sodic soils can have poor physical structure with severe impacts on soil aeration, which affects plant roots and biological activity.

Permanganate oxidizable C (also referred to as reactive C or active C) quantifies the C most readily degradable by microorganisms. This simple method uses slightly dilute alkaline KM_nO_4 that reacts with the most oxidizable forms of C (Weil et al., 2003). Culman et al. (2012) found that the permanganate oxidizable C is an easy and inexpensive method that demonstrated sensitivity to changes in management and environmental variation. Reactive C is considered a chemical indicator because of its association with soil minerals. Clay minerals can strongly bind SOM and so protect the organic matter and reactive C from rapid mineralization, whereas sand and silt are non-binding (USDA, 2014). Reactive C sources include fresh organic material, soil microbial biomass, particulate organic matter and C loosely bound to soil minerals with a short residence time of 2–5 years (USDA, 2015). Reactive C may be used as an indicator of the changes produced by cropping and soil management practices that manipulate SOM content (USDA, 2015). Patra et al. (2015) recommended microbial C and N, as well as soil respiration as biological indicators for measuring soil quality in Indian soils.

In addition to individual soil health indicators, integrative soil health tests that combine different indicators are used. The Comprehensive Assessment of Soil Health (CASH) by Cornell University includes pH and a standard soil test analysis package that measures levels of P, K, Mg, Fe, Mn and Zn as chemical indicators to assess soil health and soil management impacts (Idowu et al., 2009; Moebius-Clune et al., 2016). In addition, add-on chemical indicators including salinity, sodicity and heavy metals are used (Moebius-Clune et al., 2016). The Cornell soil health assessment integrates measurements of numerous soil attributes that are scored on a 0–100 scale, from which an un-weighted average is derived for the overall soil health score (Idowu et al., 2009). Fine et al. (2017) reported on soil health status of over 5700 samples from various locations in the United States as evaluated using CASH. Best subset regression found that active C is the best single predictor (45%) of an overall CASH score. Penetration resistance, respiration and aggregate stability data were also informative (68%). Congreves (2015) used principal component analysis to weigh individual scores and developed the Ontario soil health assessment, which was more sensitive than the original approach. The soil health tool developed by NRCS is an integrated approach to soil testing using chemical and biological soil test data (Haney, 2015). The test uses water-extractable organic C, water-extractable organic N, 1-day CO_2-C respiration in addition to certain chemical tests to determine soil health.

Finding specific soil indicators useful for certain cropping systems or soils may not be straightforward. Pattison et al. (2008) concluded that there is no single indicator of soil health for the Australian banana industry, but a set of soil health indicators which would allow for the measurement of soil improvements includes: bulk density, soil C, pH, EC, total N, extractable P, effective cation exchange capacity and soil nematode community structure. The task of establishing a specific criterion for soil quality for forest ecosystems was found to be challenging because functions and values provided by forest ecosystems are variable and rely on interplay of soil physical, chemical and biological properties and processes which often differ across spatial and temporal scales (Schoenholtz et al., 2000). Takoutsing et al. (2016) investigated the possibility of selecting a minimum data set of soil health indicators to identify key land constraints for maize production in Cameroon. Nitrogen concentration was one of the parameters that were very variable. Exchangeable bases, boron content, pH, Mn content, EC, P and clay contents formed the minimum data set of soil health indicators for the study area.

3 Nitrogen

3.1 Nitrogen functions in plants and nitrogen chemistry in soils

The concentration and cycling of C, N and P affect the soil's ability to provide the necessary soil conditions for promoting sustainable crop production (Kibblewhite et al., 2008). Nitrogen is a very transient nutrient in the soil, making it the main focus of much of soil health research. More money and effort are spent on the management of N than any other element (Weil and Brady, 2017). Nitrogen management is essential for providing plant available nutrients and for supporting soil microbial communities (Lehman et al., 2015). Balancing inputs of N and associated C is important for soil health, otherwise soil health may decline in agricultural systems due to reductions in the pool of N available to organisms supporting soil functions (Kibblewhite et al., 2008).

The main function of nitrogen (N) in the plant is to serve as a constituent of proteins and nucleic acids (Marschner, 1995). A good supply of N stimulates root growth and development. Plants contain 1–6% N and utilize absorbed ammonium (NH_4^+-N) and nitrate (NO_3^--N) into amino acids and to proteins (Havlin et al., 2013). The majority of N in the soil is organic, with no more than 1% or 2% present as inorganic matter (NH_4^+-N and NO_3^--N) in the soil solution. In fact, 95–99% of the soil N is in organic compounds that protect it from loss, but is unavailable to higher plants (Weil and Brady, 2017). Decomposition of organic matter releases N compounds into the soil solution in the form of NH_4^+-N and NO_3^--N. Both of these N forms are taken up by plant roots. Mineralization of soil organic N by soil microbes is the first step in releasing NH_4^+-N, and nitrification is the second step releasing NO_3^- into the soil solution (Fig. 1). Decomposition of organic N is regulated by factors that affect microbial activity such as temperature, moisture and oxygen contents. For farmers to increase N availability to plants and soil microbes, they must be able to effectively manage field and soil conditions to release sufficient plant available N in the soil solution without releasing excess amounts and polluting the environment.

The N mineralization capacity refers to the capability of soils to transform organic N compounds into ammonium/nitrate under optimum moisture and temperature conditions over a given period of time (Gil-Sotres et al., 2005). Soil systems that are conducive to greater plant growth are also conducive to mineralization of organic N. Predicting N mineralization needs a knowledge of two indicators: mineralization potential (assessed under laboratory conditions) and rate factor, which is affected by soil temperature, soil moisture content, soil microbiological activity and their interactions (Mullen, 2011). The mineralization potential can be estimated by the C:N ratio of residues and added animal manure (Mullen, 2011). Environmental conditions are critical to determining the extent of rate factor of N mineralization under field conditions. Lab experiments have not been identified to most accurately represent conditions that will maximize N mineralization

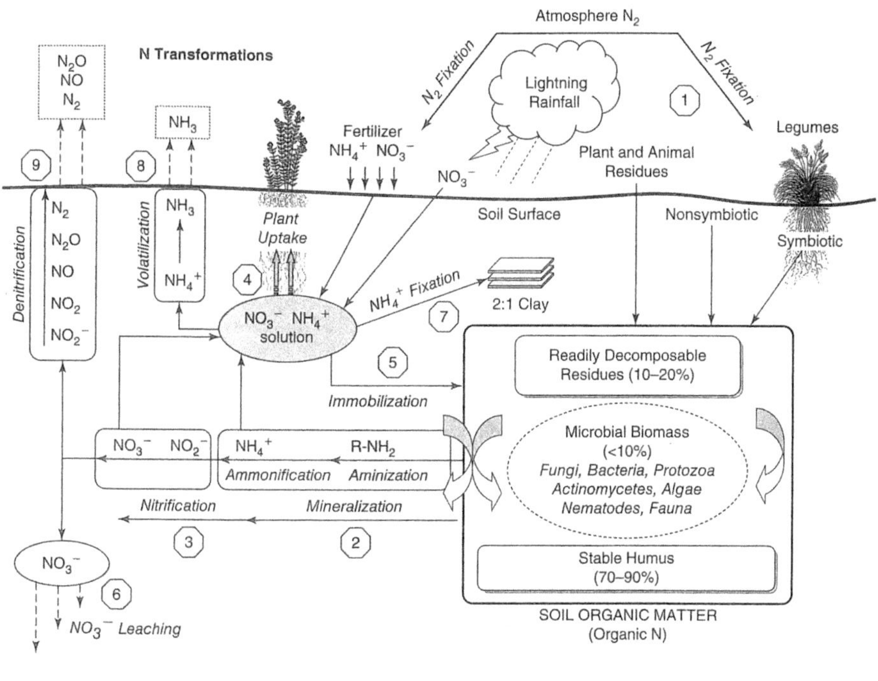

Figure 1 The N cycle showing the different reactions in the soil and the many loss mechanisms to air and water. Havlin, John L., Tisdale, Samuel L., Nelson, Werner L., Beaton, and James D. *Soil Fertility and Fertilizers*, 8th Ed., © 2013. Reprinted by permission of Pearson Education, Inc., New York.

(Mullen, 2011). Nitrogen mineralization has been used to assess the influence of soil management on soil health, but significant seasonal variation limits its value as a soil quality indicator (Gil-Sotres et al., 2005). In well-aerated soils, nitrification of NH_4^+-N is a rapid process yielding first nitrites and then nitrates. Nitrification is an acid-producing reaction carried out by two autotrophs: *Nitrosomonas* and *Nitrobacter* (Mullen, 2011; Weil and Brady, 2017). Nitrate-N fates in the soil include uptake by plant, leaching below the root zone or denitrification to gaseous forms (Cameron et al., 2013). To maintain soil health farmers need to optimize plant uptake and reduce nitrate-N leaching and denitrification.

The opposite process of mineralization is immobilization where ammonium-N and nitrate-N are converted into organic forms. This occurs in soils that have received residues with a C:N ratio greater than 30:1 and microbes decomposing these residues use inorganic N from soil solution to create new organic N compounds (Weil and Brady, 2017; Mullen, 2011). This may result in a period of plant available N suppression in the soil that depends on the type of organic materials applied or plant residues incorporated in the soil. Residue decomposition effects on N availability of corn following corn or soybean indicated less N immobilized during soybean residue decomposition and therefore less N fertilizers needed (Green and Blackmer, 1995). However, Hadas et al. (2004) found different dynamics of N immobilization that demonstrated the need to determine the decomposability of C and N compounds rather than their total contents in plant residues.

A very beneficial reaction in the N cycle that can sustainably improve soil health and subsequently sustainable crop production is biological N_2 fixation (Peoples et al., 1995). An important source of N in agricultural soils is due to the symbiotic N fixation, mutually beneficial relationship, between legumes and bacteria of the genera *Rhizobium* and *Bradyrhizobium*. These organisms infect root hairs and the cortical cells of legume plants and form root nodules that serve as the site of N_2 fixation. The legume host plant supplies the bacteria with carbohydrates for energy and the bacteria in turn supplies the plant with N compounds. The enzyme nitrogenase catalyses the reduction of N_2 gas to ammonia in an oxygen-free environment in root nodules. The ammonia is then combined with organic acids to form amino acids and proteins. The legume–*Rhizobium* associations function best on soils that are not too acidic, and are well supplied with essential nutrients excluding N (Weil and Brady, 2017). Over time, the presence of N-fixing species can significantly increase the content of N in soils and benefit non-fixing species. Lupwayi et al. (2011) concluded that grain legumes in sub-Saharan Africa have positive effects on agriculture by adding and recycling biologically fixed N_2, enhancing nutrient uptake, reducing greenhouse gas emissions, improving soil structure and breaking non-legume crop pest cycles. Grain legumes occupy about 20 million ha in Africa with cowpeas and common bean being major crops (Lupwayi et al., 2011). When coupled with P fertilization, herbaceous legume-based integrated soil fertility management sequences provide a potential entry point for reversing soil degradation, improving soil health and offering opportunities for increasing crop productivity in smallholder farming systems of Southern Africa (Nezomba et al., 2015). Indeed, biologically fixed N improves the N economy of soils and can be achieved in almost all agricultural ecosystems through the activity of many different symbiotic associations (Peoples et al., 1995). The fixed N becomes available as a result of mineralization of N-rich compounds in root exudates and in sloughed-off root and nodule tissue (Weil and Brady, 2017). Although the short-term recovery of N from these biological sources may not always match fertilizer application, there are consistent benefits to subsequent cereal crops in the rotation (Peoples et al., 1995). Typical levels of N_2 fixation from different sources are listed in Table 1.

Table 1 Typical levels of nitrogen fixation from different systems

Crop or plant	Associated organism	Typical level of nitrogen fixation, kg N ha^{-1} yr^{-1}
Symbiotic legumes (nodulated)		
Ipil-ipil tree (*Leucaena leucocephala*)	Bacteria (*Rhizobium*)	100–500
Locust tree (Rovina spp.)		75–200
Alfalfa (*Medicago sativa*)		150–250
Clover (*Trifolium pratense* L.)		100–150
Lupine (*Lupinus*)		50–100
Vetch (*Vicia villosa*)		50–150
Bean (*Phaseolus vulgaris*)		30–50
Cowpea (*Vigna unguiculata*)	Bacteria (*Bradyrhizobium*)	50–100
Peanut (*Arachis*)		40–80
Soybean (*Glycine* max L.)		50–150
Pigeon pea (*Cajanus*)		150–280
Kudzu (*Pueraria*)		100–140

Source: © Weil and Brady (2017), Adapted with permission of Pearson Education, Inc., New York.

3.2 Challenges in nitrogen management

Improving soil health involves increasing and retaining soil nutrient levels including those of N, by using sustainable agricultural practices to reduce reliance on the use of inorganic fertilizers and increase the use of organic N sources. Nitrogen lost from addition of organic sources, for example, repeated manure applications, can be substantial, especially via ammonia volatilization and nitrate leaching. There are many loss mechanisms in the N cycle including nitrate leaching, denitrification of nitrate-N to N gases and volatilization of NH_3 into the atmosphere (Fig. 1). This not only reduces considerably the efficiency of N fertilization, but also leads to water and air pollution. One of the major challenges that face farmers and scientists alike is how to increase SOM and available N in a sustainable manner with no negative economic or environmental impact. Nitrate ions being negatively charged move freely downwards in soils with drainage water and are readily leached from the soil. Nitrate leaching is more pronounced under coarse-textured soils and humid conditions. Nitrate-N becomes a water quality problem when it reaches and contaminates groundwater, causing health hazards if this water is used for drinking purposes (Mullen, 2011). Leached nitrates in groundwater can cause blue baby syndrome, methemoglobinemia, which can be fatal to humans (Pepper, 2013). Factors affecting nitrate leaching losses from soil have been reviewed recently by Cameron et al. (2013). These include season and climate, soil properties, land use, irrigation, fertilizer application, organic manure and waste additions.

Gaseous loss of N by denitrification entails a series of reduction reactions of nitrate into various N gaseous forms. Denitrification of nitrate occurs under very low oxygen conditions and in the presence of organic residues that provide energy for the denitrifiers. Generally, the end product from denitrification process is N_2 gas, but NO and N_2O gases are commonly released during nitrification under fluctuating aeration conditions (Weil and Brady, 2017). Atmospheric N_2 is inert, but N oxide gases are very reactive and have potential for causing serious environmental damage. The emission of nitrous oxide gas can represent a significant loss of N from the soil/plant system and contributes to climate change and depletion of the ozone layer (Cameron et al., 2013).

Another loss mechanism of N is ammonia (NH_3) gas volatilization from manure and fertilizers such as urea and anhydrous ammonia that can be extensive under certain conditions. Ammonia volatilization is more pronounced in high-pH soils and sandy soils (Weil and Brady, 2017). Volatilization of ammonia can also be appreciable when N fertilizers are added to flooded rice even on slightly acid soils (Weil and Brady, 2017). This is due to an increase in pH of the water as applied fertilizers stimulate algae growth that uses CO_2 from water and reduces the amount of carbonic acid formed. Loss of N through leaching, denitrification and ammonia volatilization can be soil-degrading reactions if the soil system is nitrogen limited, decreasing the soil's ability to provide N to plants and microbes.

3.3 Nitrogen: key issues and challenges being addressed

Kibblewhite et al. (2008) defines a healthy agricultural soil as one that is capable of supporting both an adequate production of food and fibre, and the continued delivery of essential ecosystem services. Strategies that include the addition of animal manure and use of mineral fertilizers counter the losses of N and other nutrients with the aim of restoring and sustaining soil health. On the other hand, excess addition of nutrients leads to loss from the soil system and pollution of air, water and unhealthy soils (Kibblewhite et al., 2008). Key challenges for N management include: 1) identifying management practices that increase SOM content; 2) accurately estimating the contribution and timing of organic N to the nutrition of plants in a cropping season and 3) reducing the many possible loss reactions in the N cycle. Adoption of practices by farmers to increase SOM can be slow due to added expense and lack of knowledge.

Increasing organic matter content of the soil has many physical, chemical and biological benefits. Not only will organic matter increase the CEC of the soil but also will the decomposition of added and residual organic compounds release available N for plant uptake. There are various management practices to increase SOM and total N in soils, including using cover crops (Kaspar and Singer, 2011) and green manures (GMs), crop rotations, crop residue management (Turmel et al., 2015; Murphy et al., 2016), addition of animal manure and compost (D'Hose et al., 2016), biochar addition (Lehmann, 2007), adopting no-till or minimum tillage systems (Mazzoncini et al., 2016; Sapkota et al., 2017) or a combination of these practices (Bulluck et al., 2002; Raphael et al., 2016; Wei et al., 2016; Ghafoor et al., 2017; Mulvaney et al., 2017).

Cover crops grown during fallow periods in cropping systems can increase total N in the soil by changing the annual patterns of N uptake and mineralization, reducing NO_3 leaching, retrieving NO_3 from deep soil layers and fixing atmospheric N_2, if the cover crops are legumes, increasing SOM and providing a living mulch and forage for livestock (Kaspar and Singer, 2011; Chatterjee and Clay, 2016). This is highly dependent on climate, planting date and termination dates with greater amounts of N_2 fixation occurring under warm conditions and earlier termination dates (Kaspar and Singer, 2011). Replacing fallow in no till winter wheat–fallow rotation with winter and spring cover crops for five years in the semi-arid central Great Plains in the United States – reduced runoff loss of sediment, total P and NO_3-N (Blanco-Canqui et al., 2013). Choosing appropriate plant species is an important consideration to obtain desired benefits as nitrate reductions due to leaching in cover crops vary greatly with cover crop species, growth of cover crop, amount of N in the soil and percolating water (Kaspar and Singer, 2011). For example, annual rye grass develops dense, shallow root systems that prevent soil erosion, while barley can have very deep roots to scavenge nutrients (Chatterjee and Clay, 2016).

One of the main factors affecting decomposition of cover crop residues and release of N is the C-to-N ratio of the residues. Residue incorporation also increases N release. Biochemical composition and environmental conditions were the primary factors affecting short-term N mineralization from warm season legume and grass cover crops in organic farming systems (O'Connell et al., 2015). Greater potential N mineralization was found in legume cover crops, but even cover crops with C:N ratio of >40:1 had net N mineralization (O'Connell et al., 2015). Use of legumes as a summer fallow crop in small grain systems resulted in higher potentially mineralizable C and N, and microbial biomass C (O'Dea et al., 2015). Legumes may increase long-term no-till system resilience and sustainability in the Northern Great Plains by increasing the available N supply by 26–50% compared to wheat-only systems, thereby reducing the need for N fertilizer application in subsequent crops. The half-life of potentially mineralizable C was shortest in intensified systems and was longest in the legume systems, potentially mitigating the negative effects of SOM losses from summer fallow (O'Dea et al., 2015). Liebman et al. (2012) showed that red clover GMs had an N fertilizer replacement value for corn of 87–184 kg N ha^{-1}, and alfalfa supplied corn with equivalent 70–121 kg N ha^{-1} in a study conducted in Iowa. Ultimately, successful management of N using cover crops requires that N availability be synchronized so that inorganic N is readily available during periods of active uptake by cash crops and minimally available during periods when cash crops are not growing to reduce losses of N to air and water (Kaspar and Singer, 2011). This may not be easily achievable and is an important area of research to focus on to be able to utilize nutrients released from cover crops efficiently.

A systems analysis reviewing studies of the benefits of residue retention was shown to be regionally variable and depended on both agro-climatic and socio-economic factors (Turmel et al., 2015). Overall, most studies from developing countries in Asia, Latin America and Africa showed positive effects of retaining crop residues on SOM, soil moisture and enhanced nutrient cycling, but variation was observed in the effect of surface retention vs incorporation in the soil, indicating the importance of taking into account abiotic factors such as climate, soil texture and agronomic practices as well as sampling methods and study duration (Turmel et al., 2015). Murphy et al. (2016) showed that crop residue retention under conventional and zero tillage resulted in a significant increase in recovery of fertilizer N applied in smallholder maize systems in Chiapas, Mexico. They concluded that residue retention is a promising option for farms to reduce fertilizer-N losses, particularly for sandy soils within sub-humid tropic regions.

Manure addition is another practice that can increase SOM and available N, reducing reliance on inorganic N fertilizers. Singh et al. (2015) concluded that substituting 25% inorganic N requirement by farmyard manure showed good performance in plant–ratoon system, indicating possibilities of saving 25% of N fertilizers without affecting the yield levels of sugarcane. The continuous use of an integrated nutrient source (farmyard manure and vermicomposting) improved soil health and vegetable production in the hill ecosystem of northeastern Himalayas with no addition of synthetic fertilizers (Patel et al., 2015). The continued use of the integrated nutrient source, after five years of organic farming, improved soil health including soil organic C, available P, K and microbial biomass C, and water-holding capacity (Patel et al., 2015). Calcium, K, Mg and Mn levels increased in soils amended with composted cotton-gin trash, composted yard waste or cattle manure compared to soils that only received synthetic fertilizers (Bulluck et al., 2002).

Emerging research is looking at biofertilizers and compost teas as a source of nutrients and microbial inoculants. Biofertilizers and compost teas are liquids anaerobically or aerobically fermented or extracted through compost and other organic materials containing microbes. Islam et al. (2016) found that N levels in compost teas depended on the nitrogen content of the compost, compost-to-water ratio, extraction days and storage time. They found the highest N level of 1.1 g L^{-1} in compost tea made with a compost containing 16.7 g kg^{-1} at a compost/water ratio of 1:2.5 with two days' extraction and zero storage days (Islam et al., 2016)

Tillage can increase mineralization potential by introducing oxygen, which increases microbial activity (Mullen, 2011). Tilling can destabilize protected organic N pools (Kristensen et al., 2000). A long-term no-tillage (NT) experiment was conducted for 28 years in Central Italy to determine impacts on soil organic carbon (SOC) and total N (Mazzoncini et al., 2016). The shallow layer (0–10 cm) under NT showed the highest SOC and total N concentration after ten years. After 28 years, continuous NT increased SOC and soil total N content in the 30-cm soil depth by 22% compared to initial values due mainly to increases in the 0–10-cm depth. For the whole profile (0–30 cm), the mean annual SOC increase under NT was +0.40 Mg ha^{-1} yr^{-1} and NT was a useful alternative management practice increasing C sequestration and soil health in the Mediterranean conditions (Mazzoncini et al., 2016). Increased biomass production was achieved through a combination of zero tillage and partial residue retention that increased SOC between 3.0 and 4.7 t C ha^{-1} after seven years of rice–wheat rotations in the eastern Indo-Gangetic Plains of India (Sapkota et al., 2017).

Conversion of tillage systems from conventional to no-till may take a long time, however, to show increases in SOC or total N. Mulvaney et al. (2017) used a combination of no-till with high-biomass crops and invasive legume mulches on a sandy loam soil in a humid tropical area in Alabama, USA. They found that forage soybean as a summer cover crop did not increase SOC over three years, whereas the inclusion of cut and carry mulches increased SOC from 6.3 to >22.6 g kg^{-1} soil over three years without impacting collard yields. Stirling et al. (2010) compared conventionally managed sugarcane soils (sugarcane monoculture, extensive tillage and occasional burning of crop residues) with a new sugarcane farming system based on crop rotation, minimum tillage and residue retention. They compared the sugarcane soils to perennial pasture soils, and their results indicated that the biological status of sugarcane soils is poor relative to soils under the pasture. After 5–7 years of implementation, they did not find any differences in total C or total N measured in the new sugarcane farming system, but these soils had lower C:N ratios and more free-living nematodes indicating a shift towards the biological properties of pasture soils (Stirling et al., 2010). Congreves et al. (2017) investigated the impact of long-term N fertilizer application, crop rotation and tillage systems on SOC and total N gains in a humid temperate climate in southeastern Canada. They reported that changes in SOC and TN storage due to N fertilizer application were highly dependent on tillage systems and the crop rotation. SOC and TN were highest with N fertilizer application in no-till and conventional tillage systems in the corn–soybean–wheat rotation compared to continuous corn and corn–soybean rotations (Congreves et al., 2017). The authors conclude that the quantity of crop residue's C input due to N fertilization may not be as important as plant species composition and quality of crop residues due to addition of wheat in the rotation. The positive influence of wheat on SOC and TN may be related to the quantity and composition of lignin that may take longer to decompose than corn or soybean (Congreves et al., 2017).

Reactive N is a term used for N compounds that support growth directly or indirectly. These species include ammonia (NH_3), nitrous oxide (N_2O), as well as nitrate (NO_3^-). Lower reactive N in organic systems minimizes the potential for N losses; however, the addition of organic manures and residues, containing mineralizable N and C, has the potential to enhance N_2O emissions. Almost 70% of global nitrous oxide (N_2O) emissions originate from fertilization of agronomic systems (Syakila and Kroeze, 2011). Cover crops have the potential to reduce gaseous losses of N through microbial reactions by reducing soil N availability (Davidson et al., 2000). Mitchell et al. (2013) found that the rye cover crop decreased N_2O emissions when no N fertilizer was applied, but it increased N_2O emissions at an N rate near the economic optimum for a maize cropping system with banded N fertilizers. Peyrard et al. (2016), however, found only limited effect of increased use of legumes and cover crops on N_2O emissions in a conventional wheat–sunflower rotation in South West France. Cumulative N_2O fluxes measured for three years were in the low range between 0.78 and 1.12 kg N_2O-N ha^{-1} y^{-1} with significantly higher flux in systems that received higher N rates of fertilizer regardless of cover crop in the rotation (Peyrard et al., 2016). Bhowmik et al. (2016) discovered that long-term organic vegetable cropping system receiving compost additions have the potential to accumulate C and immobilize excess reactive N (particularly nitrates), thereby improving soil health and reducing N_2O emissions. Research on the use of organic amendments will need to focus on the role of these amendments to release greenhouse gases including CO_2 and N_2O gases.

In the United States, N fertilizer recommendations for a corn crop takes into account N credits from a previous legume crop into the rotation as well as from manure additions [see Bundy (1994) and Kaiser et al. (2016) for examples]. Nitrogen credit is the fertilizer N replacement value obtained in a crop sequence compared with a non-continuously grown crop, and usually refers to the legume component of crop rotations (Lory et al., 1995). However, N inputs from other sources are sometimes included. The major sources of non-fertilizer N include soil residual nitrate, nitrate in the irrigation water and N mineralized during the growing season from organic material such as organic matter, plant residues and organic amendments (Geisseler, 2017). Schmidt et al. (2002) reported that variable rate N recommendations based on SOM alone did a poor job of indicating N mineralization within a given field. Their research indicates that SOM alone is too simplistic to reflect what actually occurs at the field level. Measurement of more microbially active organic matter fractions may be a better estimator of the net N mineralization potential (Schmidt et al., 2002). The development of practical methods to better predict the magnitude of soil N mineralization is critical to improving N use efficiency (Zebarth et al., 2009). Computer tools are used for N fertilizer recommendations. Corn nitrogen calculator (CNC) is used to generate N recommendations for maize (*Zea mays* L.) in New York. The CNC takes into account N mineralization from organic matter based on soil type and assumes an N use efficiency factor based on soil type (Ketterings et al., 2003). The tool accounts for previous N application such as start-up N applied with planting or N availability from manure application or sods (Sela et al., 2017). Another tool used is Adapt-N, a dynamic, in-season sidedress N recommendation, that requires multiple soil and management inputs such as soil texture, organic matter content, crop characteristics and management, previous N applications (synthetic fertilizer, manure and soil management). Adapt-N uses real-time weather data and adjusts rates for different weather conditions and SOM levels with recommended rates lower than those from CNC (Sela et al., 2017). Adapt-N sidedress rates were positively related to the cumulative early-year rainfall amount accounting for higher N losses from the soil before sidedress and inversely related to organic matter levels accounting for N gains from mineralization (Sela et al., 2017).

In summary, there is a wealth of information regarding various management practices that may increase SOM, TN and available N in soils, thus improving soil health. The impact of these practices is quite dependent on geographical location, soils, climate, type and combination of these practices. Indeed, diverse crop rotations, cover crops and integrating livestock make organic cropping systems potentially more sustainable, provide N and P and increased SOM compared to other agroecosystems (Lynch, 2015; Bhowmik et al., 2016). However, the method for estimating N credit from these sources needs to be improved taking into account weather and soil conditions that impact mineralization and losses of N. After 30 years of research in Africa, Lotter (2015) advocates integrated soil fertility management, a combination of organic (manure, legume GMs as well as integrated pest management) and green revolution methods (synthetic fertilizers, herbicides, improved varieties) to face food insecurity in Africa. He concluded that organic conservation agriculture alone is not feasible at this time, and the use of herbicides is needed to reduce tillage and soil erosion (Lotter, 2015). Increasing and maintaining soil N as part of soil health management is important for supporting sustainable crop production and soil organisms.

4 Phosphorus

4.1 Functions in plant and phosphorus chemistry in soil

A healthy soil is able to provide and maintain available P for sustainable crop production and soil biological activity without providing excess amounts that may lead to contamination of surface waters. Extensive tracts of land in the tropical and subtropical regions of Asia, Africa and Latin America contain highly weathered and inherently infertile soils. In addition to socio-economic factors, the main constraints in these soils are soil acidity and low inherent N and P fertility (Lal, 1990), leading to degraded and unhealthy soils. Phosphorus is essential to plants and microbes as it is a constituent in nucleic acids, which as units of the DNA molecule, are the carriers of genetic information and, as units of RNA, are responsible for translation of the genetic information (Marschner, 1995). Phosphorus is also essential for energy transfer; it is an essential component of adenosine triphosphate (ATP) that drives most biological processes. Adequate P availability in the soil enhances photosynthesis, N_2 fixation, flowering, fruiting, seed production and maturation (Weil and Brady, 2017).

Impact of phosphorus deficiency is two-fold, affecting soil fertility and environmental quality. Naturally occurring P compounds in soils have low solubility and availability for plant uptake. In addition, when P fertilizers are added, P interacts with various ions to form several inorganic compounds in low- and high-pH soils; these compounds become highly insoluble with time. Adsorption of P on surfaces of oxides can be very strong in acid soils. Phosphorus is also part of SOM, which depending on its structure can have a wide range of decomposition rates. Current routine soil tests do not include contribution of organic P to meet crop demands. In many developing countries, for example, sub-Saharan Africa, underuse of fertilizers is the rule, which makes P the first limiting factor in food crop production and leads to land degradation (Weil and Brady, 2017). Many soils in the tropics are prone to degradation due to lack of fertility, leading to unhealthy soils and unsustainable agriculture. Major factors that constrain tropical soil fertility include high P fixation, high acidity with Al toxicity, erosion, low nutrients and low soil biodiversity (Cardoso and Kuyper, 2006). On the other hand, in many developed countries, the overuse of P fertilizers and manure addition have led to water quality degradation. Eutrophication of surface waters due to algae blooms caused by excess P in runoff water is widely documented, for example, in the Chesapeake Bay and the Everglades ecosystem in the United States [for example, see Sims and Sharpley (2005) and Daroub et al. (2011)]. Non-point P source pollution from agriculture and urban sources can have negative ecological impacts on fresh water and estuaries. Phosphorus has been implicated as the limiting nutrient in the eutrophication and ecological changes in the Everglades ecosystem in South Florida, USA (Richardson, 2008). Research and restoration efforts have been under way since the 1990s and cost billions of dollars. Restoration includes the use of on-farm best management practices to reduce generation and transport of P off farm (Daroub et al., 2012) and the construction of storm treatment wetlands to clean up the water by the use of vegetation before the water is released into the Everglades National Park (SFWMD, 2017).

Plants take up P from the soil solution, and in any soil the soil solution P is very low, generally from 0.001 mg L^{-1} in very infertile soils to about 1 mg L^{-1} in rich, heavily fertilized soils (Weil and Brady, 2017) (Fig. 2). Phosphorus moves very slowly by diffusion in the soil solution. Mycorrhizal fungi infection of plant roots can become an important mechanism to improve P absorption by plant and these fungi are generally best developed in undisturbed soils with low P availability (Weil and Brady, 2017). Arbuscular mycorrhizal fungi develop intensively inside roots and within the soil by forming an extensive extra radical network, and this helps plants to exploit mineral nutrients and water from the soil (Jeffries et al., 2003). Beneficial plant–microbe interactions in the rhizosphere are primary determinants of plant and soil health.

The two P species most prominent in soil solution between pH 4 and 7 are the orthophosphate ions (HPO_4^{2-} and $H_2PO_4^{-}$). When P is added either through fertilization or mineralization of organic P sources, these dissolved P species undergo 'fixation' reactions in low- and high-pH soils. Fixation reactions include precipitation reactions as well as adsorption reactions of P on the surface of hydrous oxides and silicate clays. pH is a master variable that affects P fixation reactions and solubility of P compounds (Fig. 2).

Johnston and Richards (2003) argued that adsorption and absorption reactions are more common than precipitation reactions when a fertilizer containing water-soluble P is added to soil. Initially the P forms high-solubility reaction products, but these reaction products are short lived and the majority of the P becomes distributed into readily and less readily available pools by the processes of adsorption and then absorption (Fig. 2). In sub-Saharan Africa, P deficiency is one of the largest constraints to crop production due to low native soil P and high P fixation capacity due to high degree of weathering (Margenot et al., 2017).

A variety of soil test extractants for P have been developed based on a fundamental understanding of which forms of P in the soil make up the labile P for plant uptake. Al-P and Ca-P are the major fractions of labile soil P that supply P to plants in acid to neutral soils. In calcareous soils, the major fraction of labile P is adsorbed onto the surface of $CaCO_3$ in the soil and various Ca-P minerals (Beegle, 2005). A complete list of common and routine

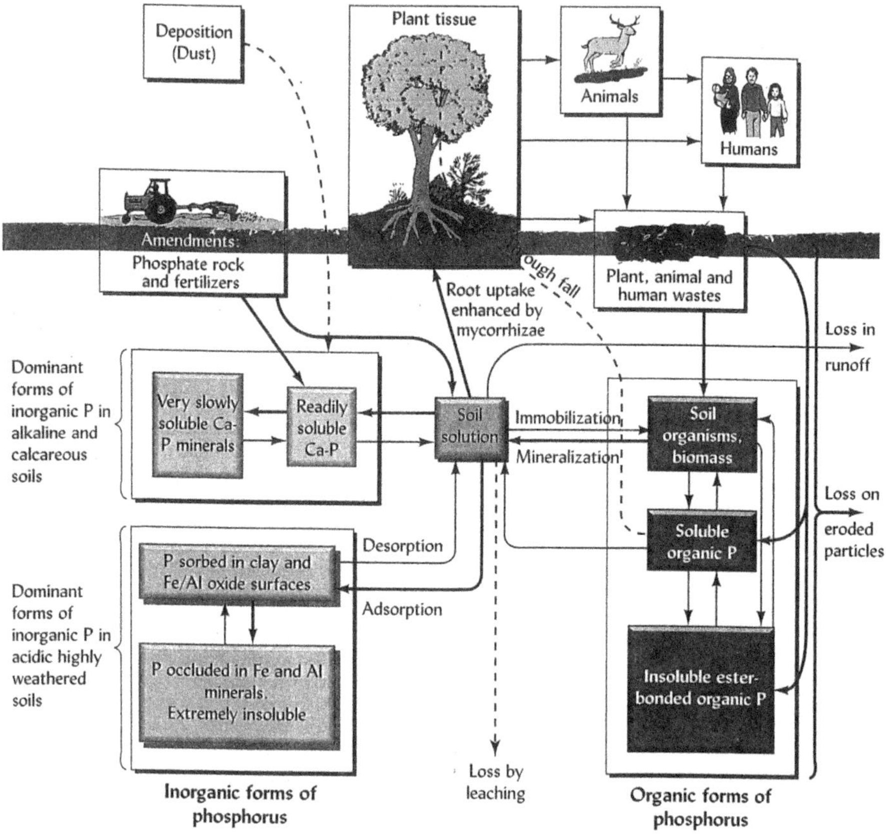

Figure 2 The phosphorus cycle in soils. Weil, Raymond R., and Brady, Nyle C. *Nature and Properties of Soils*, 15th Ed., © 2017. Reprinted by permission of Pearson Education, Inc., New York.

soil P extractants can be found in Beegle (2005). Current soil test for available P (STP) does not take into account contribution of organic P. Often more detailed fractionation schemes are used to separate various fractions of soil P. The P fractionation procedure developed by Hedley and Stewart (1982) and Tiessen et al. (1984) is one of the most commonly used procedures. It extracts both inorganic P and organic P, extracting labile P first and ending with the more resistant forms of P.

Phosphorus sorption isotherms are often used to determine the soil equilibrium P concentration (EPC) as well as the maximum sorption capacity. Phosphorus desorption can occur when fixed P is present in relatively soluble forms and most of the fixation sites are already occupied by P. There are two ways where sorption isotherms, EPC and maximum sorption are used. Equilibrium P concentration (EPC_0) can be obtained from sorption isotherms; EPC_0 indicates P solution concentration at which P is neither fixed in the soil nor released from the soil into the soil solution. The EPC_0 is an important parameter for measuring soil fertility as it indicates the capacity to replenish soil solution as P is depleted by plants. It is also an important parameter for environmental assessment as it indicates the rate at which the soil will release P into runoff and leaching water (Weil and Brady, 2017).

4.2 Factors influencing phosphorus fixation in soils

There are various factors that influence P solubility in soils, including pH, clay percent and mineralogy, redox potential and SOM. pH controls the solubility of P, and maximum P solubility in soils occurs at pH 6–7. In addition, clay per cent and clay mineralogy in soils have an equal importance in determining P solubility. In low-pH soils, the iron and aluminium hydroxy phosphate minerals form with high stability and low solubility. Common minerals include Strengite ($FePO_4{\cdot}2H_2O$) and Variscite ($AlPO_4{\cdot}2H_2O$). The solubility of these compounds increases with an increase in pH. In addition, phosphate ions in solution become adsorbed, sometimes very strongly, to the surfaces of iron, aluminium and manganese oxides and 1:1 silicate clay minerals. The $H_2PO_4^-$ ion may be attracted to positive charges of these oxides and can undergo anion exchange reactions with other anions. This P held by outer sphere complexes may be slowly available to plants. However, $H_2PO_4^-$ forms inner sphere complexes with oxides which can bind the P strongly and therefore not be readily replaced by other ions (Weil and Brady, 2017). Fe and Al oxides and 1:1 silicate clays are more abundant in highly weathered soils like Oxisols and Ultisols. P adsorption on these soils is much higher compared to temperate soils (for example, Mollisols) that have a higher percent of 2:1 minerals in the clay fraction of the soil (Sanchez and Uehara, 1980). Acid soils which fix large quantities of P are invariably medium- to fine-textured soils rich in oxides and hydroxides of Fe and Al. There are several broad soil groups which have a notoriously high P-fixation capacity: the suborder andepts, the orders Oxisols and Ultisols, and certain rhodic or oxic Alfisols and Inceptisols (Sanchez and Uehara, 1980). It is estimated that 14.7% of soils in Africa have a pH < 5.5 which favours fixation by Al and Fe (Margenot et al., 2017). As mentioned earlier and documented extensively, highly weathered acid soils in large areas of Africa, Asia and Latin America have low inherent fertility (Lal, 1990), leading to soil degradation and unhealthy, unproductive soils. This is largely due to acidity, low total N (TN) and total P, and highly unavailable P as a result of the various fixation reactions mentioned.

High-pH calcareous soils can also fix P through the formation of increasingly insoluble Ca phosphate compounds with time from tricalcium phosphate $Ca_3(PO_4)_2$ to fluorapatite $[3Ca_3(PO_4)_2]\cdot CaF_2$ (Weil and Brady, 2017). Less-soluble Ca phosphate compounds, for example, tricalcium phosphate or apatite, maintain lower soluble P in soil solution than the more soluble compounds. Solubility of Ca phosphate compounds increases with decrease in pH. Less weathered soils in Africa, mostly Entisols and Inceptisols in North Africa, have a pH above 8.5, favouring P fixation by Ca (Margenot et al., 2017). In general, P is relatively more available when it is fixed by Ca, making P fixation less of concern in alkaline than in acid soils.

Redox is another factor that controls the solubility of P as it plays a major role in the solubility of iron (Fe) phosphate compounds. Poorly crystalline and amorphous oxides and hydroxides of Fe play a primary role in P retention by flooded soils and sediments (Khalid et al., 1977). Under low redox potential in soils, P will be released due to reduction of Fe–P compounds, making this P biologically available again and improving productivity and soil health in highly weathered soils. In the highly weathered soils of humid tropical forests, Fe plays a key role in ecosystem biogeochemical cycling through its interactions with carbon (C) and P (Liptzin and Silver, 2009). Peretyazhko and Sposito (2005) showed that anoxic conditions caused significant Fe (II) production in an Ultisol sampled from a riparian valley in a humid tropical forest in Puerto Rico. Significant release of soluble P also occurred, mainly in organic form, an observation that is consistent with the accumulation of humus and organic P in highly weathered forest soils under wet moisture regimes (Peretyazhko and Sposito, 2005). In ferralitic soils characteristic of humid tropical forests, soil P availability is strongly coupled to C and Fe cycles. Microbial reduction of Fe (III) led to the release of P strongly sorbed on the surfaces of Fe oxides (Chacon et al., 2006). In a laboratory study, Liptzin and Silver (2009) found that Fe reduction was generally greater with higher C additions in upland tropical forest soils with fluctuating redox potential. Microbial biomass P increased significantly, suggesting rapid microbial uptake of P liberated from Fe. The immobilization of P by microbes provides a P conservation mechanism in these soils with fluctuating redox potential. The authors conclude that Fe cycling appears to be an important source of P for the biota; and that the periodic release and rapid immobilization of released P with Fe reduction may help maintain the high net primary productivity of wet tropical forests (Liptzin and Silver, 2009).

SOM can increase P availability not only due to P mineralization, but also due to the formation of organophosphate complexes that are more soluble, organic anion replacement of H_2PO_4 on adsorption sites and coating of Fe/Al oxides by humus to form a protective cover and reduce P adsorption (Havlin et al., 2013). Organic P represents about 50% of total soil P and varies between 15 and 80% (Havlin et al., 2013). The knowledge of the specific nature of much of the organic P in soils is limited (Cheesman et al., 2012; Havlin et al., 2013). Three board categories of common organic P compounds in soils are: 1) Inositol phosphates (IP*x*) are the most abundant and make up 10–60% of total organic P in soil under both acid and alkaline conditions. These are orthophosphate monoesters with *myo*-inositol hexakisphosphate (IP6) (phytic acid) being the most common IP*x* compound in the environment, 2) orthophosphate diesters such as nucleic acids and 3) phospholipids. Both nucleic acids and phospholipids comprise only 1–2% of organic P in most soils. Rest of the organic P compounds have not been identified; however, recent advances in spectroscopic methods, for example, ^{31}P nuclear magnetic resonance spectroscopy (NMR) and P XANES spectroscopy, are revealing more details on organic P compound levels in soils (Cade-Menun, 2005; Doolette and Smernik, 2011). Turner and Engelbrecht (2011) assessed the composition of soil organic P in 19 soils of varying pH (3.3–7) and total P (74–1650 mg kg^{-1}) under lowland tropic forest in Panama using NaOH-EDTA extraction and solution ^{31}P NMR. Soil organic P ranged between 22 and 494 mg kg^{-1} and occurred mainly as phosphate monoesters (68–96%) with small concentrations of phosphate diesters in the form of DNA (4–32% of total organic P). Most soils did not contain detectable concentrations of either myo- or scyllo-inositol hexakisphosphate, which is in marked contrast to many temperate mineral soils. Organic P identified by solution ^{31}P NMR spectroscopy in surface soils of a tropical ombrotrophic wetland included phosphomonoesters (12–17%), phosphodiesters (10–14%) and phosphonates (up to 3.3% of total P). Inositol phosphates were not detected in these acidic peats (Cheesman et al., 2012).

Organic P can play a quite important role in maintaining soil health by supplying P to plants in highly weathered and low-P fertility soils in tropical regions. This is due to the high potential turnover of organic P into plant available inorganic forms under warm climates (Margenot et al., 2017). Crops grown on degraded soils in Southern Africa show weak responses to mineral fertilization, indicating a need to use integrated soil fertility management supported through farmers' local knowledge of soils to increase organic P and subsequently human health (Nezomba et al., 2015). Short-term restoration of productivity of degraded sandy soils was achieved through high-quality organic resource application (indigenous legumes and sunn hemp) in combination with cattle manure and P fertilization. This led to an increase in soil biological activity (microbial biomass and basal respiration) and the responsiveness of the degraded soils to mineral N fertilizer (Nezomba et al., 2015). It is commonly reported that conservation tillage systems, especially NT, increase organic C levels in topsoil due to accumulation of residues. NT also increases total and available P concentrations in the topsoil compared to the lower soil horizons, as fertilizer P as well as P released from plant residues remain at the soil surface (Daroub et al., 2000; Jansa et al., 2003). The contribution of organic P cycling in tropical Oxisols of the Brazilian savanna increased after conversion to agriculture and was higher under NT systems (Rodrigues et al., 2016). Labile organic P fractions, however, were not significantly increased in three soils in Michigan that have been under NT for 4–13 years possibly due to organic P mineralization (Daroub et al., 2000).

Currently routine STP, however, does not capture the contribution of organic P to soil fertility. Mineralization of organic P occurs by the same processes that release N from organic matter. Net mineralization of P happens when plant residues or other sources of organic P added to the soil have a C:P ratio less than 200:1, while net immobilization happens when C:P ratio of added organic materials is higher than 300:1. Phosphatase enzymes catalyse the mineralization reaction of organic P. Factors that affect the quantity of P mineralization and immobilization are temperature, moisture, aeration, pH and tillage similar to organic N cycling. In addition, management practices discussed earlier with regard to N can increase organic P and its contribution to crop nutrition. STP and tools that can predict the release of P accurately from organic sources under different soils and climates are needed.

4.3 Phosphorus: key issues and challenges being addressed

Maintaining soil P levels is essential for soil health to have optimal crop production and for supporting soil biology. There is great uncertainty in the estimation of the world phosphate rock (PR) reserves that can be economically extracted, but currently estimated to last 93 years (Fixen and Johnston, 2012). Limited economically extractable deposits of PR, used to produce P fertilizers, lead to the critical need to manage P resources and adopt more sustainable P uses and recycling strategies (Oberson et al., 2001; Dodd and Sharpley, 2015). In the United States, the USDA soil health initiative promotes on-farm management practices aimed to improve soil health (Dodd and Sharpley, 2015). Practices like addition of amendments, mostly manure, for crop nutrient requirements, adoption of no-till and crop rotations can influence soil C and different soil organic P forms, which are important controls of the soil P cycle.

Organic P management is necessary for building sustainable soil health conditions and preventing further degradation in P-limited soils, as PR is a limited resource. Organic P provides a potential source of P via mineralization and stimulates biological cycling of P, chiefly via microbial drivers (Margenot et al., 2017). Organic P can then be an important source to maintain soil fertility not only in low-P acid soils but in other soils also. Key questions would be how to make sure that P is used efficiently in agriculture and how organic C and P cycling can be best utilized to increase plant uptake of soil organic P and at the same time minimize the loss of P to surface water (Weaver and Wong, 2011; Dodd and Sharpley, 2015). Various P management strategies, including soil testing, application of inorganic fertilizers and organic P solubilization, in both low- and high-pH soils are reviewed in this section, with more focus on low-pH soils.

A recurrent issue in acid soils with high P fixation capacity is how to saturate the P fixation capacity of the soil and maintain adequate levels of P in soil solution to meet crop demands. Sanchez and Uehara (1980) summarized management concepts for acid soils with high P fixation capacity: *'Over the years, two main management strategies have emerged. The first or high input option consists of a heavy initial investment followed by substantial residual effects for several years. In addition to solving the problem immediately, heavy P applications also serve to act as a soil amendment, producing beneficial changes in soil physical and chemical properties in highly weathered Oxisols. The second strategy involves a low input option. It has been traditionally based on applying P in bands to satisfy the fixation capacity in a small soil volume'*. Addition of high rates of soluble P fertilizer or amendments can lead to P saturation and leaching of P even in high sorption capacity soils (Gikonyo et al., 2010). Alleoni et al. (2014) showed an Oxisol from Brazil reached critical levels of P saturation in the top 10 cm with a high rate of biosolid application at 76 265 kg ha^{-1}. Given the high cost of fertilizers and potential loss of P, the second strategy may be more sustainable in soils with high P fixation capacity. Johnston and Richards (2003) recommended that plant available P in soil should first be raised to the critical level for the crops being grown and then maintained at this level by replacing the P removed in harvest crops. This reduced both cost and risk of environmental contamination by not increasing plant available P beyond the critical level. Another benefit of SOM inputs, for example, from plant residues, is ameliorating P fixation. This is believed to reflect decreases in binding sites and/or exchangeable acidity as a result of SOM complexation of exchangeable Al^{3+} and consumption of H^+ via decomposition processes and OM buffering (Margenot et al., 2017).

Band application of inorganic P fertilizers has been used as an important strategy in low- and high-pH soils not only to reduce the fixation reactions that occur, but also to reduce the rate of P fertilizer applied. Application of the appropriate P fertilizer, in bands near the plant roots, can change the pH of the rhizosphere, increasing the availability of P. Less P applied translates to less P transported off farm that reduces environmental pollution. In addition to band application of P fertilizers, direct application of PR, adding lime and silicate amendments and choosing species and varieties tolerant to low P are all viable options (Sanchez and Uehara, 1980). The International Fertilizer Development Center conducted research with a focus on the use of indigenous PR deposits and modified PR as a source of P for crop production in developing countries (Chien and Friesen, 1992). Adding lime to increase the pH of highly acidic soils will increase available P. Given the potential high cost of lime that many subsistence farmers may not afford, organic waste materials can be an efficient alternative to the use of lime and inorganic fertilizers. Bougnom et al. (2009) found that compost addition with and without wood ash could ameliorate acid tropical soils in a greenhouse experiment. Addition of organic waste composts increased soil pH, SOM, EC, total C, TN, P, exchangeable cations (Ca, Mg, Na and K) and micronutrients (Zn and Cu). Concentrations of exchangeable Al, Fe and Mn decreased. Ch'ng et al. (2016) tested biochar and compost produced from poultry manure and pineapple leaves with regard to production of *Z. mays* L. in a tropical acid soil. Organic amendments significantly increased soil pH, increased P availability, reduced exchangeable acidity and exchangeable forms of aluminium and iron compared with non-organic amendments (Ch'ng et al., 2016).

Use of recycled and recovered P management of organic P includes tillage, residue management and the use of GMs. Margenot et al. (2017) tested the combined effect of tillage and residue management on soil P status in an Oxisol from Kenya after nine years of adopting these practices. They found out that reduced tillage increased aggregation, labile P, maximum P sorption by a mean of 9% and decreased soil pH by 0.15 units at the

0–15-cm depth. Residue management, however, had no effect on soil P and TC that was attributed to the low residue retention rates of 2 t maize residues per year. Their work identified limited but positive impacts of conservation agriculture on soil P availability in weathered soils driven by reduced tillage (Margenot et al., 2017).

Phosphorus can be managed through use of P-efficient plants, which can be used as GMs, intercropping or in crop rotations with non P-efficient crops to increase and maintain the cycling of labile P. Phosphorus-efficient crops can solubilize sorbed P through excretion of organic anions and acid phosphatase (Liu et al., 2016). Examples of P-efficient crops include white lupin, chickpea, pigeon pea and fava bean (Simpson et al., 2011). Dissanayaka et al. (2017) reviewed literature of intercropping lupin for increased P acquisition and concluded that lupin may help increase P uptake with non-P efficient crops but only in P-limited soils when crops share rhizospheres. Continued research needs to look at the effectiveness of P-efficient GMs, intercropping and crop rotations in field studies (Simpson et al., 2011). All of these practices have the potential of increasing SOM though increased productivity, potentially improving the health of the soil.

The ^{32}P–^{33}P double isotope labelling technique was used to determine the relative contribution of GMs and P fertilizers to P uptake by Setaria grass grown in an amended tropical acid soil with a high P retention capacity in a glasshouse study (Bah et al., 2006). Because of their low P contents, the GMs (even the legume GMs) directly contributed less than 5% to total P uptake; but by mobilizing soil P, the GMs caused much higher P uptake in Setaria than the P fertilizers measured 15 weeks after establishment of grass. The GMs also increased the utilization of the reactive PR and TSP (Bah et al., 2006). In a 42-year-long term fertilizer trails on flooded tropic rice in India, grain yield was increased significantly by 74% over the control under the combined application of farm yard manure + NPK. Long-term balanced fertilization in the form of FYM + NPK for 42 years lowered the bonding energy and adsorption capacity for P in soil but increased its desorption potential, increasing P availability to the plant and leading to higher P uptake and yield maintenance (Bhattacharyya et al., 2015).

Other management strategies include addition of organic acids and phosphate-solubilizing microorganisms. These strategies may become important in highly weathered acidic soils with high P fixation capacities. Citric acid significantly enhanced the solubilization of organic P by 34.7% in native or plantation forests in subtropical and tropical Australia with 16–87% of soil total P being in organic form (Wei et al., 2010). Liu et al. (2014) isolated three efficient phosphate-solubilizing bacteria from the rhizosphere and bulk soil of betel nut. These isolates produced gluconic acid and were more abundant and had greater ability to solubilize insoluble phosphate in soils with low available P content (Liu et al., 2014). The variability in the performance of naturally abundant phosphate-solubilizing microorganisms, and poor understanding of their microbial ecology in soil, has hampered the large-scale application of phosphate-solubilizing microorganisms in sustainable agriculture (Khan et al., 2007; Sharma et al., 2013). Despite the variations in their performance, phosphate-solubilizing microorganisms are applied in agronomic practices in order to increase the productivity of crops while maintaining the health of soils (Khan et al., 2007). Several bacterial (Pseudomonads and Bacilli) and fungal strains (Aspergilli and Penicillium) have been identified as phosphate-solubilizing microorganisms (Sharma et al., 2013).

Although P is fixed or sorbed in soils, residual or legacy P can be slowly available for plant uptake over a long period of time (Johnston and Richards, 2003). Legacy P accumulates from past inputs of fertilizers or manures and can be a large secondary source of P and substitutes for P fertilizers. Legacy P has a continuum of availability that is dependent on pH and soil mineralogy (Rowe et al., 2016). Johnston and Richards (2003) demonstrated that P has not been fixed irreversibly in soils using evidence from long-term experiments at Rothamsted with no P fertilizers applied since 1856. The annual release of legacy/residual P from the Rothamsted experiments, however, was not sufficient to achieve the yield potential of currently used cultivars of barley and wheat. Highly weathered acidic soils with high Al and Fe oxides will bind inorganic P much more strongly than young base-rich temperate soils and therefore residual P is not available (Tiessen et al., 1984). Rodrigues et al. (2016) have reported that the total P content in Brazil's Cerrado soils has nearly doubled in the last 30–40 years due regular application of inorganic P fertilizers; however, P is mostly bound in moderately labile and non-labile forms associated with Fe/Al oxyhydroxides. NT agriculture had a positive albeit small impact on P availability in the topsoil and especially increased the contribution of organic P to labile P (Rodrigues et al., 2016). Withers et al. (2014) emphasized the need to quantify the value of the large amounts of legacy P present in most soils and develop ways to exploit it.

In summary, there is a critical need to reduce the reliance on inorganic P fertilizers due to limited availability of PR reserves and eutrophication problems in surface waters. Sustainable agriculture requires practices that maintain ecological balance of soil ecosystems and are environmentally friendly. Withers et al. (2014) proposed four key steps that include reduced crop P demand through breeding, quantify legacy P and develop ways to exploit it, increase use of recycled and recovered P and finally develop more efficient fertilizer formulations and application methods. Reviewing management impacts on biological P cycling in cropped soils, Oberson et al. (2001) recommended to increase the use of mineral fertilizers where soil P is being mined in conjunction with best management practices. Recommended best management practices include appropriate timing and placement of fertilizers, improved crop germplasm with high P acquisition, introduction of legume fallows or GMs to enhance P cycling, proper use of animal manure and urban solid and liquid waste, and minimizing P losses by erosion and runoff by utilizing minimal tillage and cover crops (Oberson et al., 2001). All of these practices use legacy P in soils more efficiently, reduce the reliance on chemical fertilizers and increase SOM, leading to healthy soils that are more biological active.

5 Potassium

5.1 Function in plant and potassium chemistry

To maintain soil health, crop production aims to not mine soil nutrients but manage soil fertility through integrated nutrient management. Excess N fertilizer application in crop production to push for intensification often leads to mining of N and potassium (K) from the soil (Patra et al., 2015). After N and P, K is the third most likely nutrient to limit crop productivity (Brouder, 2011). Potassium is the most limiting nutrient in 75% of rice paddies in China and half of wheat production in Australia (Zörb et al., 2014). Potassium is needed in plants as an enzyme activator for energy metabolism, starch synthesis, photosynthesis and sugar degradation (Marschner, 1995). Potassium in plants is not incorporated into the structure of organic compounds, but remains in the ionic form (K^+) in solution in the cell or acts as an activator of cellular enzymes. Potassium plays a critical role in lowering cellular osmotic potential, thereby reducing the loss of water from leaf stomata. Potassium is especially important in helping plants adapt to environmental stresses, such as drought tolerance, winder hardiness and resistance to plant diseases (Prabhu et al., 2007; Weil and Brady, 2017). Potassium is taken in by plants in large quantities similar to N, with K content in leaves ranging between 1% and 4%.

Potassium is found in comparatively high levels in most mineral soils, except those containing a large per cent of quartz sand. In soils, K is found in four main pools: the soil solution (0.1–0.2% of total K), exchangeable sites of colloids (1–2%), fixed in non-exchangeable form in 2:1 silicate clay minerals (1–10%) and lastly in primary minerals, for example, micas and feldspars (90–98%) (Weil and Brady, 2017). The readily available K includes soluble and exchangeable K on soil colloids, while non-exchangeable K is considered slowly available. Exchangeable K is readily available to plants and is in equilibrium with soil solution K. The amount of available K in unfertilized soils depends on the amount and kinds of K minerals in parent materials. Most of the K is found in relatively insoluble primary minerals such as feldspars and micas or is fixed in non-exchangeable forms in the mineral lattice of certain clays, such as hydrous mica (illite). Feldspars are quite resistant to chemical breakdown and weather slowly. The structural and unavailable K in micas exists in the interlayer space. The internal crystal-bonding differences cause trioctahedral biotite to weather more rapidly than dioctahedral muscovite (Ellis and Foth, 1996). Relatively unavailable forms, mostly micas and feldspars, supply little K during the growing season. Inoculation with K-solubilizing bacteria or choosing crops with the ability to produce K-solubilizing root execrates has been found to increase K solubility and biological availability (Zörb et al., 2014).

5.2 Factors affecting potassium availability

Potassium availability in soils depends on the weathering of primary minerals, clay mineralogy, the CEC of the soil, content of Ca and Mg in exchange sites, soil moisture, soil temperature and soil aeration (Weil and Brady, 2017; Ellis and Foth, 1996). Kaolinite fixes little K, but 2:1 clays such as vermiculite, fine-grained mica and smectite fix K readily and in large quantities. Alternate wetting/drying and freezing/thawing has been shown to enhance both the fixation of K into non-exchangeable forms and the release of previously fixed K to the soil solution. The practical importance of this is not well understood (Weil and Brady, 2017). Potassium buffering capacity or buffer power reflects the change in exchangeable or solid phase K concentration that occurs in response to change in solution phase K concentration (Brouder, 2011). A higher K buffer capacity reflects a higher soil capability for maintaining the solution K while experiencing a large decrease in exchangeable K. Well-buffered soils tend to have high CEC, large amounts of exchangeable K and K located in the wedge position in mica minerals (Ellis and Foth, 1996).

Calcareous soils often have high Ca content in solution and in exchange sites, which may decrease K uptake. Both Ca and Mg compete with K for plant uptake. Application of lime sometimes may result in an increase in K fixation of soils. Diffusion of K through soil water is the most limiting step in K uptake during the growing season. As soils dry, K diffusion decreases, resulting in less available K for uptake (Ellis and Foth, 1996). Potassium leaching can be determined by the quantity of K in soil and amount of water leaching through the soil profile. Leaching losses and extent of K movement tend to increase as soils become sandy. But K is neither thought to contribute to eutrophication of surface waters nor considered as a pollutant (Brouder, 2011).

A primary mechanism of K management for optimal crop productivity is soil testing (Brouder, 2011). Exchangeable and soil solution K are primarily extracted in routine soil test using 1 M ammonium acetate. However, fixed or non-exchangeable K is continually being released to the exchangeable forms and this may reduce the rate of K fertilizers to be applied. Current soil tests for K, however, do not capture the dynamic relationships among K forms that determine soil K-supplying power. Simple use of CEC and/or soil texture along with soil test has not proved useful in practice (Brouder, 2011). Except in very sandy soils, the problem of K fertility is not of total supply, but rather one of adequate transformation from non-available to available forms. The rate and distribution of K uptake is also highly dependent on crops and stage of development. For forage legumes such as alfalfa, K fertilization at high levels is needed, even if there are adequate supplies of K minerals. In the United States, K fertilizers were added historically to increase the supplying power of soils. Now these soils with high baseline K levels only need a smaller yearly K fertilizers maintenance application. However, the global use of K fertilizers is estimated to increase especially in row crop agriculture and in regions where sandy soils or highly weathered soils are prominent (Weil and Brady, 2017).

6 Other nutrients

6.1 Calcium and magnesium

Calcium (Ca) and magnesium (Mg) are macronutrients essential for life. Ca, Mg and K are also essential for sustaining soil physical characteristics due to their ability to aid flocculation (Zörb et al., 2014). Plants generally use Ca in amounts second only to N and K. The amount of Ca in plant foliage varies widely from as low as 0.1 to as high as 5% depending on the crop. Most grasses and other monocots have lower % of Ca in tissue. Some dicots, however, have 1–3% Ca in the leaves for optimal growth. Calcium is a major component of cell walls and is involved in cell elongation and division, membrane permeability and activation of several critical enzymes (Marschner, 1995). Calcium is usually the most abundant of non-acid cations in soils, so most plant deficiencies of Ca are quite rare except in very acid soils. Certain plants, however, show Ca deficiency even when the soil pH is adequately maintained. Such deficiencies are related to the transport of Ca within the plant. This deficiency or disorder is usually associated with unevenness in the water supply that interrupts the flow of Ca (Weil and Brady, 2017).

Magnesium concentration in plants ranges between 0.15 and 0.75% of dry matter. Magnesium is a central component of the chlorophyll molecule and is intimately involved in photosynthesis in plants. Magnesium also plays critical roles in the synthesis of oils and proteins and activation of enzymes involved in energy metabolism. Both Ca and Mg are absorbed by plants as divalent cations Ca^{2+} and Mg^{2+}. Calcium and Mg in soils are found mainly in three pools: 1) Ca- and Mg-containing minerals, for example, calcite ($CaCO_3$) and dolomite ($CaMgCO_3$). In addition to carbonates, minerals such as hornblende, augite, anorthite, olivine, biotite mica and gypsum supply soil Ca and Mg as they weather; 2) Ca and Mg complexed with organic matter; and 3) Ca and Mg held by the exchange sites on the clay and organic colloids. In most soils the principal sources that supply Ca and Mg to the soil solution for plant uptake are exchangeable Ca and Mg, and Ca and Mg in readily weathered minerals such as carbonates and apatite (Ellis and Foth, 1996). Exchangeable Mg is held less tightly than exchangeable Ca; therefore, exchangeable Mg normally saturates only 5–20% of the effective CEC as compared to the 60–90% typical for Ca in neutral to moderately sodic soils (Weil and Brady, 2017).

Numerous research studies have shown that plants grow very well and meet their Ca and Mg needs in soils that vary widely in Ca-to-Mg ratios anywhere from 1:1 to 15:1 (Weil and Brady, 2017). Sodic soils develop, however, when there is excess Na in soil solution and exchange sites relative to Ca and Mg. An ESP greater than 15 indicates a sodic soil that may lead to soil structure degradation and poor soil health. An amendment like gypsum ($CaSO_4$) is applied to ameliorate sodic soils and reduce Na saturation in exchange sites. Salt-affected soils create an extreme environment for soil microbes with an inverse relationship between soil EC, microbial biomass and activity (Yuan et al., 2007). In acidic soils, Ca and Mg are often applied as liming amendments to increase the pH of the soil and base saturation of acidic soils (Havlin et al., 2013). Maintaining a healthy quantity of Ca and Mg is important for soil health as salt-affected soils or acidic soils negatively impact various soil chemical, biological and physical growth factors.

6.2 Sulphur

Sulphur is a macronutrient element essential for plant and microbial life. It is used by plants in amounts similar to those of P. Sulphur is a constituent of the essential amino acids methionine, cysteine and cystine. Sulphur is closely associated with N in the process of protein and enzyme synthesis (Marschner, 1995). Sulphur deficiencies in crops have become increasingly common during the past several decades due to the enforcement of clean air standards in many countries. In addition, the S content of today's highly concentrated NPK fertilizers is far lower than those generations ago. The problem of maintaining adequate quantities of sulphur for nutrition of plants and soil health is becoming increasingly concerning. In some parts of the world sulphur is already the next most limiting nutrient after N (Ellis and Foth, 1996).

The three major natural sources of sulphur are: 1) organic matter, 2) soil minerals and 3) sulphur gases in the atmosphere. In surface soils in temperate humid regions, 90–98% of the sulphur is usually present in organic forms. As is the case for N, the exact forms of S in organic matter are not well known. Over time, soil microorganisms break down organic sulphur compounds into some soluble forms similar to the release of ammonium-N and nitrate-N from organic matter. Factors that affect microbial activity, that is, temperature, moisture, oxygen and pH, affect S availability. Similar to N most of the S is organic with only a very small percentage in the mineralized sulphate form (Weil and Brady, 2017). Application of gypsum as a source of sulphur at 30 kg $acre^{-1}$ and 60 kg $acre^{-1}$ saw an increase in microbial biomass between 8.2 and 12.8% (Ram et al., 2017).

Depending on the soil redox conditions, most inorganic sulphur is present either as sulphates (SO_4^{2-}) or sulphides (S^{2-}). The sulphate minerals are most easily solubilized and the sulphate ion is easily assimilated by plants. The most common sulphate mineral is gypsum. The sulphate anion can be leached from the soil, especially in humid regions. But sulphate anions are adsorptive to the surfaces of iron and aluminium oxides that are coating soil particles. Sulphate anions are attracted by the positive charges that characterize acid soils containing these clays. Most sulphates may be held in the subsoil horizons of all Ultisols and Oxisols because they contain iron oxides (Weil and Brady, 2017).

In soils with poor aeration, for example, wetlands, sulphur is present in reduced forms such as S° or as sulphides. Sulphide-rich sediments are widespread in many low-lying coastal areas worldwide and are prevalent in Australia, Asia and Europe. Extreme acidity in soils can form when the common form of reduced sulphur is pyrite (FeS_2 iron disulphide). These sulphides are stable as long as oxygen is not present. When the submerged or buried materials are drained and exposed to oxygen, the sulphides and/or the elemental sulphur oxidizes to form sulphuric acid driving pH levels as low as two. Plants cannot grow under these conditions. Thousands of streams have been seriously polluted in this manner, due to the acid drainage. Drainage of extensive areas of acid sulphate soils in Queensland, Australia, has caused acidification, metal contamination, deoxygenation and iron precipitation in the Great Barrier Reef receiving waters (Powell and Martens, 2005). The acid and metal ions (Al and toxic trace elements) draining into adjacent waterways cause great environmental damage; degrade aquatic ecosystems, large fish kills; and affect infrastructure and human health (Powell and Martens, 2005; Wu et al., 2015). Acid sulphate soils will be dominated with sulphur- and iron-oxidizing microorganisms (Wu et al., 2015) and need expensive application of lime to get them to a pH that will support plants and healthy microbial activity.

The atmosphere contains hydrogen sulphide, sulphur dioxide and carbonyl sulphide and other sulphur gases. These atmospheric forms of sulphur arise from volcanic eruptions, release from soils, ocean spray, biomass fires and industrial plants. In the atmosphere, most of the sulphur-containing materials are eventually oxidized to sulphates forming sulphuric acid, which contributes to acid rain. Sulphur emissions in North America and parts of Europe have declined by more than half since the late 1980s, due to regulatory programmes to reduce these emissions. The acid precipitation caused partly by atmospheric sulphur and N is a serious threat to the health of lakes, forest and ecosystems (Weil and Brady, 2017).

7 Case study: integrated watershed model in India

The International Crops Research Institute for the Semi-Arid Tropics (ICRISAT) developed the Integrated Watershed Model (IWM), to improve rural livelihoods with a focus on building soil health. The IWM case study was chosen to expand the discussion of soil health chemical components beyond NPK by expanding upon the impacts of three other chemical components, micronutrients, organic matter and sodicity on soil health. IWM has been adopted by more than 368 watersheds across Asia between 2001 and 2008 (Wani et al., 2008), and then implemented state wide in Karnataka, involving five million farmers in the Bhoochetana programme (Wani et al., 2017b). The goals of IWM programmes implemented by ICRISAT are to improve rural livelihoods through the dissemination of technologies such as integrated nutrient management for soil health, integrated pest management, rainwater-harvesting structures, improved cropping practices, improved crop varieties and micro-enterprises for women. To improve soil health, ICRISAT focuses on secondary and micronutrient management, and increasing SOC. Soil sodicity is a rising concern for soil health in the semi-arid tropics, as irrigation with groundwater becomes more wide spread.

ICRISAT focuses on developing and disseminating technologies to increase livelihood while reducing environmental degradation in the semi-arid tropical regions of Africa and

Asia (ICRISAT, 2014). Increasing soil health in the semi-arid region of India is important for Indian food security. The semi-arid climate is classified as having precipitation greater than or equal to evapotranspiration between 2 and 4.5 months a year. The soils are not highly weathered due to the limited precipitation between 350 and 800 mm. Soil chemical limitations to soil health in the semi-arid include nutrient management, SOC management and salt-affected soils (Juo and Franzluebbers, 2003).

7.1 Integrated nutrient management

The green revolution and government subsidies led Indian farmers to apply excess amounts of NPK for more than 50 years. After years of applying only NPK, soil degradation occurred, and yields became stagnant as secondary and micronutrients were exploited from the soil (Rego et al., 2006). There are limited soil testing facilities in India that are capable of testing secondary and micronutrients. ICRISAT is one of the few institutions able to provide complete soil health recommendations. ICRISAT soil sampling has improved the locality of soil micronutrient recommendations from blanket statewide recommendations to community-level recommendations (Wani et al., 2017b).

The ICRISAT Development Center (IDC) framework uses integrated nutrient management recommendations for soil health as an entry point activity for establishing trust with communities in new development programmes. First, when ICRISAT enters into a community, soil samples (0–15 cm) are collected through a random, stratified sampling scheme. The strata represent cropping system, size of farm and topography (Wani et al., 2003b). Farmers' interest in soil health is developed through workshops and participatory soil sampling (Chander et al., 2016). One aggregated soil health recommendation is then made for the community, and is communicated through farmer field trials, workshops, soil health card dissemination and soil fertilizer recommendations painted on walls in communities' meeting halls. Parameters shown in the soil health card include soil pH in water; electrical conductivity (dS m^{-1}); organic carbon (%); and available phosphorus, potassium, calcium, magnesium, sulphur, zinc, boron, iron, copper and manganese (mg kg^{-1}) (Wani et al., 2017a). For ease of communication, each chemical soil health component is denoted as high, medium, normal, sufficient or deficient. Also noted on the soil health card are fertilizer recommendations given in kg per hectare for urea, diammonium phosphate, potassium chloride, gypsum, $ZnSO_4$, and borax for various cropping systems. If more than 50% of samples were deficient in a micronutrient, then a full dose of fertilizer was recommended. If less than 50% of samples were deficient in a micronutrient, then a half dose of fertilizer was recommended. A full dose of fertilizer includes sulphur, boron and zinc at 30, 5 and 10 kg ha^{-1}, respectively.

In the Indian states of Telangana, Andhra Pradesh, Madhya Pradesh and Rajasthan, 80–100% of soil samples analysed by ICRISAT had critical deficiencies of zinc, boron and/or sulphur. Zinc is the most deficient micronutrient in India (Rego et al., 2006). Kothapalli was the first community where the IWM was implemented. Kothapally saw crop yields increase between 1.7 and 3.8 times compared to the pre-programme yields between 1998 and 2002 (Wani et al., 2003a). Field trials in the community attributed to between 12 and 53% of yield increases due to the applications of boron and/or sulphur for sorghum, maize and soybean (Rego et al., 2006).

Bhoochetana, an IWM development project, was implemented by ICRISAT state wide in Karnataka, involving five million farmers (Wani et al., 2017b). The state of Karnataka is 7.5 million ha. Fifty per cent of the land area was enrolled in Bhoochetana (Wani et al., 2017a).

Table 2 Grain yields of farmers participating in Bhoochetana during 2009–13

Year	Grain yield crop	Farmers' practice (kg ha^{-1})	Improved practice (kg ha^{-1})	Increase (%)	LSD*	F value
2009	Maize	5420	7800	44	356**	174.76
	Finger millet	1680	2550	52	157**	120.06
2010	Maize	5820	7560	30	441**	60.31
	Finger millet	1910	2620	37	112**	152.71
2011	Maize	3900	5110	31	440**	29.32
	Finger millet	1500	1960	31	215**	18.12
2012	Maize	3700	4850	31	437**	26.73
	Finger millet	1380	1830	32	204**	18.54
2013	Maize	4100	5340	30	220**	121.62
	Finger millet	1460	1870	28	116**	17.94

* LSD, Least significant difference.
** significant at 1% level.
Adapted with permission from Anantha et al. (2016).

From 2009 to 2012, Bhoochetana reported an increase of crop yields between 28 and 52% (Table 2). During the project, 92 904 soil samples were collected and analysed to develop soil health recommendations for communities (Wani, 2016). The majority of farms were deficient, with 55% deficient in Zn, 62% deficient in B and 52% deficient in S (Wani et al., 2017a). Over 3700 on-farm trials helped disseminate these soil health recommendations by practically showing farmers how improved practices improved crop and soil health (Wani, 2016). The soil health status for various subdivisions of a district (*taluks*) for the Raichur District in Karnataka is shown in Table 3.

Table 3 Soil health status of Raichur District, Karnataka. Percentage of farmers' fields deficient in nutrients

Taluk	Organic carbon (%)	Available phosphorus (%)	Available potassium (%)	Available sulphur (%)	Available zinc (%)	Available boron (%)
Raichur	77	45	14	73	78	63
Manvi	79	49	0	70	85	18
Sindhanur	40	31	0	26	53	12
Lingasur	64	54	4	84	86	52
Devadurga	85	54	6	70	84	46
Average for Raichur District	69	47	4	64	78	37

Adapted with permission from Sawargaonkar et al. (2016).

7.2 Soil organic carbon management

In addition in Karnataka, SOC levels were low with an average between 0.25-0.5% (Wani et al., 2017a), with 52% of them deficient and below the recommended 1% (Chander et al., 2016). To improve SOC and to reduce input of inorganic fertilizers, ICRISAT recommended the application of manure. However, because of the shift in labour force from agricultural to urban jobs there was limited accessibility and availability of manures. Instead, ICRISAT disseminated information on vermicomposting (Nagavallemma et al., 2004), aerobic compost (Wani et al., 2015) and GMs (Wani et al., 2017b). There was often limited availability of organic fertilizers in communities in India despite the high demand by farmers. To fill the gap, ICRISAT facilitated the development of women-owned vermicompost businesses, where women had their own income source and produced a high-demand product for the community (ICRISAT, 2016). The vermicompost was created through the decomposition of cow manure, weeds and agriculture waste (cotton and pigeonpea stalks) by earthworms with the addition of rock phosphate (Wani et al., 2003a). In aerobic compost decomposition, it was facilitated by adding a microbial inoculant consisting of 21 isolated bacteria (Wani et al., 2017b). A chopping machine was recommended by ICRISAT to break agricultural waste into smaller pieces, which facilitated the decomposition process (Wani et al., 2015).

In the Bhoochetana project, ICRISAT helped women-owned businesses to grow a multipurpose forage tree legume, *Gliricidia,* to be used as a GM by farmers (ICRISAT, 2016). *Gliricidia* can be a source of organic carbon and nitrogen. ICRISAT recommended *Gliricidia* grown at 75-cm spacing on farm bunds, where leaves and brunches are cut yearly and incorporated into the soil. The cuttings can provide 30 kg N ha^{-1}, while also providing a source of SOC (Wani, 2008). 50 000 *Gliricidia* saplings were planted by farmers during the IWM programme at Kothapolly (Wani et al., 2003a).

7.3 Soil sodicity

ICRISAT implemented IWM in many watersheds in the Deccan plateau of south India. Some of these watersheds have sodic Vertisols. The sodic Vertisols' cotton yield was 50% less than non-sodic soils', according to a reported field trial at ICRISAT (Pal et al., 2012). A sodic soil is one that has an electrical conductivity of less than 4 dS m^{-1}, a pH greater than 8.5 and an ESP greater than 15, indicating Na is the dominant cation in the soil's exchange sites. Sodic soils cover 3.77 Mha in India (Dalal et al., 2011).

In the Deccan plateau, sodic Vertisols developed under natural pedogenesis. The soil profiles of semi-arid soils already have a high concentration of bases, because there was limited leaching due to limited rainfall and the upward movement of base cations in the soil profile due to the upward movement of water from of evapotranspiration. Over time, micro-high areas of the Vertisols underwent a slightly more arid climate, causing a greater formation of pedogenic $CaCO_3$, creating micro-site sodic areas (Pal, 2017). The sodicity was not wide spread, but the sodicity affects the soil health of the farmers involved in IWM. See Sahrawat et al. (2016) for the range of pH and electrical conductivity of soils in each district of Karnataka. Soil sodicity was a common limitation in crop production in the semi-arid tropics (Pal et al., 2012).

ICRISAT recommended that sodic conditions in Vertisols be improved either quickly or slowly. Sodic conditions can be remediated quickly by adding gypsum; however, this only reduces sodicity at the soil surface and subsoil sodicity is not remediated (Bhattacharyya et al., 2016a). In addition, limited water often inhibits this method's effectiveness. Sodic conditions can also be remediated slowly through improved management practices, such as improved cropping systems, improved crop varieties, increasing SOC and improving soil fertility management, which leads to the slow dissolution of $CaCO_3$, slowly reducing the ESP (Bhattacharyya et al., 2016a; Pal, 2017). Pal et al. (2012) reported that improved management practices compared to traditional management practices over a 30-year time period led to a significantly higher accumulation of SOC, leading to dissolution of $CaCO_3$, thus replacing Na^+ by Ca^{2+} ions in the soil exchange sites and reversing soil sodicity. Submerged rice paddy cultivation can also increase the rate of the dissolution of $CaCO_3$ (Pal, 2017). ICRISAT also recommended improvement of groundwater quality by increasing groundwater levels and limited use of groundwater for irrigation as preventative measures to reduce the impacts of soil sodicity.

8 Summary: how can integrated nutrient management contribute to enhanced and sustainable crop production through improved soil health

In this chapter, we focused on the chemical indicators of soil health and the chemistry of nutrients, but the biological factors influencing soil health especially for N and P could not be identified. We illustrated a case study from India that showed how integrated nutrient management, using both inorganic fertilizers and practices to improve and increase SOC, can be done using site-specific research and available resources.

Several chemical properties are used as indicators for measuring soil health, including pH, salinity, reactive C and nutrient levels. It is clear that the use of these tests is site- and ecosystem-specific. The use of integrative soil health indices, for example, the CASH by Cornell University, is an approach that combines individual soil chemical tests (Moebius-Clune et al., 2016).

Integrated nutrient management systems need to be developed specific to geographical location, soils, climates, crop rotation, available resources and limitations. Finding a one-size-fits-all solution is not possible due to differences in soil properties, climate and available resources. Site-specific research in nutrient management that includes a holistic approach, taking into account the use of inorganic fertilizers as well as management practices to improve organic matter, is needed. Optimum nutrient cycling to meet the demands of plant growth and soil functions is an important concept of soil heath. Deficiencies and losses of N have profound impacts on the health and productivity of soils and the world's ecosystems. There are many loss pathways in the N cycle that lead to water and air pollution in addition to less sustainable ecosystems. Improving soil health incorporates the use of sustainable practices that may include cover crops, GMs and conservation tillage. Use of inorganic N fertilizers should be considered as part of a sustainable system that is optimized to minimize the loss of reactive N. The use of biological N_2 fixation to increase total and available N supplies in soils especially in low-input systems is crucial. The importance of biological N_2 fixation to world food security is unquestionable (Vance et al., 2000).

Phosphorus is second to N in its importance for restoring degraded soils, especially weathered acid soils. Phosphorus undergoes fixation reactions in low- and high-pH soils, rendering the applied P in less soluble forms. Major factors that constrain tropical soil fertility include high P fixation, high acidity with Al toxicity, erosion and low nutrients. Improving P fertility in these soils can lead to increased productivity and improvement of soil health. Practices to increase organic matter and P are essential. There are also innovative new approaches to improve P fertility by solubilizing insoluble P or using P-efficient plants. In more developed areas, P losses to surface water, via erosion and surface runoff, is a major concern leading to water eutrophication and degradation of ecosystems.

Through the IWM, ICRISAT was able to bring localized soil health recommendations to millions of farmers across Asia. Previously, state-level soil fertility recommendations that only provided NPK recommendations led to exploitation of secondary and micronutrients, since farmers only added NPK to their soil. By providing more holistic soil health recommendations, farmers have reduced the amount of NPK fertilizers applied, and increased crop productivity through the application of secondary and micronutrients. ICRISAT reported widespread deficiencies of boron, zinc and sulphur across the semi-arid regions of India. ICRISAT also reported a widespread deficiency in SOC across the state of Karnataka, India. IWM focused on providing recommendations for improving SOC levels by developing new composting methods that use organic carbon sources available locally to farmers. Finally, the use of improved management practices to reverse soil sodicity, instead of the generalized recommendation of application of gypsum, matches the local availability limitation of high-quality water. Disseminating soil health cards to farmers involved in IWM helped farmers identify sodic soil and the correct improved management practices. Localized soil health recommendations can help farmers identify soil health chemical components that may be limiting their production.

9 Future trends in research

The future of research in agriculture and soil health chemical components needs to look into soil health at a food system scale. Future research should not only look at traditional soil science topics but expand its view by placing soil science within the greater perspective of the entire food system.

Some traditional soil science research areas that need to be explored include (1) soil fertilizer recommendations which are still missing for non-grain crops in many soils of the world. These recommendations would help prevent further soil and environmental degradation. (2) Further research into microbial inoculants, compost teas and other biofertilizers to understand what their chemical components are; how they vary depending on storage, application and source of products used and how biofertilizers interact with microbes to make more nutrients available to the plant. (3) Research on the use of organic amendments and waste products will need to expand to include the role of these amendments in releasing greenhouse gases including CO_2 and N_2O gases. (4) Ultimately, successful management of N (and other nutrients), using cover crops and amendments requires that availability of nutrients be synchronized so that nutrients are readily available during periods of active uptake by cash crops and minimally available during periods when cash crops are not growing to reduce losses to air and water (Kaspar and Singer, 2011). This is an important area of research to focus on, to be able to utilize nutrients released from cover crops and organic amendments efficiently, and reduce the reliance on inorganic fertilizers which may not be easily available or affordable.

Soil and agriculture cannot be examined in isolation from the larger food system, which supports them. For example, soil science can be viewed from a larger, more holistic view of the food system. One example is examining the relationship between soil science and food waste. Food waste, product quality preferences and available food system infrastructures can all be linked to improve soil health and subsequently the resiliency of the food system. It is estimated that 30–40% of food produced is wasted along the supply chain (Godfray et al., 2010). In developing countries the majority of food is wasted in post-harvest and pre-retail to the consumer, whereas in developed countries the majority of food is wasted post-retail (Godfray et al., 2010). How can soil health chemical components be managed better while also eliminating food waste and/or using food waste to improve soil health chemical components? What soil management strategies can be followed to reduce post-harvest food waste?

For example, 30–40% of an individual's waste in Belgium is organic and is separated by each household for mandatory collection in each municipality. Belgium was motivated by lack of landfill space to create infrastructure that diverts organic waste from landfills (Gellynck et al., 2011). Further research needs to look at how this organic waste can be used in agricultural production and its impact on soil health. For example, Treuer et al. (2017) reported improved soil health and plant available nutrients after application of orange juice by-products on degraded pasture soil. Broader research needs to look at what policy was needed to create the organic waste infrastructure, how the public perceives its relationship to improve soil health and ecosystem services and how these policies could be implemented across the globe.

Finally, effective information transfer regarding soil health and its importance in food security is crucial. When communicating to farmers and future farmers about soil health, what are the best ways to communicate the importance of soil health chemical components? Are researchers providing relevant information to match the local soil health conditions? How can the perception of food quality be changed so that it matches the idea that high-quality food is grown in healthy soil? How can consumer preference be better aligned with products that grow best in their local soil conditions? The field of ethnopedology has begun to ask basic research related to how the general public interacts with soil, perceives soil health, values soil health (Barrera-Bassols and Zinck, 2003) and governs soil (Montanarella and Vargas, 2012). This research needs to continue and expand to include the public perception of soil within the larger scope of the food system.

10 Where to look for further information

1. Soil health assessment and soil quality indicator sheets from USDA (https://www.nrcs.usda.gov/wps/portal/nrcs/main/soils/health/assessment/)
2. The Soil Health Institute: http://soilhealthinstitue.org
3. Download an app from the University of Wisconsin to calculate legume credits for N and manure N, P, K and S credits: http://ipcm.wisc.edu/apps/npkcredits/
4. For publications on role of phosphorus in water quality and methods to measure P and to minimize phosphorus losses from agriculture: https://sera17.org/publications/
5. Use of phosphate rocks in Sustainable Agriculture: http://www.fao.org/docrep/007/y5053e/y5053e00.htm#Contents.

11 References

Alleoni, L. R. F., Fernandes, A. R. and Campos, M. de (2014), 'Degree of phosphorus saturation of an Oxisol amended with biosolids in a long-term field experiment', *Environmental Science and Pollution Research*, 21(8), pp. 5511–20. doi:10.1007/s11356-013-2469-0.

Anantha, K. H., Raju, K. V., Wani, S. P., Pardhasaradhi, G., Srinivas, K., Garg, K. K. and Krishnappa, K. (2016), 'Social and economic benefits', In: Raju, K. V. and Wani, S. P. (Eds), *Harnessing Dividends from Drylands: Innovative Scaling Up with Soil Nutrients*. Oxfordshire: CABI, pp. 259–89.

Bah, A. R., Zaharah, A. R. and Hussin, A. (2006), 'Phosphorus uptake from green manures and phosphate fertilizers applied in an acid tropical soil', *Communications in Soil Science and Plant Analysis*, 37(13–14), pp. 2077–93. doi:10.1080/00103620600770433.

Barrera-Bassols, N. and Zinck, J. A. (2003), 'Ethnopedology: A worldwide view on the soil knowledge of local people', *Geoderma*, 111(3–4), pp. 171–95. doi:10.1016/S0016-7061(02)00263-X.

Beegle, D. (2005), 'Assessing soil phosphorus for crop production by soil testing', In: *Phosphorus: Agriculture and the Environment: Monograph (46)*. Madison, WI: American Society of Agronomy, Crop Science Society of America, and Soil Science Society of America, pp. 123–43. doi:10.2134/agronmonogr46.c5.

Bhattacharyya, P., Nayak, A. K., Shahid, M., Tripathi, R., Mohanty, S., Kumar, A., Raja, R., Panda, B. B., Lal, B., Gautam, P., Swain, C. K., Roy, K. S. and Dash, P. K. (2015), 'Effects of 42-year long-term fertilizer management on soil phosphorus availability, fractionation, adsorption–desorption isotherm and plant uptake in flooded tropical rice', *The Crop Journal*, 3(5), pp. 387–95. doi:10.1016/j.cj.2015.03.009.

Bhattacharyya, T., Pal, D. K., Wani, S. P. and Sahrawat, K. L. (2016a), 'Resilience of the semi-arid tropical soils', *Current Science*, 110(09), pp. 1784–8.

Bhattacharyya, T., Wani, S. P., Pal, D., Sahrawat, K. L., Pillai, S., Nimje, A., Telpande, B. and Chandran, P. (2016b), 'ICRISAT, India soils: Yesterday, today and tomorrow', *Current Science*, 110(09), pp. 1652–70.

Bhowmik, A., Fortuna, A. M., Cihacek, L. J., Bary, A. I. and Cogger, C. G. (2016), 'Use of biological indicators of soil health to estimate reactive nitrogen dynamics in long-term organic vegetable and pasture systems', *Soil Biology and Biochemistry*, 103, pp. 308–19. doi:10.1016/j.soilbio.2016.09.004.

Blanco-Canqui, H., Holman, J. D., Schlegel, A. J., Tatarko, J. and Shaver, T. M. (2013), 'Replacing fallow with cover crops in a semiarid soil: Effects on soil properties', *Soil Science Society of America Journal*, 77(3), pp. 1026–34. doi:10.2136/sssaj2013.01.0006.

Bougnom, B. P., Mair, J., Etoa, F. X. and Insam, H. (2009), 'Composts with wood ash addition: A risk or a chance for ameliorating acid tropical soils?', *Geoderma*, 153(3), pp. 402–7. doi:10.1016/j.geoderma.2009.09.003.

Brouder, S. (2011), 'Potassium cycling', In: Hatfield, J. L. and Sauer, T. J. (Eds), *Soil Management: Building a Stable Base for Agriculture*. Madison, WI: Soil Science Society of America, pp. 79–102. doi:10.2136/2011.soilmanagement.c6.

Bulluck, L. R., Brosius, M., Evanylo, G. K. and Kistaino, J. B. (2002), 'Organic and synthetic fertility amendments influence soil microbial, physical and chemical properties on organic and conventional farms', *Applied Soil Ecology*, 19(2), pp. 147–60. doi:10.1016/S0929-1393(01)00187-1.

Bundy, L. G. (1994), *Corn Fertilization*. Madison, WI: University of Wisconsin-Extension (Cooperative Extension, AA3340).

Cade-Menun, B. (2005), 'Using phosphorus-31 nuclear magnetic resonance spectroscopy to characterize organic phosphorus in environmental samples', In: Turner, B. L. (Ed.), *Organic Phosphorus in the Environment*. Wallingford, UK: CABI, pp. 21–44.

Cameron, K. C., Di, H. J. and Moir, J. L. (2013), 'Nitrogen losses from the soil/plant system: A review', *Annals of Applied Biology*, 162(2), pp. 145–73. doi:10.1111/aab.12014.

Cardoso, I. M. and Kuyper, T. W. (2006), 'Mycorrhizas and tropical soil fertility', *Agriculture, Ecosystems & Environment (Nutrient Management in Tropical Agroecosystems)*, 116(1), pp. 72–84. doi:10.1016/j.agee.2006.03.011.

Chacon, N., Silver, W. L., Dubinsky, E. A. and Cusack, D. F. (2006), 'Iron reduction and soil phosphorus solubilization in humid tropical forests soils: The roles of labile carbon pools and an electron shuttle compound', *Biogeochemistry*, 78(1), pp. 67–84. doi:10.1007/s10533-005-2343-3.

Chander, G., Wani, S. P., Krishnappa, K., Sahrawat, K. L., Parthasaradhi, G. and Jangawad, L. S. (2016), 'Soil mapping and variety-based entry-point interventions for strengthening agriculture-based livelihoods – exemplar case of 'Bhoochetana' in India', *Current Science*, 110(09), pp. 1683–91.

Chatterjee, A. and Clay, D. E. (2016), 'Cover crops impacts on nitrogen scavenging, nitrous oxide emissions, nitrogen fertilizer replacement, erosion, and soil health', In: Chatterjee, A. and Clay, D. (Eds), *Soil Fertility Management in Agroecosystems*. Madison, WI: American Society of Agronomy, Crop Science Society of America, and Soil Science Society of America, Inc., pp. 76–89. doi:10.2134/soilfertility.2016.0012.

Cheesman, A. W., Turner, B. L. and Ramesh Reddy, K. (2012), 'Soil phosphorus forms along a strong nutrient gradient in a tropical ombrotrophic wetland', *Soil Science Society of America Journal*, 76(4), pp. 1496–506. doi:10.2136/sssaj2011.0365.

Chien, S. H. and Friesen, D. K. (1992), 'Phosphate rock for direct application', In: Sikora, F. J. (Ed.), *Future Directions for Agricultural Phosphorus Research* (Bulletin, Y-224). Muscle Shoals, USA: Valley Authority, pp. 47–52.

Ch'ng, H. Y., Ahned, O. H. and Majid, N. K. A. (2016), 'Improving phosphorus availability, nutrient uptake and dry matter production of *Zea mays* l. on a tropical acid soil using poultry manure biochar and pineapple leaves compost', *Experimental Agriculture*, 52(3), pp. 447–65. doi:10.1017/S0014479715000204.

Congreves, K. A., Hayes, A., Verhallen, E. A. and Van Eerd, L. L. (2015), 'Long-term impact of tillage and crop rotation on soil health at four temperate agroecosystems', *Soil and Tillage Research*, 152, pp. 17–28. doi:10.1016/j.still.2015.03.012.

Congreves, K. A., Hooker, D. C., Hayes, A., Verhallen, E. A. and Van Eerd, L. L. (2017), 'Interaction of long-term nitrogen fertilizer application, crop rotation, and tillage system on soil carbon and nitrogen dynamics', *Plant and Soil*, 410(1–2), pp. 113–27. doi:10.1007/s11104-016-2986-y.

Culman, S. W., Snapp, S. S., Freeman, M. A., Schipanski, M. E., Beniston, J., Lal, R., Drinkwater, L. E., Franzluebbers, A. J., Glover, J. D., Grandy, A. S., Lee, J., Siz, J., Maul, J. E., Mirksy, S. B., Spargo, J. T. and Wander, M. M. (2012), 'Permanganate oxidizable carbon reflects a processed soil fraction that is sensitive to management', *Soil Science Society of America Journal*, 76(2), pp. 494–504. doi:10.2136/sssaj2011.0286.

Dalal, R. C., Wong, V. N. L. and Sahrawat, K. L. (2011), 'Salinity and sodicity affect organic carbon dynamics in soil', *Bulletin of the Indian Society of Soil Science*, 28, pp. 95–117.

Daroub, S. H., Pierce, F. J. and Ellis, B. G. (2000), 'Phosphorus fractions and fate of phosphorus-33 in soils under plowing and no-tillage', *Soil Science Society of America Journal*, 64(1), pp. 170–6. doi:10.2136/sssaj2000.641170x.

Daroub, S. H., Lang, T. A., Josan, T. A. and Bhadha, J. H. (2011), 'Best management practices and long-term water quality trends in the Everglades Agricultural Area', *Critical Reviews in Environmental Science and Technology*, 41(suppl. 1), pp. 608–32. doi:10.1080/10643389.2010.530905.

Daroub, S. H., Diaz, O. H., Lang, T. A. and Chen, M. (2012), *Best Management Practices in the Everglades Agricultural Area: Soil Testing*. Gainesville, FL: Soil and water Science Department, Florida Cooperative Extension Service, Institute of Food and Agricultural Sciences, University of Florida. Available at: http://edis.ifas.ufl.edu/ss445.

Davidson, E. A., Keller, M., Erickson, H. E., Verchot, L. V. and Veldkamp, E. (2000), 'Testing a conceptual model of soil emissions of nitrous and nitric oxides: Using two functions based on soil nitrogen availability and soil water content, the hole-in-the-pipe model characterizes a large fraction of the observed variation of nitric oxide and nitrous oxide emissions from soils', *BioScience*, 50(8), pp. 667–80.

D'Hose, T., Ruysschaert, G., Viaene, N., Debode, J., Vanden Nest, T., Van Vaerenbergh, J., Cornelis, W., Willekens, K. and Vandecasteele, B. (2016), 'Farm compost amendment and non-inversion tillage improve soil quality without increasing the risk for N and P leaching', *Agriculture, Ecosystems & Environment*, 225, pp. 126–39. doi:10.1016/j.agee.2016.03.035.

Dissanayaka, D. M. S. B., Wickramasinghe, W. M. K. R., Marambe, B. and Wasaki, J. (2017), 'Phosphorous-mobilization strategy based on carboxylate exudation in lupins (Lioinus, favaceae): A mechanism facilitating the growth and phosphorus acquisition of neighbouring plants under phosphorus-limited conditions', *Experimental Agriculture*, 53(2), pp. 308–19. doi:10.1017/S0014479716000351.

Dodd, R. J. and Sharpley, A. N. (2015), 'Recognizing the role of soil organic phosphorus in soil fertility and water quality', *Resources, Conservation and Recycling*, 105, pp. 282–93. doi:10.1016/j.resconrec.2015.10.001.

Doolette, A. L. and Smernik, R. J. (2011), 'Soil organic phosphorus speciation using spectroscopic techniques', In: Bünemann, E., Oberson, A. and Frossard, E. (Eds), *Phosphorus in Action*. Berlin, Heidelberg: Springer, pp. 3–36. doi:10.1007/978-3-642-15271-9_1.

Doran, J. W. and Zeiss, M. R. (2000), 'Soil health and sustainability: managing the biotic component of soil quality', *Applied Soil Ecology*, 15(1), pp. 3–11. doi:10.1016/S0929-1393(00)00067-6.

Ellis, B. and Foth, H. (1996), *Soil Fertility*, Second Edition. Boca Raton: CRC Press.

Fine, A. K., van Es, H. M. and Schindelbeack, R. R. (2017), 'Statistics, scoring functions, and regional analysis of a comprehensive soil health database', *Soil Science Society of America Journal*, 81(3), pp. 589–601. doi:10.2136/sssaj2016.09.0286.

Fixen, P. E. and Johnston, A. M. (2012), 'World fertilizer nutrient reserves: A view to the future', *Journal of the Science of Food and Agriculture*, 92(5), pp. 1001–5. doi:10.1002/jsfa.4532.

Geisseler, D. (2017), 'Estimating non-fertilizer nitrogen contributions', In: *Western Nutrient Management Conference*, Reno, NV: International Plant Nutrition Workshop, pp. 32–6.

Gellynck, X., Jacobsen, R. and Verhelst, P. (2011), 'Identifying the key factors in increasing recycling and reducing residual household waste: A case study of the Flemish region of Belgium', *Journal of Environmental Management*, 92(10), pp. 2683–90. doi:10.1016/j.jenvman.2011.06.006.

Ghafoor, A., Poeplau, C. and Kätterer, T. (2017), 'Fate of straw- and root-derived carbon in a Swedish agricultural soil', *Biology and Fertility of Soils*, 53(2), pp. 257–67. doi:10.1007/s00374-016-1168-7.

Gikonyo, E., Zaharah, A. R., Hanafi, M. M. and Anuar, R. (2010), 'Phosphorus leaching in an acid tropical soil 'recapitalized' with phosphate rock and triple superphosphate', *The Scientific World Journal*, 10, pp. 1498–508. doi:10.1100/tsw.2010.156.

Gil-Sotres, F., Trasar-Cepeda, C., Leirós, M. C. and Seoane, S. (2005), 'Different approaches to evaluating soil quality using biochemical properties', *Soil Biology and Biochemistry*, 37(5), pp. 877–87.

Godfray, H. C. J., Beddington, J. R., Crute, I. R., Haddad, L., Lawrence, D., Muir, J. F., Pretty, J., Robinson, S., Thomas, S. M. and Toulmin, C. (2010), 'Food security: The challenge of feeding 9 billion people', *Science*, 327(5967), pp. 812–18. doi:10.1126/science.1185383.

Green, C. J. and Blackmer, A. M. (1995), 'Residue decomposition effects on nitrogen availability to corn following corn or soybean', *Soil Science Society of America Journal*, 59(4), pp. 1065–70. doi:10.2136/sssaj1995.03615995005900040016x.

Hadas, A., Kautsky, L., Goek, M. and Erman Kara, E. (2004), 'Rates of decomposition of plant residues and available nitrogen in soil, related to residue composition through simulation of carbon and nitrogen turnover', *Soil Biology and Biochemistry*, 36(2), pp. 255–66. doi:10.1016/j.soilbio.2003.09.012.

Haney, R. (2015), *Soil Health Tool (SHT) ver 4.3: An Integrated Approach to Soil Testing, Soil Testing in Nature's Image*. Washington, DC: USDA.

Havlin, J. L., Tisdale, S. L., Nelson, W. L. and Beaton, J. D. (2013), *Soil Fertility and Fertilizers*. 8th Edition. Upper Saddle River, N.J.: Pearson.

Hedley, M. J. and Stewart, J. W. B. (1982), 'Method to measure microbial phosphate in soils', *Soil Biology and Biochemistry*, 14(4), pp. 377–85. doi:10.1016/0038-0717(82)90009-8.

ICRISAT (2014), *ICRISAT Development Center*. Patancheru: International Crops Research Institute for the Semi-Arid Tropics. Available at: http://idc.icrisat.org/wp-content/uploads/2016/02/Flyer-on-IDC.pdf (Accessed: 29 May 2017).

ICRISAT (2016), *Strengthening Bhoochetana: A Sustainable Agriculture Mission for Improved Livelihoods in Karnataka*. Annual Report 2015–2016. Patancheru: International Crops Research Institute for the Semi-Arid Tropics, p. 72.

Idowu, O. J., van Es, H. M., Abawi, D. W., Schindelbeck, R. R., Moebius-Clune, B. N. and Gugino, B. K. (2009), 'Use of an integrative soil health test for evaluation of soil management impacts', *Renewable Agriculture and Food Systems*, 24(3), pp. 214–24. doi:10.1017/S1742170509990068.

Islam, M. K., Yaseen, T., Traversa, A., Ben Kheder, M., Brunetti, G. and Cocozza, C. (2016), 'Effects of the main extraction parameters on chemical and microbial characteristics of compost tea', *Waste Management*, 52, pp. 62–8. doi:10.1016/j.wasman.2016.03.042.

Jansa, J., Mozafar, A., Kuhn, G., Anken, T., Ruh, R., Sanders, I. R. and Frossard, E. (2003), 'Soil tillage affects the community structure of mycorrhizal fungi in maize roots', *Ecological Applications*, 13(4), pp. 1164–76. doi:10.1890/1051-0761(2003)13[1164: STATCS]2.0.CO;2.

Jeffries, P., Gianinazzi, S., Perotto, S., Turnau, K. and Barea, J.-M. (2003), 'The contribution of arbuscular mycorrhizal fungi in sustainable maintenance of plant health and soil fertility', *Biology and Fertility of Soils*, 37(1), pp. 1–16. doi:10.1007/s00374-002-0546-5.

Johnston, A. E. and Richards, I. R. (2003), 'Effectiveness of different precipitated phosphates as phosphorus sources for plants', *Soil Use and Management*, 19(1), pp. 45–9. doi:10.1111/j.1475-2743.2003.tb00278.x.

Juo, A. S. R. and Franzluebbers, K. (2003), *Tropical Soils - Properties and Management for Sustainable Agriculture*. Oxford: Oxford University Press. Available at: http://app.knovel.com/hotlink/toc/id:kpTSPMSA01/tropical-soils-properties/tropical-soils-properties.

Kaiser, D. E., Fernandez, F., Lamb, J. A., Coulter, J. A. and Barber, B. (2016), *Fertilizing Corn in Minnesota*. AG-FO-3790-D. St. Paul, MN: University of Minnesota Extension. Available at: https://www.extension.umn.edu/agriculture/nutrient-management/nutrient-lime-guidelines/fertilizing-corn-in-minnesota/docs/fertilizing-corn.pdf.

Kaspar, T. and Singer, J. (2011), 'The use of cover crops to manage soil', Publications from USDA-ARS/UNL Faculty. Available at: http://digitalcommons.unl.edu/usdaarsfacpub/1382.

Ketterings, Q. M., Klausner, S. D. and Czymmek, K. J. (2003), *Nitrogen Guidelines for Field Crops in New York*. Ithaca, NY: Cornell University (Department of Crop and Soil Extension Series, E03-16).

Khalid, R. A., Patrick, W. H. and DeLaune, R. D. (1977), 'Phosphorus sorption characteristics of flooded soils', *Soil Science Society of America Journal*, 41(2), pp. 305–10. doi:10.2136/sssaj1977.03615995004100020026x.

Khan, M. S., Zaidi, A. and Wani, P. A. (2007), 'Role of phosphate-solubilizing microorganisms in sustainable agriculture – A review', *Agronomy for Sustainable Development*, 27(1), pp. 29–43. doi:10.1051/agro:2006011.

Kibblewhite, M. G., Ritz, K. and Swift, M. J. (2008), 'Soil health in agricultural systems', *Philosophical Transactions of the Royal Society of London B: Biological Sciences*, 363(1492), pp. 685–701. doi:10.1098/rstb.2007.2178.

Kristensen, H. L., McCarty, G. W. and Meisinger, J. J. (2000), 'Effects of soil structure disturbance on mineralization of organic soil nitrogen', *Soil Science Society of America Journal*, 64(1), pp. 371–8. doi:10.2136/sssaj2000.641371x.

Lal, R. (1990), 'Soil Erosion and Land Degradation: The Global Risks', In: *Advances in Soil Science*. New York, NY: Springer, pp. 129–72. doi:10.1007/978-1-4612-3322-0_4.

Lehmann, J. (2007), 'A handful of carbon', *Nature*, 447(7141), pp. 143–4. doi:10.1038/447143a.

Lehman, R. M., Cambardella, C. A., Stott, C. A., Acosta-Martinez, V., Manter, D. K., Buyer, J. S., Maul, J. E., Smith, J. L., Collins, H. P., Halvorson, J. J., Kremer, R. J., Lundgren, J. G., Ducey, T. F., Jin, V. L. and Karlen, V. L. (2015), 'Understanding and enhancing soil biological health: The solution for reversing soil degradation', *Sustainability*, 7(1), pp. 988–1027. doi:10.3390/su7010988.

Liebman, M., Graef, R. L., Nettleton, D. and Cambardella, C. A. (2012), 'Use of legume green manures as nitrogen sources for corn production', *Renewable Agriculture and Food Systems*, 27(3), pp. 180–91. doi:10.1017/S1742170511000299.

Liptzin, D. and Silver, W. L. (2009), 'Effects of carbon additions on iron reduction and phosphorus availability in a humid tropical forest soil', *Soil Biology and Biochemistry*, 41(8), pp. 1696–702. doi:10.1016/j.soilbio.2009.05.013.

Liu, F.-P., Liu, H.-Q., Zhou, H.-L., Dong, Z.-G., Bai, X.-H., Bai, P. and Qiao, J.-J. (2014), 'Isolation and characterization of phosphate-solubilizing bacteria from betel nut (*Areca catechu*) and their effects on plant growth and phosphorus mobilization in tropical soils', *Biology and Fertility of Soils*, 50(6), pp. 927–37. doi:10.1007/s00374-014-0913-z.

Liu, H., White, P. J. and Li, C. (2016), 'Biomass partitioning and rhizosphere responses of maize and faba bean to phosphorus deficiency', *Crop and Pasture Science*, 67(8), pp. 847–56. doi:10.1071/CP16015.

Lory, J. A., Russelle, M. P. and Peterson, T. A. (1995), 'A comparison of two nitrogen credit methods: Traditional vs. difference', *Agronomy Journal*, 87(4), pp. 648–51. doi:10.2134/agronj1995.00021962008700040007x.

Lotter, D. (2015), 'Facing food insecurity in Africa: Why, after 30 years of work in organic agriculture, I am promoting the use of synthetic fertilizers and herbicides in small-scale staple crop production', *Agriculture and Human Values*, 32(1), pp. 111–18. doi:10.1007/s10460-014-9547-x.

Lupwayi, N. Z., Kennedy, A. C., and Chirwa, M. (2011), 'Grain legume impacts on soil biological processes in sub-Saharan Africa', *African Journal of Plant Science*, 5(1), pp. 1–7.

Lynch, D. H. (2015), 'Nutrient cycling and soil health in organic cropping systems - Importance of Management strategies and soil resilience', *Sustainable Agriculture Research*, 4(3), p. 80. doi:10.5539/sar.v4n3p80.

Margenot, A. J., Paul, B. K., Sommer, R. R., Pulleman, M. M., Parikh, S. J., Jackson, L. E. and Font, S. J. (2017), 'Can conservation agriculture improve phosphorus (P) availability in weathered soils? Effects of tillage and residue management on soil P status after 9 years in a Kenyan Oxisol', *Soil and Tillage Research*, 166, pp. 157–66. doi:10.1016/j.still.2016.09.003.

Marschner, H. (Ed.) (1995), *Mineral Nutrition of Higher Plants*. New York: Academic Press, pp. 299–312.

Mazzoncini, M., Antichi, D., Di Bene, C., Risaliti, R., Petri, M. and Bonar, E. (2016), 'Soil carbon and nitrogen changes after 28 years of no-tillage management under Mediterranean conditions', *European Journal of Agronomy*, 77, pp. 156–65. doi:10.1016/j.eja.2016.02.011.

Mitchell, D. C., Castellano, M. J., Sawyer, J. E. and Pantoja, J. (2013), 'Cover crop effects on nitrous oxide emissions: Role of mineralizable carbon', *Soil Science Society of America Journal*, 77(5), pp. 1765–73. doi:10.2136/sssaj2013.02.0074.

Moebius-Clune, B. N., Moebius-Clune, D. J., Gugino, D. J., Idowu, O. J., Schindelbeck, R. R., Ristow, A. J., van Es, H. M., Thies, J. E., Shayler, H. A., McBride, M. B., Wolfe, D. W. and Abawi, G. S. (2016), *Comprehensive Assessment of Soil Health –The Cornell Framework Manual*, Volume 3.1. Geneva, NY: Cornell University.

Montanarella, L. and Vargas, R. (2012), 'Global governance of soil resources as a necessary condition for sustainable development', *Current Opinion in Environmental Sustainability. (Terrestrial systems)*, 4(5), pp. 559–64. doi:10.1016/j.cosust.2012.06.007.

Mullen, R. W. (2011), 'Nutrient cycling in soils: Nitrogen', In: Hatfield, J. L. and Sauer, T. J. (Eds), *Soil Management: Building a Stable Base for Agriculture*. Madison, WI: Soil Science Society of America, pp. 67–78. doi:10.2136/2011.soilmanagement.c5.

Mulvaney, M. J., Wood, C. W., Balkcom, K. S., Kemble, J. and Shannon, D. A. (2017), 'No-till with high biomass cover crops and invasive legume mulches increased total soil carbon after three years of collard production', *Agroecology and Sustainable Food Systems*, 41(1), pp. 30–45. doi:10.1080/21683565.2016.1236766.

Murphy, R. P., Montes-Molina, J. A., Govaerts, B., Six, J., van Kessel, C. and Fonte, S. J. (2016), 'Crop residue retention enhances soil properties and nitrogen cycling in smallholder maize systems of Chiapas, Mexico', *Applied Soil Ecology*, 103, pp. 110–16. doi:10.1016/j.apsoil.2016.03.014.

Nezomba, H., Mtambanengwe, F., Tittonell, P. and Mapfumo, P. (2004), *Vermicomposting: Recycling Wastes into Valuable Organic Fertilizer*, Volume 8. Patancheru: International Crops Research Institute for the Semi-Arid Tropics.
Nezomba, H., Mtambanengwe, F., Tittonell, P. and Mapfumo, P. (2015), 'Point of no return? Rehabilitating degraded soils for increased crop productivity on smallholder farms in eastern Zimbabwe', *Geoderma*, 239, pp. 143–55. doi:10.1016/j.geoderma.2014.10.006.
Oberson, A., Friesen, I. M., Rao, S., Bühler, D. K. and Frossard, E. (2001), 'Phosphorus transformations in an Oxisol under contrasting land-use systems: The role of the soil microbial biomass', *Plant and Soil*, 237(2), pp. 197–210. doi:10.1023/A:1013301716913.
O'Connell, S., Grossman, J.M., Hoyt, G. D., Shi, W., Bowen, S., Marticorena, D. C., Fager, K. L. and Creamer, N. G. (2015), 'A survey of cover crop practices and perceptions of sustainable farmers in North Carolina and the surrounding region', *Renewable Agriculture and Food Systems*, 30(6), pp. 550–62. doi:10.1017/S1742170514000398.
O'Dea, J. K., Jones, C. A., Zabinski, C. A., Miller, P. R. and Keren, I. N. (2015), 'Legume, cropping intensity, and N-fertilization effects on soil attributes and processes from an eight-year-old semiarid wheat system', *Nutrient Cycling in Agroecosystems*, 102(2), pp. 179–94. doi:10.1007/s10705-015-9687-4.
Pal, D. K. (2017), *A Treatise of Indian and Tropical Soils*. Cham, Switzerland: Springer.
Pal, D. K., Wani, S. P. and Sahrawat, K. L. (2012), 'Role of calcium carbonate minerals in improving sustainability of cracking clay soils (Sodic Haplusterts) by improved management: An appraisal of results from the semi arid zone of India', *Clay Research*, 31(2), pp. 94–108.
Patel, D. P., Das, A., Kumar, M., Munda, G. C., Ngachan, S. V., Ramkrushna, G. I., Layek, J., Pongla, N., Buragohain, J. and Somireddy, U. (2015), 'Continuous application of organic amendments enhances soil health, produce quality and system productivity of vegetable-based cropping systems in subtropical eastern Himalayas', *Experimental Agriculture*, 51(1), pp. 85–106. doi:10.1017/S0014479714000167.
Patra, A. K., Lenka, N. K. and Biswas, A. K. (2015), 'Soil health assessment and management: Issues and strategies', *Indian Journal of Fertility*, 11(12), pp. 16–25.
Pattison, A. B., Moody, P. W., Badcock, K. A., Smith, L. J., Armour, J. A., Rasiah, V., Cobon, J. A., Gulino, L.-M. and Mayer, R. (2008), 'Development of key soil health indicators for the Australian banana industry', *Applied Soil Ecology*, 40(1), pp. 155–64. doi:10.1016/j.apsoil.2008.04.002.
Peoples, M. B., Herridge, D. F. and Ladha, J. K. (1995), 'Biological nitrogen fixation: An efficient source of nitrogen for sustainable agricultural production?', *Plant and Soil*, 174(1–2), pp. 3–28. doi:10.1007/BF00032239.
Pepper, I. L. (2013), 'The Soil Health-Human Health Nexus', *Critical Reviews in Environmental Science and Technology*, 43(24), pp. 2617–52. doi:10.1080/10643389.2012.694330.
Peretyazhko, T. and Sposito, G. (2005), 'Iron(III) reduction and phosphorus solubilization in humid tropical forest soils', *Geochimica et Cosmochimica Acta*, 69(14), pp. 3643–52. doi:10.1016/j.gca.2005.03.045.
Peyrard, C., Mary, B., Perrin, P., Véricel, G., Gréhan, E., Justes, E. and Léonard, J. (2016), 'N2O emissions of low input cropping systems as affected by legume and cover crops use', *Agriculture, Ecosystems & Environment*, 224, pp. 145–56. doi:10.1016/j.agee.2016.03.028.
Powell, B. and Martens, M. (2005), 'A review of acid sulfate soil impacts, actions and policies that impact on water quality in Great Barrier Reef catchments, including a case study on remediation at East Trinity', *Marine Pollution Bulletin*, 51(1), pp. 149–64. doi:10.1016/j.marpolbul.2004.10.047.
Prabhu, A. S., Fageria, N. K., Huber, D. M. and Rodrigues, F. A. (2007), 'Potassium and plant disease', In: Datnoff, L. E., Elmer, W. H. and Huber, D. M. (Eds), *Mineral Nutrition and Plant Disease*. Saint Paul: The American Phytopathological Society Press, pp. 57–78.
Ram, A., Kumar, D., Babu, S., Prasad, D. and Dev, I. (2017), 'Effect of sulphur on soil biological properties, residual fertility and yield of aerobic rice grown under aerobic rice-wheat cropping system in Inceptisols', *Journal of Environmental Biology; Lucknow*, 38(4), pp. 587–93. doi: http://dx.doi.org/10.22438/jeb/38/4/MS-275.

Raphael, J. P. A., Calonego, J. C., Milori, D. M. B. P. and Rosolem, C. A. (2016), 'Soil organic matter in crop rotations under no-till', *Soil and Tillage Research*, 155, pp. 45–53. doi:10.1016/j.still.2015.07.020.

Rego, T. J., Wani, S. P., Sahrawat, K. L. and Pardhasaradhi, G. (2006), 'Macro-benefits from boron, zinc and sulfur application in Indian SAT: A step for grey to green revolution in agriculture', *Journal of SAT Agricultural Research*, 2(1), pp. 1–21.

Reicosky, D. C. (2015), 'Conservation tillage is not conservation agriculture', *Journal of Soil and Water Conservation*, 70(5), p. 103A–108A. doi:10.2489/jswc.70.5.103A.

Richardson, C. J. (2008), *The Everglades Experiments: Lessons for Ecosystem Restoration*. New York: Springer (Ecological Studies).

Rodrigues, M., Pavinato, P. S., Withers, P. J. A., Teles, A. P. B. and Herrera, W. F. B. (2016), 'Legacy phosphorus and no tillage agriculture in tropical oxisols of the Brazilian savanna', *Science of The Total Environment*, 542, pp. 1050–61. doi:10.1016/j.scitotenv.2015.08.118.

Rowe, H., Withers, P. J. A., Baas, P., Chan, N. I., Doody, D., Holiman, J., Jacobs, B., Li, H., MacDonald, G. K., McDowell, R., Sharpley, A. N., Shen, J., Taheri, W., Wallenstein, M. and Weintraub, M. N. (2016), 'Integrating legacy soil phosphorus into sustainable nutrient management strategies for future food, bioenergy and water security', *Nutrient Cycling in Agroecosystems*, 104(3), pp. 393–412. doi:10.1007/s10705-015-9726-1.

Sahrawat, K. L. (2016), 'How fertile are semi-arid tropical soils?', *Current Science*, 110(9), pp. 1671–4.

Sanchez, P. A. and Uehara, G. (1980), 'Management considerations for acid soils with high phosphorus fixation capacity', In: Knasawneh, F. E., Sample, E. C. and Kamrath, E. J. (Eds), *The Role of Phosphorus in Agriculture*. Madison, WI: ASA-CSSA-SSSA, pp. 471–514. doi:10.2134/1980.role of phosphorus.c18.

Sapkota, T. B., Jat, R. K., Singh, R. G., Jat, M. L., Stirling, C. M., Jat, M. K., Bijarniya, D., Kumar, M., Yadvinder-Singh, Saharawat, Y. S. and Gupta, R. K. (2017), 'Soil organic carbon changes after seven years of conservation agriculture in a rice–wheat system of the eastern Indo-Gangetic Plains', *Soil Use and Management*, 33(1), pp. 81–9. doi:10.1111/sum.12331.

Sawargaonkar, G., Raghavendra Rao, S. and Wani, S. P. (2016), 'An integrated approach for productivity enhancement', In: Raju, K. V. and Wani, S. P. (Eds), *Harnessing Dividends from Drylands: Innovative Scaling up with Soil Nutrients*. Oxfordshire: CABI, pp. 236–58.

Schmidt, J. P., DeJoia, A. J., Ferguson, R. B., Taylor, R. K., Young, R. K. and Havlin, J. L. (2002), 'Corn yield response to nitrogen at multiple in-field locations', *Agronomy Journal*, 94(4), pp. 798–806. doi:10.2134/agronj2002.7980.

Schoenholtz, S. H., Miegroet, H. V. and Burger, J. A. (2000), 'A review of chemical and physical properties as indicators of forest soil quality: Challenges and opportunities', *Forest Ecology and Management*, 138(1), pp. 335–56. doi:10.1016/S0378-1127(00)00423-0.

Sela, S., van Es, H. M., Moebius-Clune, B. N., Marjerison, R., Moebius-Clune, D., Schindelbeck, R., Severson, K. and Young, E. (2017), 'Dynamic model improves agronomic and environmental outcomes for maize nitrogen management over static approach', *Journal of Environmental Quality*, 46(2), pp. 311–19. doi:10.2134/jeq2016.05.0182.

SFWMD (2017), *2017 South Florida Environmental Report*. South Florida Environmental Report. South Florida Water Management District, p. 8. Available at: https://www.sfwmd.gov/science-data/sfer.

Sharma, S. B., Sayyed, R. Z., Trivedi, M. H. and Gobi, T. A. (2013), 'Phosphate solubilizing microbes: sustainable approach for managing phosphorus deficiency in agricultural soils', *SpringerPlus*, 2, p. 587. doi:10.1186/2193-1801-2-587.

Shrivastava, P. and Kumar, R. (2015), 'Soil salinity: A serious environmental issue and plant growth promoting bacteria as one of the tools for its alleviation', *Saudi Journal of Biological Sciences*, 22(2), pp. 123–31. doi:10.1016/j.sjbs.2014.12.001.

Simpson, R. J., Oberson, A., Culvenor, R. A., Ryan, M. H., Veneklaas, E. J., Lambers, H., Lynch, J. P., Ryan, P. R., Delhaize, E., Smith, F. A., Smith, S. E., Harvey, P. R. and Richardson, A. E. (2011), 'Strategies and agronomic interventions to improve the phosphorus-use efficiency of farming systems', *Plant and Soil*, 349(1–2), pp. 89–120. doi:10.1007/s11104-011-0880-1.

Sims, J. T. and Sharpley, A. N. (Eds) (2005), *Phosphorus: Agriculture and the Environment.* Madison, WI: American Society of Agronomy. Available at: https://www.cabdirect.org/cabdirect/abstract/20073096530 (Accessed: 15 July 2017).

Singh, K., Chaudhary, O. P., Singh, K., Singh, H., Singh, S. and Singh, R. S. (2015), 'Identification of integrated nutrient management for sustaining soil health and sugarcane yield in South Western Punjab', *Journal of Environmental Biology*, 36(3), pp. 551–5.

Smith, J. L. and Doran, J. W. (1996), 'Measurement and use of pH and electrical conductivity for soil quality analysis', In: Doran, J. W. and Jones, A. J. (Eds), *Methods for Assessing Soil Quality: SSSA special publication No. 49*. Madison, WI: SSSA, pp. 169–85. doi:10.2136/sssaspecpub49.c10.

Stirling, G. R., Moody, P. W. and Stirling, A. M. (2010), 'The impact of an improved sugarcane farming system on chemical, biochemical and biological properties associated with soil health', *Applied Soil Ecology*, 46(3), pp. 470–7. doi:10.1016/j.apsoil.2010.08.015.

Syakila, A. and Kroeze, C. (2011), 'The global nitrous oxide budget revisited', *Greenhouse Gas Measurement and Management*, 1(1), pp. 17–26. doi:10.3763/ghgmm.2010.0007.

Takoutsing, B., Weber, J., Aynekulu, E., Rodríguez Martín, J. A., Shepherd, K., Sila, A., Tchoundjeu, Z. and Diby, L. (2016), 'Assessment of soil health indicators for sustainable production of maize in smallholder farming systems in the highlands of Cameroon', *Geoderma*, 276, pp. 64–73. doi:10.1016/j.geoderma.2016.04.027.

Tiessen, H., Stewart, J. W. B. and Cole, C. V. (1984), 'Pathways of phosphorus transformations in soils of differing pedogenesis', *Soil Science Society of America Journal*, 48(4), pp. 853–8. doi:10.2136/sssaj1984.03615995004800040031x.

Treuer, T. L. H., Choi, J. J., Janzen, D. H., Hallwachs, W., Peréz-Aviles, D., Dobson, A. P., Powers, J. S., Shanks, L. C., Werden, L. K. and Wilcove, D. S. (2017), 'Low-cost agricultural waste accelerates tropical forest regeneration', *Restoration Ecology*. doi:10.1111/rec.12565.

Turmel, M.-S., Speratti, A., Baudron, F., Verhulst, N. and Govaerts, B. (2015), 'Crop residue management and soil health: A systems analysis', *Agricultural Systems*, 134, pp. 6–16. doi:10.1016/j.agsy.2014.05.009.

Turner, B. L. and Engelbrecht, B. M. J. (2011), 'Soil organic phosphorus in lowland tropical rain forests', *Biogeochemistry*, 103(1–3), pp. 297–315. doi:10.1007/s10533-010-9466-x.

USDA (1954), *Diagnosis and Improvement of Saline and Alkali Soils* (Richard, L. A. (Ed.)). Washington DC: United States Department of Agriculture (Agriculture Handbook, 60). Available at: https://www.ars.usda.gov/ARSUserFiles/20360500/hb60_pdf/hb60intro.pdf.

USDA (1996), *Indicators for Soil Quality Evaluation*. Washington DC: United States Department of Agriculture. Available at: https://www.nrcs.usda.gov/Internet/FSE_DOCUMENTS/nrcs142p2_053149.pdf.

USDA (2014), *Soil Quality Indicators: Reactive Carbon*. Washington DC: United States Department of Agriculture. Available at: https://www.nrcs.usda.gov/wps/portal/nrcs/detail/soils/health/assessment/?cid=stelprdb1237387.

USDA (2015), *Soil Quality Indicators. Physical, Chemical, Biological Indicators for Soil Quality Assessment and Management*. Washington DC: USDA. Available at: https://www.nrcs.usda.gov/wps/portal/nrcs/detail/soils/health/assessment/?cid=stelprdb1237387.

Vance, C. P., Graham, P. H. and Allan, D. L. (2000), 'Biological nitrogen fixation: Phosphorus - A critical future need?', In: *Nitrogen Fixation: From Molecules to Crop Productivity*. Dordrecht: Springer (Current Plant Science and Biotechnology in Agriculture), pp. 509–14. doi:10.1007/0-306-47615-0_291.

Wani, S. P. (2008), 'Taking soil science to farmers' doorsteps through community watershed management', *Journal of the Indian Society of Soil Science*, 56(4), pp. 367–77.

Wani, S. P. (2016), 'Evolution of bhoochetana', In: Raju, K. V. and Wani, S. P. (Eds), *Harnessing Dividends from Drylands: Innovative Scaling Up with Soil Nutrients*. Oxfordshire, UK: CABI, p. 34.

Wani, S. P., Singh, H. P., Sreedevi, T. K., Pathak, P., Rego, T. J., Shiferaw, B. and Iyer, S. R. (2003a), 'Farmer-participatory integrated watershed management: Adarsha watershed, Kothapally India - an innovative and upscalable approach', *Journal of SAT Agricultural Research*, 2(1), pp. 1–27.

Wani, S. P., Anantha, K. H., Dharmarajan, B. K. and Krishnappa, K. (2015), *Bhoo Samruddhi: A Compendium of Success Stories*. Research Report IDC-2. Patancheru: International Crops Research Institute for the Semi-Arid Tropics, p. 49.
Wani, S. P., Anantha, K. H. and Garg, K. K. (2017a), 'Soil properties, crop yield, and economics under integrated crop management practices in Karnataka, Southern India', *World Development*, 93, pp. 43–61. doi:10.1016/j.worlddev.2016.12.012.
Wani, S. P., Chander, G. and Anantha, K. H. (2017b), 'Enhancing resource use efficiency through soil management for improving livelihoods', In: Rakshit, A. et al. (Eds), *Adaptive Soil Management: From Theory to Practices*. Singapore: Springer Singapore, pp. 413–47.
Wani, S. P., Sreedevi, T. K., Pathak, P., Rego, T. J., Ranga Rao, G. V., Jangawad, L. S., Pardhasaradhi, G. and Iyer, S. R. (2003b), 'Minimizing land degradation and sustaining productivity by integrated watershed management: Adarsha Watershed, Kothapally, India', In: Wani, S. P., Maglinao, A. R., Ramakrishna, A. and Rego, T. J. (Eds), *Proceedings of the ADB-ICRISAT-IWMI Project Review and Planning Meeting. Integrated watershed management for land and water conservation and sustainable agricultural production in Asia*, Hanoi, Vietnam: International Crops Research Institute for the Semi-Arid Tropics, p. 268.
Wani, S. P., Sreedevi, T. K., Vamsidhar Reddy, T. S., Venkateshvarlu, B. and Shambhu Prasad, C. (2008), 'Community watersheds for improved livelihoods through consortium approach in drought prone rainfed areas', *Journal of Hydrological Research and Development*, 23, pp. 55–77.
Weaver, D. M. and Wong, M. T. F. (2011), 'Scope to improve phosphorus (P) management and balance efficiency of crop and pasture soils with contrasting P status and buffering indices', *Plant and Soil*, 349(1–2), pp. 37–54. doi:10.1007/s11104-011-0996-3.
Wei, L., Chen, C. and Xu, Z. (2010), 'Citric acid enhances the mobilization of organic phosphorus in subtropical and tropical forest soils', *Biology and Fertility of Soils*, 46(7), pp. 765–9. doi:10.1007/s00374-010-0464-x.
Wei, W., Yan, Y., Cao, J., Christie, P., Zhang, F. and Fan, M. (2016), 'Effects of combined application of organic amendments and fertilizers on crop yield and soil organic matter: An integrated analysis of long-term experiments', *Agriculture, Ecosystems & Environment*, 225, pp. 86–92. doi:10.1016/j.agee.2016.04.004.
Weil, R. R., Islam, K. R., Stine, M. A., Gruver, J. B. and Samson-Liebig, S. E. (2003), 'Estimating active carbon for soil quality assessment: A simplified method for laboratory and field use', *American Journal of Alternative Agriculture*, 18(1), pp. 3–17. doi:10.1079/AJAA200228.
Weil, R. R. and Brady, N. C. (2017), *Nature and Properties of Soils*. 15th Edition. New York, NY: Pearson.
Withers, P. J. A., Neal, C., Jarvie, H. P. and Doody, D. G. (2014), 'Agriculture and eutrophication: Where do we go from here?', *Sustainability*, 6(9), pp. 5853–75. doi:10.3390/su6095853.
Wu, X., Sten, P., Engblom, S., Nowak, P., Österholm, P. and Dopson, M. (2015), 'Impact of mitigation strategies on acid sulfate soil chemistry and microbial community', *Science of The Total Environment*, 526(Suppl. C), pp. 215–21. doi:10.1016/j.scitotenv.2015.04.049.
Yuan, B.-C., Li, Z.-Z., Liu, H., Gao, M. and Zhang, Y.-Y. (2007), 'Microbial biomass and activity in salt affected soils under arid conditions', *Applied Soil Ecology*, 35(2), pp. 319–28. doi:10.1016/j.apsoil.2006.07.004.
Zebarth, B. J., Drury, C. F., Tremblay, N. and Cambouris, A. N.. (2009), 'Opportunities for improved fertilizer nitrogen management in production of arable crops in eastern Canada: A review', *Canadian Journal of Soil Science*, 89(2), pp. 113–32. doi:10.4141/CJSS07102.
Zörb, C., Senbayram, M. and Peiter, E. (2014), 'Potassium in agriculture – Status and perspectives', *Journal of Plant Physiology*, 171(9), pp. 656–69. doi:10.1016/j.jplph.2013.08.008.

Chapter 8

Soil microorganisms: role in soil health

Penny R. Hirsch, Rothamsted Research, UK

1 Introduction

Without life, soil would remain an assemblage of mineral particles produced by weathering of rocks. Soil is itself essential for supporting terrestrial life, providing a substrate for plants and hosting the complex food web of soil biota. This includes microorganisms and micro- and mesofauna, most of which are reliant ultimately on nutrients supplied directly or indirectly by plants. The microfauna range in size from the smallest single-celled protozoans (protists) of around 5 μm to small nematodes, unsegmented worms and tardigrades (eight-legged arthropods) up to 1 mm but not readily visible by eye; mesofauna are larger soil invertebrates which move and shred plant residues or predate on other members of the soil biota (Nannipieri et al., 2003). However, the focus of this chapter is on microorganisms: prokaryotic bacteria and archaea (smaller than 5 μm), and eukaryotic fungi with hyphae up to 50 μm diameter but potentially many metres long. Residues from plant, animal and microbial activity provide organic components, making soils fertile and binding together mineral particles into aggregates that, with the associated pore spaces, confer structure. Soil microorganisms are pivotal for nutrient provision in soil systems through mineralization, where simple and complex organic materials are oxidized to labile inorganic compounds, a large proportion of which is recycled via plants.

Soil microbial communities are both extraordinarily large and diverse with an estimated 10^9 bacterial and archaeal cells g^{-1} comprising 10^4 to 10^6 species g^{-1} in temperate soils (Bent and Forney, 2008). Species are defined in bacteria and archaea as strains that

http://dx.doi.org/10.19103/AS.2017.0033.10

show at least 70% DNA homology over the whole genome and at least 97% in the 16S ribosomal RNA (rRNA) gene (Cohan, 2002). Equivalent information on soil fungi and microfauna is not yet available as there is much less sequence information available in the DNA databases. To date (2016), there are >230 000 bacterial and >2000 archaeal genomes available on the US Joint Genome Institute website https://gold.jgi.doe.gov/index (Mukherjee et al., 2017); the fungal portal http://jgi.doe.gov/fungi is aiming to provide 1000 fungal genomes (Grigoriev et al., 2014). The micro-scale structural and physico-chemical variability of soil creates multiple microenvironments providing the basis of the biodiversity, a measure which includes the number of different operational taxonomic units (OTUs) or species (richness) and the relative frequency with which they occur (evenness). The spatial separation of micro-sites enables co-evolution of multiple lineages in any particular soil, over time. This endows the soil with resilience, the ability to continue to function when challenged by stress or perturbation (Allison and Martiny, 2008; Girvan et al., 2005), which is related to functional redundancy, that is, due to the number of different species of different ecotypes that can perform a given function (Konopka, 2009). These properties are essential to maintain microbial activity in the face of periodic or catastrophic changes.

The earth has finite resources, with the elements essential for life partitioned in different reservoirs: terrestrial, aquatic, atmospheric and biotic. Ultimately, solar and geothermal energy drives the geochemical cycles, converting elements between different forms; microorganisms have an essential role in circulating them between abiotic and biotic pools. This includes fixation of carbon and nitrogen, the solubilization and uptake of minerals, and decomposition of organic residues with the associated release of mineral nutrients. Other essential elements including P, K, S, Mg, Ca, Fe and a range of trace element micronutrients are derived initially from minerals and may be released by volcanism or solubilized by weathering and enzymes secreted by microorganisms and plant roots (Falkowski et al., 2008).

Agricultural practice has continually influenced soil properties from its origin as small family settlements cultivating crops and domesticated animals selected for desirable traits, to modern intensive tillage agriculture, over the past 10 000 years (Montgomery, 2007a,b). The impact of human activity on soil microbiology is apparent in ancient sites: Amazonian terra preta, dark coloured soils rich in charred organic material and phosphate, some thousands of years old and thought to be derived from ancient human habitation, have distinct microbial communities with more species than pristine forest (Grossman et al., 2010). Similarly, woodland sites of Roman farms in France are reported to contain mycorrhizal fungi that differ from adjoining areas (Diedhiou et al., 2010). Changes are not necessarily detrimental, however, problems arising from agricultural mismanagement are a recurrent theme in social history, linked to the downfall of civilizations from Europe and Asia to Mesoamerica (Tainter, 1990). A combination of deforestation and intensification to provide food for increasing urban populations and stochastic natural disasters can lead to loss of soil structure and fertility resulting in soil erosion. Problems associated with soil erosion and degradation persist and may be exacerbated by climate change (Quinton et al., 2010; Banwart, 2011). Other activities such as metal mining and refining released toxic compounds creating polluted, infertile soils on a small scale until a massive expansion following the industrial revolution.

Modern agricultural tillage practices including mechanization, chemical fertilizers, pesticides and modern high-yielding crop varieties have increased productivity. Despite its success in sustaining the global human population (the FAO reports that over the past decade cereal production has risen in line with demand), modern agricultural practice raises many environmental concerns. These include annexing of previously 'natural' environments, soil compaction and erosion due to agricultural vehicle traffic, over-cropping, overgrazing and a reduction in soil organic matter. There may also be depletion of surface and groundwater supplies diverted to crop production, pollution due to run-off of excess applied nutrients and pesticide residues, and generation of greenhouse gasses (GHG). Loss of natural habitats to agriculture, from <10 % of land globally in 1700 to ~40% in 2008 (FAOSTAT), has been detrimental to many wild plant and animal species, raising concerns about the less visible life below-ground. The global human population increased 10-fold between 1700 and 2000 to exceed 6 billion and reached 7 billion in 2011. The UN predicts that this increase in population will rise to ~9 billion by 2050. Pressure on land for food production and aggravated climate change will also increase in the future.

Whilst feedback mechanisms have been identified and mitigation options in soil have been proposed (Singh et al., 2010), there will undoubtedly be stress from broader extremes of temperature and precipitation associated with anthropogenic climate change. To ensure new practices do not compromise food security, it is essential to establish the likely environmental impacts on soil microbial communities and the geochemical cycle functions they perform, especially those where national decisions on land management have global significance.

A practical definition of soil health for any soil, whether nature reserve or intensively managed cropland, is 'the continued capacity of soil to function as a vital living system, within ecosystem and land use boundaries, to sustain biological productivity, promote the quality of air and water environments, and maintain plant, animal, and human health' (Pankhurst et al., 1997). This chapter reviews the contribution of soil microorganisms to sustainable agriculture and considers their role in both crop plant productivity and in major geochemical cycles, how this may have varied over time and how it may be influenced by changes in land use now and in the future. In this context, recent advances in the understanding of microbial community function arising from the application of high-throughput molecular methods for microbial ecology are discussed.

2 Methods for investigating microorganisms in soil

Historically, measurement of soil properties such as C and N content, bulk density, pH, cation exchange capacity (CEC) and microbial biomass and functions including aerobic respiration, nitrification and denitrification rates indicated the consequences of changes in land use. However, such 'black-box' approaches cannot describe soil microbial community structure, nor indicate reduced biodiversity and functional redundancy, early indicators of loss of resilience. Prior to development of methods based on the direct extraction of DNA and RNA from soil, termed metagenomics and metatranscriptomics, respectively, results were biased towards a potentially unrepresentative minority of microorganisms: currently, around 99% cannot be grown in the laboratory using standard methods (Hirsch et al., 2010). This may improve following development of the isolation chip (iChip), where soil organisms are incubated in discrete diffusion chambers incubated *in situ* in soil, reported to increase culture success to 50% (Nichols et al., 2010).

Metagenomic approaches that analyse community DNA extracted directly from environments provide data on the diversity and relative abundance of functional and phylogenetic (mostly based on rRNA) gene sequences. This is a major advance despite limitations on the depth of sequencing that is practical, albeit that costs are falling and efficiency is increasing. Potential bias arises from DNA extraction methods (Delmont et al., 2011) and poor gene assignment due to incomplete knowledge of functional genes (Konopka, 2009). Currently, all methods assessing relative numbers and diversity of genes, whether based on PCR amplification, direct sequencing or hybridization to microarrays, are subject to technical and practical constraints limiting representation of less abundant groups (Bent and Forney, 2008). Other limitations include the fact that DNA can originate from inactive and lysed organisms, so it provides information on the potential rather than the actual activity of the community, and non-functional pseudogenes could skew estimates of redundancy. As the microorganisms represented in the DNA sequence database increase (see above), identification of metagenomic reads is constrained by the bioinformatics methods available (Thomas et al., 2012). The analysis of metagenomic DNA from soil for bacteria and archaea using universal prokaryotic primers to amplify 16S rRNA gene, sometimes referred to as 'barcoding', has provided invaluable information: many studies compare the relative abundance of 16S amplicons in different soils. The equivalent analysis for fungi is usually based on the spacer region that separates ribosomal genes, and for other eukaryotes the 18S rRNA gene or a mitochondrial gene is used but identification of these is limited by a paucity of sequence data from environmental isolates (Pawlowski et al., 2012).

However, information provided by soil metagenomes has proved invaluable for revealing community structure and sequence data to facilitate the design of molecular tools for more detailed analysis of particular genes including probes for microarrays and primers for quantitative PCR (qPCR) and targeted sequencing (Hirsch et al., 2010, 2013). Metatranscriptomics identifies active functional genes if messenger RNA (mRNA) is converted to DNA using reverse transcriptase and subjected to qPCR and methods based on *in situ* labelling of DNA denote actively growing organisms (Hirsch et al., 2010, 2013). Following metagenomics/transcriptomics, soil proteomics is beginning to identify the gene products present at the time of sampling (Keiblinger et al., 2016). A combination of these methods is anticipated to provide increasingly detailed insight of the soil system, illuminating the black box. As the cost of sequencing is reduced, and the length of each sequence is increased, we get closer to the ultimate aim of obtaining full genomes of all the microorganisms present in a soil sample and demonstrating which genes are being expressed (Thomas et al., 2012).

A frequent question in soil ecology concerns the relative abundance and activity of eukaryotic and prokaryotic microorganisms. Metagenomic data is now providing answers to the first part: on average, bacterial DNA comprises 97%, archaeal DNA 1%, fungal DNA 0.3% and other eukaryotic DNA 1.2% (the remainder is assigned to viruses or unknown groups) in eight publically available soil metagenomes, shown in Table 1. If the proportion of fungal DNA is lower than expected from estimates based on other biomarkers, it may be because the ratio of genetic material to total fungal biomass is lower than for the prokaryotes (Strickland and Rousk, 2010). Also, the hyphal and spore walls of many fungi are more difficult to lyse than those of prokaryotic cells and may be under-represented (Mauchline et al., 2002). Modern agricultural tillage disrupts fungal hyphae in soil but evidence that it reduces the proportion of fungal biomass compared to that of bacteria is contradictory (Helgason et al., 2009; Strickland and Rousk, 2010). Our results (unpublished) show that the proportion of fungal DNA is 0.05% of the total extracted in regularly tilled arable soil but 0.7% in undisturbed grassland. This is in marked

Table 1 A range of soil metagenomes publically available on MG-RAST (http://metagenomics.anl.gov/) unless indicated otherwise, analysed for the percentage of DNA sequences assigned to the Bacteria, Archaea, Fungi or other Eukaryotes

Biome	Sequencing platform	Bacteria % DNA	Archaea% DNA	Fungi % DNA	% DNA other eukaryotes	Location (reference)
Grassland	454	97.23	0.72	0.16	1.43	Park Grass, Rothamsted Research, Harpenden, Hertfordshire, UK (Delmont et al., 2012)
Grassland	454	97.56	1.17	0.51	0.71	Kellogg Biological Station, Hickory Corners, MI (Ramirez et al., 2010)
Grassland	454	97.01	1.17	0.18	1.60	Cedar Creek Ecosystem Science Reserve, Bethel, MN (Ramirez et al., 2010)
Tropical rainforest	454	96.99	1.22	0.52	1.05	Luquillo, Puerto Rico (Deangelis et al., 2010)
Tallgrass prairie	Illumina	98.07	0.38	0.27	0.35	Fricke Cemetery, NE (Fierer et al., 2013)
Arid soil	Illumina	95.75	2.48	0.27	0.21	Uluru, Northern Territories, Australia (www.bioplatforms.com/soil-biodiversity)
Rice paddy soil	Illumina	97.35	1.27	0.22	0.01	Typical paddy at unspecified location, South China (http://trace.ddbj.nig.ac.jp/DRASearch/study?acc=SRP039858)
Rain-fed bog	Illumina	97.35	1.27	0.22	0.01	Marcell Experimental Forest, MN (Lin et al., 2014)
Mean (+/− se)		**97.20** (0.43)	**1.07** (0.21)	**0.34** (0.34)	**1.20** (0.21)	

contrast to a previous study on the same grassland soil using direct measurement of biovolume and the reduction in soil respiration in response to fungal inhibitors, which estimated a fungal:bacterial ratio of 4:1 (Lin and Brookes, 1999). This demonstrates that measurement of DNA (genetic potential), RNA (gene expression), biomarkers (PLFA or ergosterol), biovolume and activity all evaluate different aspects and cannot alone provide a measure of the functional importance of fungi in soil (Strickland and Rousk, 2010). Based on the proportion of metagenomic DNA, fungi were found to be more abundant in the rhizosphere than in bulk soil, 2.54% and 0.39%, respectively, in a recent study (Guo et al., 2016). Currently, there are few metatranscriptomic studies on soil but estimates of fungal activity, according to the proportion of fungal mRNA, are much higher: up to 13% reported in garden and agricultural soils (Mcgrath et al., 2008); up to 30% in forest soil where soil is not tilled (Baldrian et al., 2012).

3 The soil environment

The soil food web recycles nutrients; respiration returns C to the atmosphere as CO_2, and other gasses and solutes are also lost from soil due to microbial activity. Nutrients can be relocated by large-scale removal of plant material (roaming herds of herbivores, agricultural practices) and released from plant residues in soil by grazing megafauna above ground and the activity of mesofauna and microorganisms in soil. The major sources, sinks and outflows for C and N in soils, including GHG emissions, are shown in Fig. 1 and in recent reviews (Singh et al., 2010; Gruber and Galloway, 2008).

3.1 Sources of natural variation in soil

Most soils developed over millennia and contain a mixture of minerals classified as sand, silt and clay according to their particle size. The relative proportion of these and their parent minerals, together with soil organic matter, determines the physico-chemical properties including pH and CEC. Clays and organic matter contribute to soil CEC and water-holding capacity and are considered beneficial in agricultural soils. Soils are subject to seasonal climatic changes, as well as natural catastrophic events such as freezing, glaciation, flooding, drought, fire and volcanic activity. Since soil is a porous solid matrix,

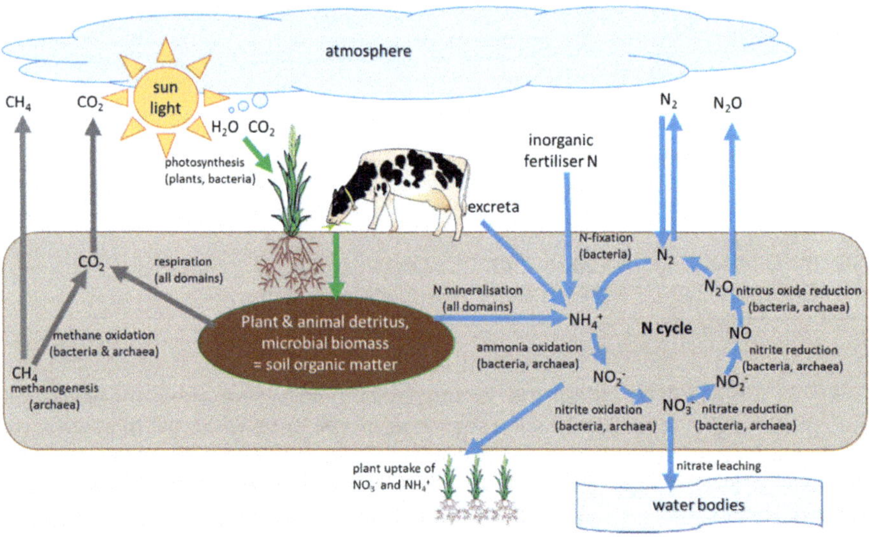

Figure 1 The major sources, sinks and outflows for C and N in soils, including GHG emissions: carbon dioxide (CO_2), methane (CH_4) and nitrous oxide (N_2O). Plants provide a major sink for inorganic N in soil; in agricultural systems this is removed during harvest or grazing. For more detail and references see Sections 4.1 and 4.2.

the microbiota is relatively immobile, although wind, water flow, root growth and faunal activity all contribute to the dispersion of microbial communities in soil particles. The soil environment varies at the landscape scale according to climatic zone and altitude, underlying geology and the plant and animal communities present. Soil pH, a key factor in mineral nutrient availability, has been shown to exert a major influence on microbial community structure (Fierer and Jackson, 2006). At the field scale, local variation influences plant, animal and hence soil communities (Garbeva et al., 2004); individual plants can have a direct influence due to quantitative and qualitative differences in root exudates and residues; and indirect effects through the herbivores, pathogens and symbionts they support (Bais et al., 2006; Haichar et al., 2008). On a micro-scale, there is extreme heterogeneity due to the variable mineral and organic matter components of soil, pore structure, the presence or absence of particular sources of nutrients (roots and plant or animal detritus) and associated differences in pH (Bronick and Lal, 2005). This is overlaid by gradients of oxygen, CO_2 and water, influenced by the frequency of perturbation by animals, freeze-thawing and agricultural cultivation.

3.2 Microbial diversity

Bacteria are the dominant microorganisms in temperate soils, archaea comprising <2% of cells (Buckley et al., 1998; Bates et al., 2011). There is strong circumstantial evidence that many archaea are ammonia oxidizers as the archaeal ammonia monooxygenase gene is relatively abundant compared to the bacterial equivalent (Leininger et al., 2006). In anaerobic soils that generate methane (bogs, paddy fields), methanogenic archaea are responsible (Conrad et al., 2006); they are present in most soils but active only in anoxic conditions (Angel et al., 2012). Soil metagenomic sequence data from different biomes shows that the well-known bacterial phyla Proteobacteria, Firmicutes and Actinomycetes are abundant in soil. Some members of these groups are associated with specific functions including nitrogen fixation, nitrification and denitrification. Other more recently described bacterial phyla including the Acidobacteria, Bacteroidetes, Gemmatimonadetes and Verrucomicrobia are also ubiquitous in soil but their role is uncertain other than a heterotrophic contribution to carbon cycling and enhancing the overall resilience of the soil (Girvan et al., 2005).

A comparison of the relative abundance of phyla using 16S rRNA gene amplicon analysis in plots with different long-term treatments on the Rothamsted farm (grassland, arable and a permanent bare fallow without plants) showed that Verrucomicrobia were significantly more abundant in the grassland, Gemmatimonadetes in the bare fallow and Nitrospirae in the arable soil (Hirsch et al., 2016). When all the prokaryotic phyla are displayed on a ternary plot (Fig. 2), it appears that most cluster in the centre indicating no overall preference for one of the treatments but there are some notable exceptions; phyla with known functions. The Thaumarchaeota, an archaeal phylum that can oxidize ammonia, is relatively more abundant in the arable plots that receive ammonium nitrate fertilizer. The photosynthetic cyanobacteria are relatively more abundant in the bare fallow where they are not shaded by plants, and N levels are low. The majority of phyla, those most abundant overall, will provide very broad functions such as CO_2 emission. However, when the 16S rRNA gene amplicons are re-analysed at the highest possible resolution (to genus for some, others can only be placed at order or family level), the distribution appears to become more biased to one or another of the treatments (Fig. 3). For example, one Verrucomicrobia family, one α-proteobacteria genus and several groups of Actinobacteria and Acidobacteria are

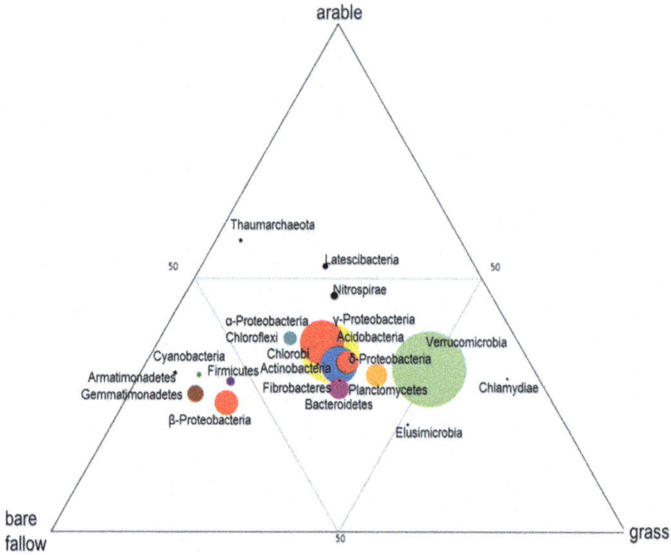

Figure 2 Ternary plot showing the overall relative abundance indicated by the size of the bubbles for prokaryotic phyla (subphyla for Proteobacteria) in three differently managed soils. The Thaumarchaeota are Archaea; other phyla are Bacteria. The position of each group on the plot shows the relative distribution between treatments. Data from Hirsch et al., 2016.

Each phylum is colour coded, in order of overall abundance in all plots: Verrucomicrobia – light green; Proteobacteria subphyla – red; Gemmatimonadetes – brown; Actinobacteria – blue; Acidobacteria – yellow; Firmicutes – violet; Chloroflexi – greenblue; Bacteroidetes – pink; Planctomycetes – orange; Nitrospirae – dark grey; Cyanobacteria – green; Latescibacteria, Thaumarchaeota, Armatimonadetes, Elusimicrobia, Fibrobacteres, Chlorobi, Chlamydiae – black.

more abundant in bare fallow plots; some Gemmatimonadetes groups favour grassland, in contrast to the overall affinities of their respective phyla shown in Fig. 2.

Thus, the relative abundance of phyla and species varies between biomes, although the implications of this are unclear with the exception of cases where phyla are associated with specific functions. Differences may not be apparent unless extreme stress is applied, such as in highly polluted soil. However, for processes performed by a restricted range of soil species, the stability and efficiency of the function is predicted to be limited when species richness is low, rising as species numbers increase, demonstrated in the case of methane oxidation (Levine et al., 2011).

3.3 Microbial interactions with plants

Microorganisms are an essential component of soil: without them, there would be no recycling of nutrients including N, no biological fixation of N_2 and no protection for crop plants against adventitious pathogens and pests. The role of specific groups is detailed below, but it is assumed that many different groups mineralize organic N and C. The rhizosphere (area of soil influenced by the root) has numerous active bacteria than bulk soil, but fewer different OTU (Dennis et al., 2010). Many bacteria and fungi possess functions beneficial to plant

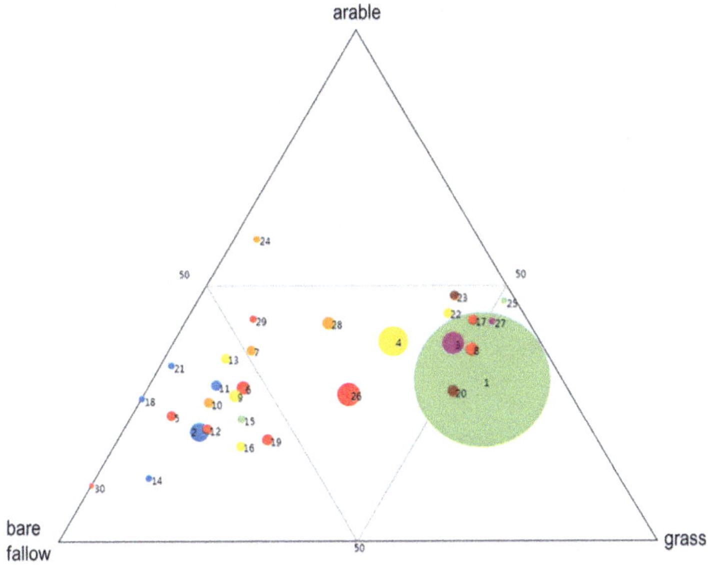

Figure 3 Ternary plot showing the relative abundance indicated by the size of the bubbles and the distribution between treatments for prokaryotic OTU defined by >97% DNA sequence similarity in the 16S rRNA gene sequence. The taxonomic identification to genus is not possible in many cases: the best match is given. Only the 30 OTU contributing more than 0.8% of the overall differences between plots are shown. Data from Hirsch et al., 2016. Colour coding for the phyla to which OTU belong is similar to the one used in Fig. 2 (Verrucomicrobia – light green; Proteobacteria – red; Gemmatimonadetes – brown; Actinobacteria – blue; Acidobacteria – yellow; Bacteroidetes – pink; Planctomycetes – orange). Key below:

1 Verrucomicrobia; genus DA101
2 Actinobacteria; family Gaiellaceae
3 Bacteroidetes; family Chitinophagaceae
4 Acidobacteria; order iii1–15
5 β-Proteobacteria; family EB1003
6 β-Proteobacteria; order SC-I-84
7 Planctomycetes; genus *Gemmata*
8 δ-Proteobacteria; family Syntrophobacteraceae
9 Acidobacteria; family Koribacteraceae
10 Planctomycetes; family Gemmataceae
11 Actinobacteria; order 0319-7L14
12 β-Proteobacteria; genus *Methylibium*
13 Acidobacteria; order Solibacterales
14 Actinobacteria; family Micrococcaceae
15 Verrucomicrobia; family Chthoniobacteraceae
16 Acidobacteria; order RB41
17 γ-Proteobacteria; family Sinobacteraceae
18 Actinobacteria; genus *Mycobacterium*
19 α-Proteobacteria; genus *Bradyrhizobium*
20 Gemmatimonadetes; class Gemm-1
21 Actinobacteria; order Acidimicrobiales
22 Acidobacteria; genus *Candidatus Koribacter*
23 Gemmatimonadetes; order Ellin 5301
24 Planctomycetes; family Pirellulaceae
25 Verrucomicrobia; family auto67_4W
26 α-Proteobacteria; genus *Rhodoplanes*
27 Bacteroidetes; order Sphingobacteriales
28 Planctomycetes; order WD2101
29 α-Proteobacteria; family Hyphomicrobiaceae
30 β-Proteobacteria; family Oxalobacteraceae

growth in addition to increasing nutrient availability, including phytohormone production and the ability to protect against pasts and diseases. There are specific associations that are mutually beneficial, including leguminous plants with rhizobia and mycorrhizal fungi with their host plants. This topic is too wide to cover in detail in this chapter, but there exist many reviews on the rhizosphere microbiome (Lakshmanan et al., 2014, Philippot et al., 2013, Mcnear Jr., 2013). Exploitation of the rhizosphere microbiome for sustainable agriculture

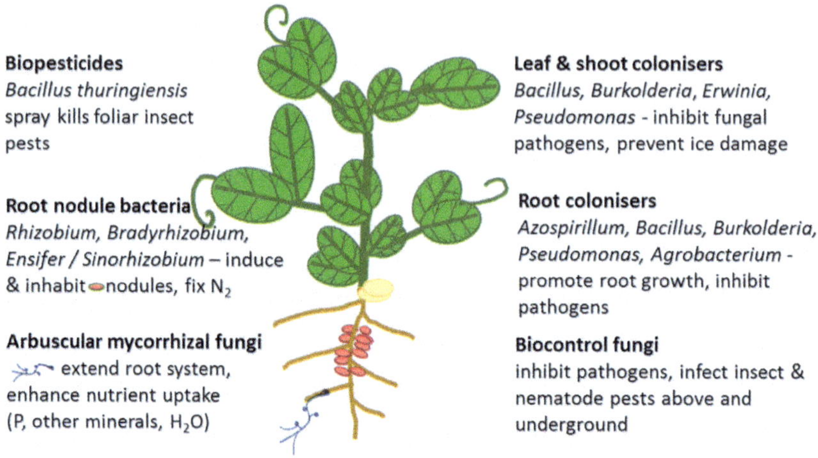

Figure 4 Examples of beneficial bacteria and fungi that may be exploited to improve plant health and crop yields. For more details and references see Sections 3.3 and 5.4.

has a long history: rhizobial inoculation of legumes on a commercial scale began in the early twentieth century (Hirsch, 2004). Some of the beneficial interactions are summarized in Fig. 4 and there is further discussion in Section 5.4.

4 Microbial inputs to geochemical cycles

4.1 Carbon cycling

Microorganisms are a reservoir of organic carbon, in addition to breaking down plant and animal residues (Fig. 1). Many different groups of microorganisms contribute to the turnover of the large reservoir of carbon-rich organic matter in soil (Nannipieri et al., 2003), some of which is lost as CO_2 and some of which is degraded and condenses over time to form humus (Martin and Haider, 1971). Inversion tillage (ploughing) makes organic matter more available for degradation; in undisturbed soils where anoxia can develop, it tends to accumulate (Bronick and Lal, 2005, Martin and Haider, 1971). Methanogenic archaea generate CH_4 in wet, organic C-rich soils (see above). Organic matter incorporation into the soil can be managed and CH_4 emission rates can be reduced when rice paddies are drained, but global warming and thawing permafrost are likely to increase methanogenesis in the future (Davidson and Janssens, 2006).

4.2 Nitrogen cycling

Soil microorganisms maintain supplies of bioavailable N for plants, whether inputs are from N fixation, manure, compost or N fertilizers based on urea or ammonia. In contrast, nitrate in fertilizers is soluble, mobile in soil and readily available for plants. Excess N, whether from animal excreta or fertilizer application, can be lost from soil by physico-chemical routes (volatilization, leaching) or microbially mediated nitrification and denitrification (Fig. 1).

Nitrogen fixation is an energy-dependent process, often performed by bacteria associated with plants, notably symbiotic endophytic bacteria (Dixon and Kahn, 2004). There are two major functional groups: rhizobia – Proteobacteria that nodulate leguminous plants; and *Frankia* – Actinobacteria that form nodules on some shrubs and trees (Franche et al., 2009). Specific rhizobial symbionts multiply when applied as inoculants (e.g. *Bradyrhizobium* for soya beans, typically applied to large tracts of land) or where host plants are nodulated by indigenous strains such as when clover is cultivated (Hirsch, 1996). Together, free-living and endophytic bacteria are estimated to contribute a similar amount of N to soils as N fertilizers, leguminous crops adding an additional 30%.

Fig. 1 shows how denitrification returns most of the N deposited on, applied to or fixed in soil to the atmosphere and is a major cause of fertilizer N losses from soil and GHG emissions. The many groups of bacteria and archaea capable of denitrification use nitrite and nitrous oxides as respiratory terminal electron acceptors, ultimately producing N_2 to complete the N cycle (Zumft, 1997).

Many different soil bacteria and archaea hydrolyse urea from fertilizers or animal excreta to produce ammonia, the substrate for nitrification (Lu et al., 2012, Lu and Jia, 2013). Relatively few groups of these oxidize ammonia to nitrite via hydroxylamine (Poth and Focht, 1985); further oxidation of nitrite to nitrate is performed by a narrow group of Proteobacteria and a deep-branching phylum, Nitrospirae (Spieck et al., 2006). These groups are thought to provide most of the nitrifying activity observed in soil. The nitrite oxidizers do not seem to be rate-limiting, as nitrite is not observed to accumulate in ammonia-fertilized soils unless specific inhibitors are added (Belser and Mays, 1980). Recently, some Nitrospirae have been shown to perform complete ammonia oxidation (comammox) with nitrite as a product (Van Kessel et al., 2015, Daims et al., 2015). These bacteria are associated with aquatic systems and there is evidence for their presence in soil (Pinto et al., 2016), although any role in nitrification in soil is unknown at present. Anaerobic ammonia oxidation, anammox, is important in aquatic systems and sludges. Anammox bacteria from a deep-branching, monophyletic group within the Planctomycetes that oxidize ammonium to dinitrogen gas anaerobically, with nitrite as an electron acceptor (Kuenen, 2008). However, these bacteria are not common in soil and it is unlikely that they are major contributors to soil nitrification (Humbert et al., 2009).

The interaction between different factors favouring ammonia-oxidizing archaea and bacteria (AOA and AOB) is complex and we do not yet have a full understanding of soil nitrifiers. The AOB are autotrophs, obtaining energy from the oxidation of inorganic electron donors in their environments rather than reduced organic carbon compounds whilst some AOA can use organic C (Tourna et al., 2011). AOA are more abundant than AOB in most soils (Leininger et al., 2006). However, there is conflicting evidence concerning the relative contribution of each to soil nitrification and it is uncertain if they are functionally interchangeable (Jia and Conrad, 2009; Xia et al., 2011). AOA require less energy and ammonia to survive than AOB and are likely to be more successful in low-input unfertilized and forest soils (Valentine, 2007; Martens-Habbena et al., 2009) despite having lower cellular rates of ammonium oxidation. The equilibrium between ammonia (NH_3) and ammonium (NH_4^+) shifts to NH_4^+ at low pH, limiting substrate availability and growth of AOB even when N inputs are relatively high; this might favour AOA in low pH soils (Zhang et al., 2012). The first AOA isolated from soil, *Nitrososphaera viennensis*, has mixotrophic growth and can use pyruvate as a carbon source (Tourna et al., 2011). In contrast, *Candidatus Nitrosotalea devanaterra* is an acidophilic chemoautotroph that grows in very low ammonia concentrations (Lehtovirta-Morley et al., 2011). There is further discussion on nitrification and denitrification in Section 6.1.

4.3 Other minerals

Whilst most essential elements are recycled in the soil food web, key nutrients may be depleted over time, in particular N and P (Wardle et al., 2004), depending on the underlying soil mineralogy, climatic factors and previous plant cover. The importance of these elements is emphasized in agricultural systems where plant growth is regularly removed and nutrient limitation becomes apparent. Fertilizers containing N, P, K, Mg, Ca, S (macronutrients) and other elements required in smaller amounts (micronutrients) are usually required to maintain yields. Unlike N which is readily recycled, phosphates released from organic debris or mineral fertilizers become firmly bound to soil minerals (Al and Fe oxides) in acidic soils and precipitated as Ca-, Al- or Fe-phosphate in alkaline soils and are not available to plants (Gyaneshwar et al., 2002). This sequestered P is only very slowly released by microbial, root exudate and chemical activity. Nevertheless, pollution from excess soluble phosphates in animal wastes and fertilizers, and soil particles carrying sequestered P, can cause severe eutrophication in receiving water (Tilman, 2001). Although soil microorganisms take up (immobilize) phosphate from mineral and organic sources, they may also make it available to plants. Plants require phosphate rather than organic forms of P and recycling organic residues is important in natural ecosystems as well as for sustainable agriculture. Plants and microorganisms produce organic anions and phosphatases that solubilize mineral phosphates (Jones, 1998; Richardson et al., 2009). Microorganisms can also secrete nucleases, releasing phosphate from nucleic acids, and phytases that degrade phytates, relatively stable plant-derived inositol phosphates (Jones, 1998). Inoculation of plants and amendment of soil with phosphate-solubilizing bacteria and fungi has been proposed since the 1930s, but the extent to which their application has any substantial effects on P nutrition of plants in field conditions is controversial (Gyaneshwar et al., 2002). There is also considerable disagreement over the contribution of microorganisms to plant P nutrition in natural ecosystems, with reports in the literature of estimates ranging from 0 to 90% (Van Der Heijden et al., 2008). By contrast, the importance of mycorrhizal fungi to plant health and P nutrition is widely accepted: they form symbiotic associations with roots in >80% of angiosperm genera, their hyphae extending into soil, effectively increasing the root surface available for nutrient and water uptake, in return obtaining plant sugars (Brundrett, 2002; Finlay, 2008). In natural ecosystems, most plants are found to be mycorrhizal but agricultural cultivation reduces infection levels, in part because tillage breaks up mycelial networks in soil, and plants replete in N and P may be less receptive to colonization by arbuscular mycorrhizal fungi (AMF), although there is considerable variation between species (Treseder and Allen, 2002).

Sulphur is a component of many minerals and it enters the soil S cycle where microorganisms release sulphates that plants can absorb; there are also natural sources of atmospheric S from volcanic eruptions, H_2S, dimethyl sulphide in sea spray and oxidation of elemental S by certain bacteria and archaea (Falkowski et al., 2008). Crop growth was not limited by S in most industrialized countries until the end of the twentieth century due to SO_2 released from fossil fuels, and serendipitous applications as part of fertilizer formulations (Zhao et al., 1999) but S fertilizer is now recommended for most crops. Indeed, changing SO_2 emissions over the past 160 y have been linked to changes in the dominant fungal pathogens causing wheat leaf blotch, detected in DNA extracted from material collected and archived (Bearchell et al., 2005).

5 Anthropogenic impacts on soil: land management and crop selection

Traditional agriculture incorporated crop rotation to reduce buildup of pests and diseases and fallow periods to prevent nutrient exhaustion, with ploughing to bury weeds and improve structure in heavy clay soils. N was supplied as human waste, animal manures or by exploiting symbiotic N_2-fixing associations of leguminous plants. Annual river flooding provided nutrient-rich sediments in some areas and irrigation systems were constructed to water crops in dry periods. Some developing countries still rely on these methods, but elsewhere they have been replaced by modern agricultural practices including chemical fertilizers, pesticides and modern high-yielding crop varieties driven by improved mechanization.

5.1 Deforestation and afforestation

There are distinct differences in microbial assemblages following inter-conversion between forest, pasture and arable land: besides differences in plant cover, forest soils tend to sequester more soil organic carbon (SOC) than pastures, arable land sequesters even less (Cookson et al., 2006). The higher G+C content observed in microbial community DNA isolated from pasture compared to forest soil may be attributed to a predominance of α- and β-proteobacteria in the former, although the reasons are as yet unknown (Nusslein and Tiedje, 1999). Functional gene microarray comparison of ancient forest, spruce plantation and cropped land identified the spruce plantation soil as having the greatest number of carbon-cycling genes, cropped land the fewest, and that genetic diversity seemed to increase with total SOC (Zhang et al., 2007). Total metagenomic analysis of forest soils indicated differences between the organic and mineral layers with relatively more carbon-cycling genes in the former, and long-term residual effects following harvesting of trees (Cardenas et al., 2015).

5.2 Tillage

Tilling or ploughing reduces weed growth, improves soil workability and mixes in manures, fertilizers and crop residues. Some soils are vulnerable to damage and erosion caused by excessive tillage. A long-term comparison of conventional, reduced- and no-till in an arable crop rotation in Switzerland indicated that the number of AMF spores was greater in untilled soil with a trend to increased species diversity (determined by ribosomal gene sequencing), although this was not statistically significant (Jansa et al., 2002). Comparison of 16S rRNA gene sequences in long-term corn and soya bean rotations in Kansas showed the overall abundance of Proteobacteria in all treatments. Tillage altered the Acidobacteria and Gemmatimonadetes community structure compared to no-till, and there was significantly higher diversity in the untilled sites in one of two years tested (Yin et al., 2010). Similarly, increased species richness was associated with reduced tillage in a study comparing 16S rRNA genes in a long-term arable rotation in Mexico (Ceja-Navarro et al., 2010). The functional significance of these changes is unclear, with ambiguity over which of several factors are implicated: soil type, climate, crop and management practice all affect microbial communities and their activity (Yin et al., 2010). However, a recent meta-analysis of 62 studies on the effects of tillage from around the world showed a statistically significant overall reduction in soil microbial biomass and enzyme activities and an increase in CO_2 evolution (Zuber and Villamil, 2016).

5.3 Flooding

Flooding in rice-growing areas helps control weeds and soil-borne pests. Flooded soils rapidly become anaerobic and soil microorganisms adopt alternative electron acceptors to oxygen (nitrate, Mn (IV) and Fe (III) ions, sulphate, acetate and CO_2), resulting in changes in soil chemistry and the generation of nitrous oxides, H_2S and CH_4 (Liesack et al., 2000). In a study comparing soil microcosms sampled at time and depth intervals, the active soil microbial community changed substantially within two days of flooding. Species diversity was decreased, with β-proteobacteria dominating the upper oxic zone and Bacteroidetes dominating the lower anoxic zone (Noll et al., 2005).

5.4 Crops

Although compositional differences in root exudates influence rhizosphere microorganisms colonizing different plant species, soil properties and cultivation practices associated with particular crops may have more impact on soil microbial communities (Berg and Smalla, 2009; Dennis et al., 2010; Bais et al., 2006). However, pathogens and symbionts build up in the presence of compatible plant hosts (Berg and Smalla, 2009) and crop rotation was developed in part to deplete crop-specific soil-borne pathogens, although this may take several years. Similarly, symbiotic rhizobia have been shown to increase in number during cultivation of their host legume crop and persist for many years afterwards (Hirsch, 1996). There are many reports of microbial community structure (i.e. which species are present or dominant) and its activity reflecting different crop types (Berg and Smalla, 2009), although in recent years the use of culture-independent community studies indicates that most are dominated by Proteobacteria, followed by Firmicutes and Actinobacteria (Deangelis et al., 2009; Mendes et al., 2011; Teixeira et al., 2010; Weinert et al., 2011). Composition of rhizosphere communities may change during plant growth (Deangelis et al., 2009) or vary in different cultivars of a single crop species (Weinert et al., 2011). Apart from pathogens and their specific antagonists (Garbeva et al., 2004; Deangelis et al., 2009; Mendes et al., 2011; Sanguin et al., 2009) the functional implications of differences in rhizosphere community structure are unclear. Some bacteria that promote plant growth have been shown to produce phytohormones which may improve root growth and thus nutrient uptake, drought tolerance and crop yields but those tested by necessity are from well-known groups (Proteobacteria, Firmicutes) and can be grown in lab culture (Ngumbi and Kloepper, 2016). Despite the large numbers of different bacterial species and types present on all plant roots, stable isotope probing shows that particular groups are active on certain plants' roots, for example, *Variovorax* and oilseed rape, Sphingomonadaceae and cereals (Haichar et al., 2008).

Bare fallow soil from which weeds have been removed regularly over a 50-year period seemed to have maintained a similar level of microbial diversity compared to plots under continuous wheat or grass cultivation for the same period, despite having a substantially less abundant soil community (Hirsch et al., 2016). The permanent grassland from which both treatments were derived has a mixed plant community and similar microbial diversity to the bare fallow, albeit more numerous. However, community structure was different in all three plots (Hirsch et al., 2009, 2016). A comparison using 16S rRNA gene pyrosequencing indicated higher species diversity in a wheat-cotton-corn rotation than pasture, with substantially lower diversity in cotton monoculture (Acosta-Martinez et al., 2008). Similarly, functional gene microarray analysis indicated less diversity in functional genes in soils under monoculture than those with mixed plant species (He et al., 2010).

6 Anthropogenic impacts on soil: fertilizers, agrochemicals, soil pH and pollution

6.1 Fertilizers

Fertilizers are used to improve plant growth and increase C inputs to soil, but can have indirect effects on the microbial community, in addition to specific impacts of N cycling organisms. Indirect effects arise from soil acidification by ammonium-based fertilizers – this includes urea, which is hydrolysed to ammonia by microbial activity (Bremner, 1995). Organic manures often contain substantial plant residues which complicate direct comparison with mineral fertilizers, particularly when the former are part of 'organic' agriculture regimes that avoid using 'artificial' manufactured fertilizers and pesticides. Like humus, farmyard manure (FYM) and compost improve soil structure, water and nutrient retention, whether part of a traditional, modern intensive or 'organic' agricultural systems. Animal manures in particular carry a distinctive microbial load, some of which will survive in soil and amend the microbiome of the host soil. However, both soils receiving regular FYM or green manures appear to maintain more diverse bacterial and fungal communities based on ribosomal gene diversity (Sun et al., 2004; Sugiyama et al., 2010). Functional diversity is also greater than that in soils receiving only mineral fertilizers, indicated by both enzyme activity and microarray analysis (Reeve et al., 2010; Mader et al., 2002). Nevertheless, fully 'organic' management generally results in lower crop yields than modern farming systems (Mader et al., 2002).

Fertilizers containing N might be anticipated to influence microorganisms involved in N cycling, removing selective pressure for biological N_2 fixation. There was no discernible effect on the diversity of the nitrogenase gene *nifH* attributable to N fertilizer inputs in the Broadbalk experiment at Rothamsted (Ogilvie et al., 2008). However, a negative association between *nifH* abundance and SOC was reported at a long-term ecological research site in the United States (Morales et al., 2010). In a comparison of multiple soil types and management regimes in southeast Australia, land use had no significant influence and the amount of soil microbial biomass C was the primary factor influencing *nifH* abundance (Hayden et al., 2010).

Impacts on denitrification are more complex, in part because denitrifiers are more numerous than any other functional group in the N cycle, comprising up to 5% of all soil bacteria (Philippot et al., 2007). Factors influencing denitrification such as anoxia and nitrate concentration are well established. Also, manures and organic fertilizers tend to increase denitrification activity relative to mineral fertilizers (Philippot et al., 2007; Hallin et al., 2009), although changes in denitrifier communities are less clear. The cytochrome cd_1 variant of nitrite reductase encoded by *nirS* is reported to be more common than

Cu-dependent *nirK* in cultured environmental bacterial isolates (Coyne et al., 1989). The preponderance of as-yet-uncultured soil bacterial genotypes renders the actual relative abundance uncertain. Environmental factors influencing the prevalence of alternate nitrite reductases *nirK* and *nirS* remain unclear and there is little consensus in the published literature: *nirK* is reportedly more abundant in arable systems (Coyne et al., 1989) and with increasing N fertilizer inputs (Philippot et al., 2007) whilst *nirS* is more abundant in intensively grazed plots with high soil ammonia and nitrate relative to less intensively grazed plots with lower N inputs (Dandie et al., 2011). In a Swedish long-term arable rotation comparing different fertilizer inputs, microbial functions were measured and the abundance of different genes assessed using qPCR. Denitrification activity correlated with the overall bacterial community size, crop yield and *nosZ* abundance whilst pH was the major soil property driving microbial community structure (Hallin et al., 2009). The distribution of *nirK* and *nirS* indicated that the forms are functional analogues (Hallin et al., 2009) and the overall abundance of denitrifiers together with available N and soil conditions determines the denitrification rate.

Continuing the discussion of the AOA and bacteria from Section 4.2, although overall AOA are more abundant than AOB in soil, in arable soils at neutral pH AOB proliferate in response to N fertilizer application and seem responsible for increased nitrification rates in agricultural systems (Wessen et al., 2011; Mendum et al., 1999; Jia and Conrad, 2009; Bates et al., 2011). Similarly, AOB, but not AOA, increased with N in grazed grasslands (Di et al., 2009) whilst in two arable soils, AOA declined with increasing N (Wessen et al., 2011; Bates et al., 2011). However, in acid soils, AOA seem to respond to N application (Gubry-Rangin et al., 2015). This reflects availability of the substrate NH_3 and the different pH optima for the two groups: AOB nitrification rate decreases below pH 7; AOA are more abundant in soils below pH 6.0 and nitrification activity decreases as soil pH increases (Nicol et al., 2008). Furthermore, community structure is influenced by soil pH: distinctly different groups of both AOA and AOB (determined by DGGE analysis of PCR products) were found in arable plots maintained from pH 4.5 to 7.5 (Nicol et al., 2008). This may explain why significant farm-scale heterogeneity in AOA and AOB abundance, community structure and function seems to be related to soil properties rather than land management by organic or conventional regimes (Wessen et al., 2011). To summarize, nitrification occurs in most soils: AOA may be responsible for most activity in nutrient-poor acidic conditions and AOB contribute more in fertilized soils at neutral pH. In consequence, there may be a delay in nitrification in response to intensive animal grazing, fertilizer or lime application on previously nutrient-poor soils, until AOB populations increase. There may also be differential responses to nitrification inhibitors: 3,4-dimethylpyrazole phosphate reduces abundance of AOB but not AOA (Kleineidam et al., 2011). Similarly, AOB were inhibited to a much greater degree than AOA by sulphadiazine residues in pig manure (Schauss et al., 2009).

6.2 Agrochemicals

As well as fertilizers, other agrochemicals are applied to most food and feed crops in the developed world to increase yields by improving growth, reducing damage from pests and pathogens, and decreasing competition with weeds, thus indirectly affecting diversity. For example, in the United Kingdom, approximately 25% of land receives agrochemical applications (Defra statistics 2009). Some agrochemicals have been reported to show transient effects either because of selective toxicity or because they are selective substrates metabolized by certain microbial groups (Hussain et al., 2009). For example, glyphosate is considered to be a relatively safe herbicide with limited non-target effects (Duke and Powles, 2008). At high concentrations the formulations used in agriculture are toxic for many microorganisms in laboratory culture, but when applied directly to soil glyphosate they provide a growth substrate, stimulating both microbial activity and functional diversity (Mijangos et al., 2009). When applied to plants it has inconsistent effects possibly due to herbicide-induced changes in root exudation (Mijangos et al., 2009) and had no significant effect on the diversity of bacteria in the maize rhizosphere, assessed by 16S rRNA gene sequencing (Barriuso et al., 2010). Similarly, no effects on microbial processes were detected in soil where spring barley received applications of five pesticides, including glyphosate, over a 20-year period (Bromilow et al., 1996).

6.3 Soil pH

The pH of soil is relatively easy to manage, influences nutrient availability and is reported to have the largest overall influence on microbial communities, greater than the effect of soil mineralogy or of plants (Fierer and Jackson, 2006). Natural soils tend to become acid: plant roots respire CO_2 forming the weak carbonic acid and also exude protons and organic acids (Jones, 1998). Soil microorganisms contribute by respiration and degradation of plant residues (Catt, 1985; Goulding and Annis, 1998); also rain is naturally acid, historically exacerbated by 'acid rain' arising from industrial processes (Goulding and Poulton, 1985). In agricultural systems acidification occurs, resulting in Ca^{2+} losses (Gasser, 1985) exacerbated by the use of elemental sulphur and ammonia-based fertilizers (Goulding and Poulton, 1985). The Romans used 'lime' ($CaCO_3$ in calcite or chalk, $CaMg(CO_3)_2$ in dolomite) to 'sweeten' soil if a soil suspension tasted acidic (Barber, 1967). The practice was widespread around 100 BCE according to the Roman agronomist Varro. The Romans probably introduced liming to Britain and marl (chalk) pits have been common since Norman times (1066–1154 CE). Generally, lime was applied every five years, but since the nineteenth century, improved understanding and monitoring of soil chemical processes has enabled maintenance of soil pH at levels optimal for particular crops (Goulding and Annis, 1998). Soil pH substantially altered the trajectory of microbial community structure development in a grassland over time, assessed by 16S rRNA gene microarray and qPCR study (Kuramae et al., 2011), and microbial diversity in a pH gradient in arable soil assessed by ribosomal gene sequencing (Rousk et al., 2010). In the latter study, the number of bacterial species doubled between pH 4 and 8, although effects on fungal diversity were not significant. In a landscape-scale study in Burgundy, France, soil pH was the main driver of differences in the diversity and abundance of microbial communities, including those involved in N cycling (Bru et al., 2011).

6.4 Pollution

Pollution from natural or anthropogenic sources can have major effects on soil microbial functioning. Soils polluted heavily with toxic metals or organic chemicals from industrial or natural sources may support only a limited number of dominant specialist organisms. However, these organisms may alleviate some of the most harmful effects of acute contamination. For example, they may reduce metal ions to less toxic (Cr(VI) to Cr(III)) or mobile (U(VI) to U(IV)) forms in anaerobic sediments, generating energy for their growth in the process; or by catabolizing polyaromatic hydrocarbons. The number and diversity of functional genes detected using a functional gene microarray decreased along an oil field contamination gradient, although the abundance of specific organic contaminant degradation genes increased (Liang et al., 2009). Plants are frequently early casualties of contamination; their absence will have a secondary effect on the soil food web. Planting soils with pollution-resistant plants is often a first step in reclamation of mine spoil heaps, providing substrates that stimulate bacterial growth.

Contamination of land with pesticides or animal wastes containing antibiotics frequently has transient effects on soil microorganisms. Metals used in fungicide preparations can be more problematical as they are not degraded and accumulate in soil. Chronic metal pollution can reduce plant productivity, making crops unfit for consumption. In productive soils, the commonest sources of pollution are through the use of copper-based fungicides (commonly applied to vineyards), atmospheric deposition and addition of sewage-sludge or municipal waste to arable soils (Nicholson et al., 2006). Sludges are rich sources of organic matter, N and P but typically contain potentially harmful metal pollutants such as Zn, Cu and Cd. Numerous studies describe the general response (species diversity, population respiration, enzyme activity) of soil microbial populations to metal pollutants in soils, for example, reporting reductions in diversity based on 16S rRNA gene analysis (Moffett et al., 2003; Ranjard et al., 2006; Macdonald et al., 2011b) or abundance of rhizobia based on nodulation gene frequency (Macdonald et al., 2011a). Chronic exposure to metals results in the soil community acquiring tolerance to the metal in a phenomenon termed pollution-induced community tolerance (Davis et al., 2004). Specific changes in functional diversity have been reported in soil microcosms, for example, Cu associated with fungicide application resulted in altered nitrification gene diversity and increased nitrification potential (Demanou et al., 2006). Ag addition also changed denitrification gene diversity, resulting in reduced activity (Throback et al., 2007) and changes in microbial communities in response to Cu addition were associated with an increase in the mineralization of soil organic matter (Bernard et al., 2009). However, short-term studies can be misleading, interactions between metal ions, soil organic matter and minerals, and microorganisms are complex, and few studies compare different soil types contaminated with the same material. Comparison of microbial communities using T-RFLP in five arable and two grassland soils treated 11 years previously with the same sludge containing Cu and Zn (at levels close to the EU maximum) and the untreated soils indicated that the geographic site had an overwhelming influence on soil community structure but some community shifts could be attributed to metals (Macdonald et al., 2011b).

7 Future perspectives

Whether for agriculture, forestry, industry or amenity, land use is driven by social, economic and political factors. Consequently, the most sustainable systems do not always prevail and there is a balance between using less land for more intensive food production, or lower input agriculture over a much wider area: 'land sparing' versus 'land sharing'. However, since land is a finite resource, calls for greater intensification are irresistible (Godfray et al., 2010). Eschewing synthetic chemicals cannot replace farming practices in industrialized countries and achieve continually increasing yields (FAO; http://www.fao.org/organicag/oa-faq/oa-faq7/en/). The next 30 years will see inevitable pressure to improve the efficiency of use of both water and fertilizers, whether synthetic, derived from organic wastes or minerals (Tilman et al., 2002). Continued development of chemicals for crop protection will be essential to provide maximum efficacy against pests with minimal impact on non-target organisms, including consumers (Beddington, 2010). An increase was predicted in the exploitation of higher-yielding or more pest resistant crop varieties with a broader genetic base, including plants modified using GM technology to incorporate novel traits (Godfray et al., 2010) and with targeted mutagenesis using CRISPR-Cas methods (Doudna and Charpentier, 2014). Designing plants to select for the most beneficial microbial colonists, or to modify the soil microbiome to optimize nutrient availability whilst minimizing losses is one potential strategy (Hirsch and Mauchline, 2012; Lareen et al., 2016).

Managing land for environmental improvement is subjective: natural mixed vegetation and their associated microbiota may thrive in soils too acidic and low in nutrients for optimal crop growth; incorporation of organic matter benefits arable soils, sequestering carbon and improving soil structure – but the same treatment may encourage denitrifying and methane-generating bacteria that increase nitrous oxide and methane emissions. Understanding the response of soil microorganisms to natural or anthropogenic perturbation including differences in land use management should enable optimal soil management, to maximize crop production, improve structure or regenerate 'natural' ecosystems.

Recent technological advances are opening the soil 'black box' and providing insight into how communities respond to changing land use (Hirsch et al, 2010). Although the rarest species may remain undiscovered, and scant information gained on less abundant phyla, high-throughput sequencing is providing increasingly deep sequencing, revealing improved details of complexity and greater knowledge of community structure and dynamics. This is complemented by new culture methods such as the iChip (Nichols et al., 2010), discussed above (Section 2), that enable isolation of previously obscure soil microorganisms.

High species diversity and substantial functional redundancy both confer resilience to soil communities, making it unlikely that essential functions will be lost permanently even if the environment is not conducive to survival of organisms and expression of genes. Mankind has manipulated soils deliberately to improve crop yields since the emergence of agriculture, often producing concomitant changes in microbial communities. Sometimes this has led to inadvertent loss of soil structure and chemistry, occasionally causing major environmental problems, but mitigation methods have developed alongside agriculture, in particular, management of soil nutrients and pH. Crop varieties and agrochemicals seem less likely to influence soil microorganisms than associated major changes in land management that affect the physico-chemical structure of soil (Griffiths and Philippot, 2013) and it is these, together with climate change, that could potentially destabilize biogeochemical cycles.

8 Where to look for further information

Additional information can be found online, in books and research papers. A comprehensive collection of methods for rhizosphere research is provided by *Molecular Microbial Ecology of the Rhizosphere* (de Bruijn, 2013). Molecular methods evolve rapidly and journal articles provide recent advances in technology, but the US Department of Energy Argonne National Laboratory hosts the Earth Microbiome Project website which provides current methods for DNA extraction and preparation of samples for amplicon sequencing to measure microbial diversity (http://www.earthmicrobiome.org/protocols-and-standards/).

A useful online overview of soil biology with links to related information on soil health is provided by the US Department of Agriculture Soil Biology Primer (https://www.nrcs.usda.gov/wps/portal/nrcs/main/soils/health/biology/). Further information on soil biology, with a large number of illustrations including soil mesofauna is available from the Global Soil Biodiversity Initiative which has produced an online atlas (https://globalsoilbiodiversity.org/). The FAO Soils Portal provides overviews of soil biodiversity and its importance for nutrient cycling and other aspects of soil health (http://www.fao.org/soils-portal/soil-biodiversity/en/).

9 Acknowledgements

Rothamsted Research receives strategic funding from the Biotechnology and Biological Research Council, United Kingdom; this work was supported by BBS/E/C/00005196 and BB/P01268X/1. I thank my colleagues Ian Clark, Andrew Neal and Tim Mauchline for their help in preparing this chapter.

10 References

Acosta-Martinez, V., Dowd, S., Sun, Y. and Allen, V. (2008), Tag-encoded pyrosequencing analysis of bacterial diversity in a single soil type as affected by management and land use, *Soil Biol. Biochem.*, 40, 2762–70.

Allison, S. D. and Martiny, J. B. H. (2008), Resistance, resilience, and redundancy in microbial communities, *Proc. Natl. Acad. Sci. U. S. A.*, 105, 11512–19.

Angel, R., Claus, P. and Conrad, R. (2012), Methanogenic archaea are globally ubiquitous in aerated soils and become active under wet anoxic conditions, *ISME J.*, 6, 847–62.

Bais, H. P., Weir, T. L., Perry, L. G., Gilroy, S. and Vivanco, J. M. (2006), The role of root exudates in rhizosphere interations with plants and other organisms, *Annu. Rev. Plant Biol.*, 57, 233–66.

Baldrian, P., Kolarik, M., Stursova, M., Kopecky, J., Valaskova, V., Vetrovsky, T., Zifcakova, L., Snajdr, J., Ridl, J., Vlcek, C. and Voriskova, J. (2012), Active and total microbial communities in forest soil are largely different and highly stratified during decomposition, *ISME J.*, 6, 248–58.

Banwart, S. (2011), Save our soils, *Nature*, 474, 151–2.

Barber, S. (1967), *Liming materials and practices*. Madison: American Society of Agronomy.

Barriuso, J., Marin, S. and Mellado, R. P. (2010), Effect of the herbicide glyphosate on glyphosate-tolerant maize rhizobacterial communities: a comparison with pre-emergency applied herbicide consisting of a combination of acetochlor and terbuthylazine, *Environ. Microbiol.*, 12, 1021–30.

Bates, S. T., Berg-Lyons, D., Caporaso, J. G., Walters, W. A., Knight, R. and Fierer, N. (2011), Examining the global distribution of dominant archaeal populations in soil, *ISME J.*, 5, 908–17.

Bearchell, S. J., Fraaije, B. A., Shaw, M. W. and Fitt, B. D. L. (2005), Wheat archive links long-term fungal pathogen population dynamics to air pollution, *Proc. Natl. Acad. Sci. U. S. A.*, 102, 5438–42.

Beddington, J. (2010), Food security: contributions from science to a new and greener revolution, *Phil. Trans. Roy. Soc. B-Biol. Sci.*, 365, 61–71.

Belser, L. W. and Mays, E. L. (1980), Specific-inhibition of nitrite oxidation by chlorate and its use in assessing nitrification in soils and sediments, *Appl. Environ. Microbiol.*, 39, 505–10.

Bent, S. J. and Forney, L. J. (2008), The tragedy of the uncommon: understanding limitations in the analysis of microbial diversity, *ISME J.*, 2, 689–95.

Berg, G. and Smalla, K. (2009), Plant species and soil type cooperatively shape the structure and function of microbial communities in the rhizosphere, *FEMS Microbiol. Ecol.*, 68, 1–13.

Bernard, L., Maron, P. A., Mougel, C., Nowak, V., Leveque, J., Marol, C., Balesdent, J., Gibiat, F. and Ranjard, L. (2009), Contamination of soil by copper affects the dynamics, diversity, and activity of soil bacterial communities involved in wheat decomposition and carbon storage, *Appl. Environ. Microbiol.*, 75, 7565–9.

Bremner, J. M. (1995), Recent research on problems in the use of urea as a nitrogen fertilizer, *Fertil. Res.*, 42, 321–9.

Bromilow, R. H., Evans, A. A., Nicholls, P. H., Todd, A. D. and Briggs, G. G. (1996), The effect on soil fertility of repeated applications of pesticides over 20 years, *Pestic. Sci.*, 48, 63–72.

Bronick, C. J. and Lal, R. (2005), Soil structure and management: a review, *Geoderma*, 124, 3–22.

Bru, D., Ramette, A., Saby, N. P. A., Dequiedt, S., Ranjard, L., Jolivet, C., Arrouays, D. and Philippot, L. (2011), Determinants of the distribution of nitrogen-cycling microbial communities at the landscape scale, *ISME J.*, 5, 532–42.

Brundrett, M. C. (2002), Coevolution of roots and mycorrhizas of land plants, *New Phytol.*, 154, 275–304.

Buckley, D. H., Graber, J. R. and Schmidt, T. M. (1998), Phylogenetic analysis of nonthermophilic members of the kingdom Crenarchaeota and their diversity and abundance in soils, *Appl. Environ. Microbiol.*, 64, 4333–9.

Cardenas, E., Kranabetter, J. M., Hope, G., Maas, K. R., Hallam, S. and Mohn, W. W. (2015), Forest harvesting reduces the soil metagenomic potential for biomass decomposition, *ISME J.*, 9, 2465–76.

Catt, J. (1985), Natural soil acidity, *Soil Use Manag.*, 1, 8–10.

Ceja-Navarro, J. A., Rivera-Orduna, F. N., Patino-Zuniga, L., Vila-Sanjurjo, A., Crossa, J., Govaerts, B. and Dendooven, L. (2010), Phylogenetic and multivariate analyses to determine the effects of different tillage and residue management practices on soil bacterial communities, *Appl. Environ. Microbiol.*, 76, 3685–91.

Cohan, F. M. (2002), What are bacterial species?, *Annu. Rev. Microbiol.*, 56, 457–87.

Conrad, R., Erkel, C. and Liesack, W. (2006), Rice Cluster I methanogens, an important group of Archaea producing greenhouse gas in soil, *Curr. Opin. Biotechnol.*, 17, 262–7.

Cookson, W. R., Marschner, P., Clark, I. M., Milton, N., Smirk, M. N., Murphy, D. V., Osman, M., Stockdale, E. A. and Hirsch, P. R. (2006), The influence of season, agricultural management, and soil properties on gross nitrogen transformations and bacterial community structure, *Aust. J. Soil Res.*, 44, 453–65.

Coyne, M. S., Arunakumari, A., Averill, B. A. and Tiedje, J. M. (1989), Immunological identification and distribution of dissimilatory heme cd1 and nonheme copper nitrite reductases in denitrifying bacteria, *Appl. Environ. Microbiol.*, 55, 2924–31.

Daims, H., Lebedeva, E. V., Pjevac, P., Han, P., Herbold, C., Albertsen, M., Jehmlich, N., Palatinszky, M., Vierheilig, J., Bulaev, A., Kirkegaard, R. H., Von Bergen, M., Rattei, T., Bendinger, B., Nielsen, P. H. and Wagner, M. (2015), Complete nitrification by Nitrospira bacteria, *Nature*, 528, 504–9.

Dandie, C. E., Wertz, S., Leclair, C. L., Goyer, C., Burton, D. L., Patten, C. L., Zebarth, B. J. and Trevors, J. T. (2011), Abundance, diversity and functional gene expression of denitrifier communities in adjacent riparian and agricultural zones, *FEMS Microbiol. Ecol.*, 77, 69–82.

Davidson, E. A. and Janssens, I. A. (2006), Temperature sensitivity of soil carbon decomposition and feedbacks to climate change, *Nature*, 440, 165–73.

Davis, M. R. H., Zhao, F. J. and McGrath, S. P. (2004), Pollution-induced community tolerance of soil microbes in response to a zinc gradient, *Environ. Toxicol. Chem.*, 23, 2665–72.

DeAngelis, K. M., Brodie, E. L., Desantis, T. Z., Andersen, G. L., Lindow, S. E. and Firestone, M. K. (2009), Selective progressive response of soil microbial community to wild oat roots, *ISME J.*, 3, 168–78.

DeAngelis, K. M., Gladden, J. M., Allgaier, M., D'haeseleer, P., Fortney, J. L., Reddy, A., Hugenholtz, P., Singer, S. W., Vander Gheynst, J. S., Silver, W. L., Simmons, B. A. and Hazen, T. C. (2010), Strategies for enhancing the effectiveness of metagenomic-based enzyme discovery in lignocellulolytic microbial communities, *Bioenergy Res.*, 3, 146–58.
de Bruijn, F.J., ed (2013) *Molecular Microbial Ecology of the Rhizosphere*. Volume 1 & 2. John Wiley & and Sons, Inc.
Delmont, T. O., Prestat, E., Keegan, K. P., Faubladier, M., Robe, P., Clark, I. M., Pelletier, E., Hirsch, P. R., Meyer, F., Gilbert, J. A., Le Paslier, D., Simonet, P. and Vogel, T. M. (2012), Structure, fluctuation and magnitude of a natural grassland soil metagenome, *ISME J.*, 6, 1677–87.
Delmont, T. O., Robe, P., Cecillon, S., Clark, I. M., Constancias, F., Simonet, P., Hirsch, P. R. and Vogel, T. M. (2011), Accessing the soil metagenome for studies of microbial diversity, *Appl. Environ. Microbiol.*, 77, 1315–24.
Demanou, J., Sharma, S., Weber, A., Wilke, B. M., Njine, T., Monkiedje, A., Munch, J. C. and Schloter, M. (2006), Shifts in microbial community functions and nitrifying communities as a result of combined application of copper and mefenoxam, *FEMS Microbiol. Lett.*, 260, 55–62.
Dennis, P., Miller, A. and Hirsch, P. (2010), Are root exudates more important than other sources of rhizodeposits in structuring rhizosphere bacterial communities?, *FEMS Microbiol. Ecol.*, 72, 313–27.
Di, H. J., Cameron, K. C., Shen, J. P., Winefield, C. S., O'callaghan, M., Bowatte, S. and He, J. Z. (2009), Nitrification driven by bacteria and not archaea in nitrogen-rich grassland soils, *Nature Geoscience*, 2, 621–4.
Diedhiou, A. G., Dupouey, J.-L., Buee, M., Dambrine, E., Lauet, L. and Garbaye, J. (2010), The functional structure of ectomycorrhizal communities in an oak forest in central France witnesses ancient Gallo-Roman farming practices, *Soil Biol. Biochem.*, 42, 860–2.
Dixon, R. and Kahn, D. (2004), Genetic regulation of biological nitrogen fixation, *Nat. Rev. Microbiol.*, 2, 621–31.
Doudna, J. A. and Charpentier, E. (2014), The new frontier of genome engineering with CRISPR-Cas9, *Science*, 346, 1258096.
Duke, S. O. and Powles, S. B. (2008), Glyphosate: a once-in-a-century herbicide, *Pest Manag. Sci.*, 64, 319–25.
Falkowski, P. G., Fenchel, T. and Delong, E. F. (2008), The microbial engines that drive Earth's biogeochemical cycles, *Science*, 320, 1034–9.
Fierer, N. and Jackson, R. B. (2006), The diversity and biogeography of soil bacterial communities, *Proc. Natl. Acad. Sci. U. S. A.*, 103, 626–31.
Fierer, N., Ladau, J., Clemente, J. C., Leff, J. W., Owens, S. M., Pollard, K. S., Knight, R., Gilbert, J. A. and Mcculley, R. L. (2013), Reconstructing the microbial diversity and function of pre-agricultural tallgrass prairie soils in the United States, *Science*, 342, 621–4.
Finlay, R. D. (2008), Ecological aspects of mycorrhizal symbiosis: With special emphasis on the functional diversity of interactions involving the extraradical mycelium, *J. Exp. Bot.*, 59, 1115–26.
Franche, C., Lindstrom, K. and Elmerich, C. (2009), Nitrogen-fixing bacteria associated with leguminous and non-leguminous plants, *Plant Soil*, 321, 35–59.
Garbeva, P., Van Veen, J. A. and Van Elsas, J. D. (2004), Microbial diversity in soil: Selection of microbial populations by plant and soil type and implications for disease suppressiveness, *Annu. Rev. Phytopathol.*, 42, 243–70.
Gasser, J. K. R. (1985), Processes causing loss of calcium from agricultural soils, *Soil Use Manag.*, 1, 14–17.
Girvan, M. S., Campbell, C. D., Killham, K., Prosser, J. I. and Glover, L. A. (2005), Bacterial diversity promotes community stability and functional resilience after perturbation, *Environ. Microbiol.*, 7, 301–13.
Goulding, K. W. T. and Annis, B. (1998), *Lime, Liming and the Management of Soil Acidity. Proceedings no. 410*. York, UK: Fertiliser Society.

Goulding, K. W. T. and Poulton, P. R. (1985), Acid deposition at Rothamsted, Saxmundham and Woburn, 1969–83, *Soil Use Manag.*, 1, 6–8.
Godfray, H. C. J., Beddington, J. R., Crute, I. R., Haddad, L., Lawrence, D., Muir, J. F., Pretty, J., Robinson, S., Thomas, S. M. and Toulmin, C. (2010), Food security: The challenge of feeding 9 Billion people, *Science*, 327, 812–18.
Griffiths, B. S. and Philippot, L. (2013), Insights into the resistance and resilience of the soil microbial community, *FEMS Microbiol. Rev.*, 37, 112–29.
Grigoriev, I. V., Nikitin, R., Haridas, S., Kuo, A., Ohm, R., Otillar, R., Riley, R., Salamov, A., Zhao, X. L., Korzeniewski, F., Smirnova, T., Nordberg, H., Dubchak, I. and Shabalov, I. (2014), MycoCosm portal: gearing up for 1000 fungal genomes, *Nucleic Acids Res.*, 42, D699–704.
Grossman, J. M., O'neill, B. E., Tsai, S. M., Liang, B., Neves, E., Lehmann, J. and Thies, J. E. (2010), Amazonian anthrosols support similar microbial communities that differ distinctly from those extant in adjacent, unmodified soils of the same mineralogy, *Microb. Ecol.*, 60, 192–205.
Gruber, N. and Galloway, J. N. (2008), An earth-system perspective of the global nitrogen cycle, *Nature*, 451, 293–6.
Gubry-Rangin, C., Kratsch, C., Williams, T. A., Mchardy, A. C., Embley, T. M., Prosser, J. I. and Macqueen, D. J. (2015), Coupling of diversification and pH adaptation during the evolution of terrestrial Thaumarchaeota, *Proc. Natl. Acad. Sci. U. S. A.*, 112, 9370–5.
Guo, J. R., Cole, J. R., Zhang, Q. P., Brown, C. T. and Tiedje, J. M. (2016), Microbial community analysis with ribosomal gene fragments from shotgun metagenomes, *Appl. Environ. Microbiol.*, 82, 157–66.
Gyaneshwar, P., Kumar, G. N., Parekh, L. J. and Poole, P. S. (2002), Role of soil microorganisms in improving P nutrition of plants, *Plant Soil*, 245, 83–93.
Haichar, F. E. Z., Marol, C., Berge, O., Rangel-Castro, J. I., Prosser, J. I., Balesdent, J., Heulin, T. and Achouak, W. (2008), Plant host habitat and root exudates shape soil bacterial community structure, *ISME J.*, 2, 1221–30.
Hallin, S., Jones, C. M., Schloter, M. and Philippot, L. (2009), Relationship between N-cycling communities and ecosystem functioning in a 50-year-old fertilization experiment, *ISME J.*, 3, 597–605.
Hayden, H. L., Drake, J., Imhof, M., Oxley, A. P. A., Norng, S. and Mele, P. M. (2010), The abundance of nitrogen cycle genes amoA and nifH depends on land-uses and soil types in South-Eastern Australia, *Soil Biol. Biochem.*, 42, 1774–83.
He, Z., Deng, Y., Van Nostrand, J. D., Tu, Q., Xu, M., Hemme, C. L., Li, X., Wu, L., Gentry, T. J., Yin, Y., Liebich, J., Hazen, T. C. and Zhou, J. (2010), GeoChip 3.0 as a high-throughput tool for analyzing microbial community composition, structure and functional activity, *ISME J.*, 4, 1167–79.
Helgason, B. L., Walley, F. L. and Germida, J. J. (2009), Fungal and bacterial abundance in long-term no-till and intensive-till soils of the northern great plains, *Soil Sci. Soc. Am. J.*, 73, 120–7.
Hirsch, P. R. (1996), Population dynamics of indigenous and genetically modified rhizobia in the field, *New Phytol.*, 133, 159–71.
Hirsch, P. R. (2004), Release of transgenic bacterial inoculants - rhizobia as a case study, *Plant Soil*, 266, 1–10.
Hirsch, P. R., Gilliam, L. M., Sohi, S. P., Williams, J. K., Clark, I. M. and Murray, P. J. (2009), Starving the soil of plant inputs for 50 years reduces abundance but not diversity of soil bacterial communities, *Soil Biol. Biochem.*, 41, 2021–4.
Hirsch, P. R., Jhurreea, D., Williams, J. K., Murray, P. J., Scott, T., Misselbrook, T. H., Goulding, K. W. T. and Clark, I. M. (2016), Soil resilience and recovery: rapid community responses to management changes, *Plant Soil*, 1–15.
Hirsch, P. R., Mauchline, T. H. and Clark, I. M. (2010), Culture-independent molecular techniques for soil microbial ecology, *Soil Biol. Biochem.*, 42, 878–87.
Hirsch, P. R., Mauchline, T. H. and Clark, I. M. (2013), Culture-independent molecular approaches to microbial ecology in soil and the rhizosphere. *Molecular Microbial Ecology of the Rhizosphere*. John Wiley and Sons, Inc.

Hirsch, P. R. and Mauchline, T. H. (2012), Who's who in the plant root microbiome?, *Nat Biotech*, 30, 961–2.
Humbert, S., Tarnawski, S., Fromin, N., Mallet, M.-P., Aragno, M. and Zopfi, J. (2009), Molecular detection of anammox bacteria in terrestrial ecosystems: distribution and diversity, *ISME J.*, 4, 450–4.
Hussain, S., Siddique, T., Saleem, M., Arshad, M. and Khalid, A. (2009), Impact of pesticides on soil microbial diversity, enzymes, and biochemical reactions, *Adv. Agron.*, 102, 159–200.
Jansa, J., Mozafar, A., Anken, T., Ruh, R., Sanders, I. R. and Frossard, E. (2002), Diversity and structure of AMF communities as affected by tillage in a temperate soil, *Mycorrhiza*, 12, 225–34.
Jia, Z. and Conrad, R. (2009), Bacteria rather than Archaea dominate microbial ammonia oxidation in an agricultural soil, *Environ. Microbiol.*, 11, 1658–71.
Jones, D. L. (1998), Organic acids in the rhizosphere - a critical review, *Plant Soil*, 205, 25–44.
Keiblinger, K. M., Fuchs, S., Zechmeister-Boltenstern, S. and Riedel, K. (2016), Soil and leaf litter metaproteomics—a brief guideline from sampling to understanding, *FEMS Microbiol. Ecol.*, 92, fiw180.
Kleineidam, K., Kosmrlj, K., Kublik, S., Palmer, I., Pfab, H., Ruser, R., Fiedler, S. and Schloter, M. (2011), Influence of the nitrification inhibitor 3,4-dimethylpyrazole phosphate (DMPP) on ammonia-oxidizing bacteria and archaea in rhizosphere and bulk soil, *Chemosphere*, 84, 182–6.
Konopka, A. (2009), What is microbial community ecology?, *ISME J.*, 3, 1223–30.
Kuenen, J. G. (2008), Anammox bacteria: from discovery to application, *Nat. Rev. Microbiol.*, 6, 320–6.
Kuramae, E., Gamper, H., Van Veen, J. and Kowalchuk, G. (2011), Soil and plant factors driving the community of soil-borne microorganisms across chronosequences of secondary succession of chalk grasslands with a neutral pH, *FEMS Microbiol. Ecol.*, 77, 285–94.
Lakshmanan, V., Selvaraj, G. and Bais, H. P. (2014), Functional soil microbiome: belowground solutions to an aboveground problem, *Plant Physiol.*, 166, 689–700.
Lareen, A., Burton, F. and Schafer, P. (2016), Plant root-microbe communication in shaping root microbiomes, *Plant Mol. Biol.*, 90, 575–87.
Lehtovirta-Morley, L. E., Stoecker, K., Vilcinskas, A., Prosser, J. I. and Nicol, G. W. (2011), Cultivation of an obligate acidophilic ammonia oxidizer from a nitrifying acid soil, *Proc. Natl. Acad. Sci. U. S. A.*, 108, 15892–7.
Leininger, S., Urich, T., Schloter, M., Schwark, L., Qi, J., Nicol, G. W., Prosser, J. I., Schuster, S. C. and Schleper, C. (2006), Archaea predominate among ammonia-oxidizing prokaryotes in soils, *Nature*, 442, 806–9.
Levine, U. Y., Teal, T. K., Robertson, G. P. and Schmidt, T. M. (2011), Agriculture's impact on microbial diversity and associated fluxes of carbon dioxide and methane, *ISME J.*, 5, 1683–91.
Liang, Y., Li, G., Van Nostrand, J. D., He, Z., Wu, L., Den, Y., Zhang, X. and Zhou, J. (2009), Microarray-based analysis of microbial functional diversity along an oil contamination gradient in oil field, *FEMS Microbiol. Ecol.*, 70, 324–33.
Liesack, W., Schnell, S. and Revsbech, N. P. (2000), Microbiology of flooded rice paddies, *FEMS Microbiol. Rev.*, 24, 625–45.
Lin, Q. and Brookes, P. C. (1999), Comparison of substrate induced respiration, selective inhibition and biovolume measurements of microbial biomass and its community structure in unamended, ryegrass-amended, fumigated and pesticide-treated soils, *Soil Biol. Biochem.*, 31, 1999–2014.
Lin, X., Tfaily, M. M., Green, S. J., Steinweg, J. M., Chanton, P., Imvittaya, A., Chanton, J. P., Cooper, W., Schadt, C. and Kostka, J. E. (2014), Microbial metabolic potential for carbon degradation and nutrient (Nitrogen and Phosphorus) acquisition in an ombrotrophic peatland, *Appl. Environ. Microbiol.*, 80, 3531–40.
Lu, L., Han, W. Y., Zhang, J. B., Wu, Y. C., Wang, B. Z., Lin, X. G., Zhu, J. G., Cai, Z. C. and Jia, Z. J. (2012), Nitrification of archaeal ammonia oxidizers in acid soils is supported by hydrolysis of urea, *ISME J.*, 6, 1978–84.

Lu, L. and Jia, Z. J. (2013), Urease gene-containing Archaea dominate autotrophic ammonia oxidation in two acid soils, *Environ. Microbiol.*, 15, 1795–809.
Macdonald, C., Clark, I., Hirsch, P., Zhao, F. and McGrath, S. (2011a), 'Development of a Real-time PCR assay for detection and quantification of rhizobium leguminosarum bacteria and discrimination between different biovars in Zinc-contaminated soil', *Appl. Environ. Microbiol.*, 77, 4626–33.
Macdonald, C., Clark, I., Zhao, F., Hirsch, P., Singh, B. and McGrath, S. (2011b), Long-term impacts of zinc and copper enriched sewage sludge additions on bacterial, archaeal and fungal communities in arable and grassland soils, *Soil Biol. Biochem.*, 43, 932–41.
Mader, P., Fliessbach, A., Dubois, D., Gunst, L., Fried, P. and Niggli, U. (2002), Soil fertility and biodiversity in organic farming, *Science*, 296, 1694–7.
Martens-Habbena, W., Berube, P. M., Urakawa, H., De La Torre, J. R. and Stahl, D. A. (2009), Ammonia oxidation kinetics determine niche separation of nitrifying Archaea and Bacteria, *Nature*, 461, 976–U234.
Martin, J. P. and Haider, K. (1971), Microbial activity in relation to soil humus formation, *Soil Sci.*, 111, 54–63.
Mauchline, T. H., Kerry, B. R. and Hirsch, P. R. (2002), Quantification in soil and the rhizosphere of the nematophagous fungus Verticillium chlamydospotium by competitive PCR and comparison with selective plating, *Appl. Environ. Microbiol.*, 68, 1846–53.
McGrath, K. C., Thomas-Hall, S. R., Cheng, C. T., Leo, L., Alexa, A., Schmidt, S. and Schenk, P. M. (2008), Isolation and analysis of mRNA from environmental microbial communities, *J. Microbiol. Methods*, 75, 172–6.
McNear Jr., D. H. (2013), The Rhizosphere — roots, soil and everything in between, *Nat. Educ. Knowl.*, 4, 1.
Mendes, R., Kruijt, M., De Bruijn, I., Dekkers, E., Van Der Voort, M., Schneider, J. H. M., Piceno, Y. M., Desantis, T. Z., Andersen, G. L., Bakker, P.a.H. M. and Raaijmakers, J. M. (2011), Deciphering the rhizosphere microbiome for disease-suppressive bacteria, *Science*, 332, 1097–100.
Mendum, T. A., Sockett, R. E. and Hirsch, P. R. (1999), Use of molecular and isotopic techniques to monitor the response of autotrophic ammonia-oxidizing populations of the beta subdivision of the class Proteobacteria in arable soils to nitrogen fertilizer, *Appl. Environ. Microbiol.*, 65, 4155–62.
Mijangos, I., Becerril, J. M., Albizu, I., Epelde, L. and Garbisu, C. (2009), Effects of glyphosate on rhizosphere soil microbial communities under two different plant compositions by cultivation-dependent and -independent methodologies, *Soil Biol. Biochem.*, 41, 505–13.
Moffett, B. F., Nicholson, F. A., Uwakwe, N. C., Chambers, B. J., Harris, J. A. and Hill, T. C. J. (2003), Zinc contamination decreases the bacterial diversity of agricultural soil, *FEMS Microbiol. Ecol.*, 43, 13–19.
Montgomery, D. R. (2007a), *Dirt: The Erosion of Civilizations*. University of California Press.
Montgomery, D. R. (2007b), Soil erosion and agricultural sustainability, *Proc. Natl. Acad. Sci. U. S. A.*, 104, 13268–72.
Morales, S. E., Cosart, T. and Holben, W. E. (2010), Bacterial gene abundances as indicators of greenhouse gas emission in soils, *ISME J.*, 4, 799–808.
Mukherjee, S., Stamatis, D., Bertsch, J., Ovchinnikova, G., Verezemska, O., Isbandi, M., Thomas, A. D., Ali, R., Sharma, K., Kyrpides, N. C. and Reddy, T. B. K. (2017), Genomes OnLine Database (GOLD) v.6: data updates and feature enhancements, *Nucleic Acids Res.*, 45, D446–56.
Nannipieri, P., Ascher, J., Ceccherini, M. T., Landi, L., Pietramellara, G. and Renella, G. (2003), Microbial diversity and soil functions, *Eur. J. Soil Sci.*, 54, 655–70.
Ngumbi, E. and Kloepper, J. (2016), Bacterial-mediated drought tolerance: current and future prospects, *Appl. Soil Ecol.*, 105, 109–25.
Nicholson, F. A., Smith, S. R., Alloway, B. J., Carlton-Smith, C. and Chambers, B. J. (2006), Quantifying heavy metal inputs to agricultural soils in England and Wales, *Water Environ. J.*, 20, 87–95.

Nicol, G. W., Leininger, S., Schleper, C. and Prosser, J. I. (2008), The influence of soil pH on the diversity, abundance and transcriptional activity of ammonia oxidizing archaea and bacteria, *Environ. Microbiol.*, 10, 2966–78.
Noll, M., Matthies, D., Frenzel, P., Derakshani, M. and Liesack, W. (2005), Succession of bacterial community structure and diversity in a paddy soil oxygen gradient, *Environ. Microbiol.*, 7, 382–95.
Nusslein, K. and Tiedje, J. M. (1999), Soil bacterial community shift correlated with change from forest to pasture vegetation in a tropical soil, *Appl. Environ. Microbiol.*, 65, 3622–6.
Ogilvie, L. A., Hirsch, P. R. and Johnston, A. W. B. (2008), Bacterial diversity of the Broadbalk 'classical' winter wheat experiment in relation to long-term fertilizer inputs, *Microb. Ecol.*, 56, 525–37.
Pankhurst, C., Doube, B. M. and Gupta, V. V. S. R. (Eds), (1997), *Biological Indicators of Soil Health.* CABI: Wallingford, UK.
Pawlowski, J., Audic, S., Adl, S., Bass, D., Belbahri, L., Berney, C., Bowser, S. S., Cepicka, I., Decelle, J., Dunthorn, M., Fiore-Donno, A. M., Gile, G. H., Holzmann, M., Jahn, R., Jirku, M., Keeling, P. J., Kostka, M., Kudryavtsev, A., Lara, E., Lukes, J., Mann, D. G., Mitchell, E.a.D., Nitsche, F., Romeralo, M., Saunders, G. W., Simpson, A. G. B., Smirnov, A. V., Spouge, J. L., Stern, R. F., Stoeck, T., Zimmermann, J., Schindel, D. and De Vargas, C. (2012), CBOL protist working group: barcoding eukaryotic richness beyond the animal, plant, and fungal kingdoms, *PLoS Biol.*, 10.
Philippot, L., Hallin, S. and Schloter, M. (2007), Ecology of denitrifying prokaryotes in agricultural soil, *Adv. Agron.*, 96, 249–305.
Philippot, L., Raaijmakers, J. M., Lemanceau, P. and Van Der Putten, W. H. (2013), Going back to the roots: the microbial ecology of the rhizosphere, *Nat. Rev. Microbiol.*, 11, 789–99.
Pinto, A. J., Marcus, D. N., Ijaz, U. Z., Bautista-De Lose Santos, Q. M., Dick, G. J. and Raskin, L. (2016), Metagenomic evidence for the presence of comammox Nitrospira-like bacteria in a drinking water system, *mSphere*, 1.
Poth, M. and Focht, D. D. (1985), ^{15}N Kinetic analysis of N_2O production by *Nitrosomonas europea* - an examination of nitrifier denitrification, *Appl. Environ. Microbiol.*, 49, 1134–41.
Quinton, J. N., Govers, G., Van Oost, K. and Bardgett, R. D. (2010), The impact of agricultural soil erosion on biogeochemical cycling, *Nat. Geosci.*, 3, 311–14.
Ramirez, K. S., Lauber, C. L., Knight, R., Bradford, M. A. and Fierer, N. (2010), Consistent effects of nitrogen fertilization on soil bacterial communities in contrasting systems, *Ecology*, 91, 3463–70.
Ranjard, L., Echairi, A., Nowak, V., Lejon, D. P. H., Nouaim, R. and Chaussod, R. (2006), Field and microcosm experiments to evaluate the effects of agricultural Cu treatment on the density and genetic structure of microbial communities in two different soils, *FEMS Microbiol. Ecol.*, 58, 303–15.
Reeve, J. R., Schadt, C. W., Carpenter-Boggs, L., Kang, S., Zhou, J. and Reganold, J. P. (2010), Effects of soil type and farm management on soil ecological functional genes and microbial activities, *ISME J.*, 4, 1099–107.
Richardson, A. E., Barea, J.-M., Mcneill, A. M. and Prigent-Combaret, C. (2009), Acquisition of phosphorus and nitrogen in the rhizosphere and plant growth promotion by microorganisms, *Plant Soil*, 321, 305–39.
Rousk, J., Baath, E., Brookes, P. C., Lauber, C. L., Lozupone, C., Caporaso, J. G., Knight, R. and Fierer, N. (2010), Soil bacterial and fungal communities across a pH gradient in an arable soil, *ISME J.*, 4, 1340–51.
Sanguin, H., Sarniguet, A., Gazengel, K., Moenne-Loccoz, Y. and Grundmann, G. L. (2009), Rhizosphere bacterial communities associated with disease suppressiveness stages of take-all decline in wheat monoculture, *New Phytol.*, 184, 694–707.
Schauss, K., Focks, A., Leininger, S., Kotzerke, A., Heuer, H., Thiele-Bruhn, S., Sharma, S., Wilke, B.-M., Matthies, M., Smalla, K., Munch, J. C., Amelung, W., Kaupenjohann, M., Schloter, M. and Schleper, C. (2009), Dynamics and functional relevance of ammonia-oxidizing archaea in two agricultural soils, *Environ. Microbiol.*, 11, 446–56.

Singh, B. K., Bardgett, R. D., Smith, P. and Reay, D. S. (2010), Microorganisms and climate change: terrestrial feedbacks and mitigation options, *Nat. Rev. Microbiol.*, 8, 779–90.
Spieck, E., Hartwig, C., Mccormack, I., Maixner, F., Wagner, M., Lipski, A. and Daims, H. (2006), Selective enrichment and molecular characterization of a previously uncultured Nitrospira-like bacterium from activated sludge, *Environ. Microbiol.*, 8, 405–15.
Strickland, M. S. and Rousk, J. (2010), Considering fungal: bacterial dominance in soils - Methods, controls, and ecosystem implications, *Soil Biol. Biochem.*, 42, 1385–95.
Sugiyama, A., Vivanco, J. M., Jayanty, S. S. and Manter, D. K. (2010), Pyrosequencing assessment of soil microbial communities in organic and conventional potato farms, *Plant Dis.*, 94, 1329–35.
Sun, H. Y., Deng, S. P. and Raun, W. R. (2004), Bacterial community structure and diversity in a century-old manure-treated agroecosystem, *Appl. Environ. Microbiol.*, 70, 5868–74.
Tainter, J. A. 1990. *The Collapse of Complex Societies (New studies in Archaeology).* Cambridge University Press.
Teixeira, L. C. R. S., Peixoto, R. S., Cury, J. C., Sul, W. J., Pellizari, V. H., Tiedje, J. and Rosado, A. S. (2010), Bacterial diversity in rhizosphere soil from Antarctic vascular plants of Admiralty Bay, maritime Antarctica, *ISME J.*, 4, 989–1001.
Thomas, T., Gilbert, J. and Meyer, F. (2012), Metagenomics - a guide from sampling to data analysis, *Microb.Inform. Exp.*, 2, 3–3.
Throback, I. N., Johansson, M., Rosenquist, M., Pell, M., Hansson, M. and Hallin, S. (2007), Silver (Ag+) reduces denitrification and induces enrichment of novel nirK genotypes in soil, *FEMS Microbiol. Lett.*, 270, 189–94.
Tilman, D. (2001), Forecasting agriculturally driven global environmental change, *Science*, 292, 281–4.
Tilman, D., Cassman, K. G., Matson, P. A., Naylor, R. and Polasky, S. (2002), Agricultural sustainability and intensive production practices, *Nature*, 418, 671–7.
Tourna, M., Stieglmeier, M., Spang, A., Koenneke, M., Schintlmeister, A., Urich, T., Engel, M., Schloter, M., Wagner, M., Richter, A. and Schleper, C. (2011), Nitrososphaera viennensis, an ammonia oxidizing archaeon from soil, *Proc. Natl. Acad. Sci. U. S. A.*, 108, 8420–5.
Treseder, K. K. and Allen, M. F. (2002), Direct nitrogen and phosphorus limitation of arbuscular mycorrhizal fungi: a model and field test, *New Phytol.*, 155, 507–15.
Valentine, D. L. (2007), Adaptations to energy stress dictate the ecology and evolution of the Archaea, *Nat. Rev. Microbiol.*, 5, 316–23.
Van Der Heijden, M. G. A., Bardgett, R. D. and Van Straalen, N. M. (2008), The unseen majority: soil microbes as drivers of plant diversity and productivity in terrestrial ecosystems, *Ecol. Lett.*, 11, 296–310.
Van Kessel, M.a.H. J., Speth, D. R., Albertsen, M., Nielsen, P. H., Op Den Camp, H. J. M., Kartal, B., Jetten, M. S. M. and Lucker, S. (2015), Complete nitrification by a single microorganism, *Nature*, 528, 555–9.
Wardle, D. A., Bardgett, R. D., Klironomos, J. N., Setala, H., Van Der Putten, W. H. and Wall, D. H. (2004), Ecological linkages between aboveground and belowground biota, *Science*, 304, 1629–33.
Weinert, N., Piceno, Y., Ding, G.-C., Meincke, R., Heuer, H., Berg, G., Schloter, M., Andersen, G. and Smalla, K. (2011), PhyloChip hybridization uncovered an enormous bacterial diversity in the rhizosphere of different potato cultivars: many common and few cultivar-dependent taxa, *FEMS Microbiol. Ecol.*, 75, 497–506.
Wessen, E., Soderstrom, M., Stenberg, M., Bru, D., Hellman, M., Welsh, A., Thomsen, F., Klemedtson, L., Philippot, L. and Hallin, S. (2011), Spatial distribution of ammonia-oxidizing bacteria and archaea across a 44-hectare farm related to ecosystem functioning, *ISME J.*, 5, 1213–25.
Xia, W., Zhang, C., Zeng, X., Feng, Y., Weng, J., Lin, X., Zhu, J., Xiong, Z., Xu, J., Cai, Z. and Jia, Z. (2011), Autotrophic growth of nitrifying community in an agricultural soil, *ISME J.*, 5, 1226–36.
Yin, C., Jones, K. L., Peterson, D. E., Garrett, K. A., Hulbert, S. H. and Paulitz, T. C. (2010), Members of soil bacterial communities sensitive to tillage and crop rotation, *Soil Biol. Biochem.*, 42, 2111–18.

Zhang, L.-M., Hu, H.-W., Shen, J.-P. and He, J.-Z. (2012), Ammonia-oxidizing archaea have more important role than ammonia-oxidizing bacteria in ammonia oxidation of strongly acidic soils, *ISME J.*, 6, 1032–45.
Zhang, Y., Zhang, X., Liu, X., Xiao, Y., Qu, L., Wu, L. and Zhou, J. (2007), Microarray-based analysis of changes in diversity of microbial genes involved in organic carbon decomposition following land use/cover changes, *FEMS Microbiol. Lett.*, 266, 144–51.
Zhao, F. J., Hawkesford, M. J. and McGrath, S. P. (1999), Sulphur assimilation and effects on yield and quality of wheat, *J. Cereal Sci.*, 30, 1–17.
Zuber, S. M. and Villamil, M. B. (2016), Meta-analysis approach to assess effect of tillage on microbial biomass and enzyme activities, *Soil Biol. Biochem.*, 97, 176–87.
Zumft, W. G. (1997), Cell biology and molecular basis of denitrification, *Microbiol. Mol. Biol. Rev.*, 61, 533–616.

Chapter 9

The role of soil fauna in soil health and delivery of ecosystem services[1]

George G. Brown, Elodie da Silva and Marcílio J. Thomazini, Embrapa Forestry, Brazil; Cíntia C. Niva, Embrapa Cerrados, Brazil; Thibaud Decaëns, Université de Montpellier, France; Luís F. N. Cunha, Cardiff University, UK; Herlon S. Nadolny, Wilian C. Demetrio, Alessandra Santos, Talita Ferreira, Lilianne S. Maia, Ana Caroline Conrado, Rodrigo F. Segalla and Alexandre Casadei Ferreira, Universidade Federal do Paraná, Brazil; Amarildo Pasini, Universidade Estadual de Londrina, Brazil; Marie L. C. Bartz and Klaus D. Sautter, Universidade Positivo, Brazil; Samuel W. James, Maharishi University of Management, USA; Dilmar Baretta, Universidade do Estado de Santa Catarina, Brazil; Zaida Inês Antoniolli, Universidade Federal de Santa Maria, Brazil; Maria Jesus Iglesias Briones, Universidad de Vigo, Spain; José Paulo Sousa, University of Coimbra, Portugal; Jörg Römbke, ECT Oekotoxikologie GmbH, Germany; and Patrick Lavelle, Institut de Recherche pour le Développement, France

1. Although some parts were updated and/or expanded, most of the text in this chapter was published previously in Portuguese as a book chapter: Brown, G. G.; Niva, C. C.; Zagatto, M. R. G.; Ferreira, S. A.; Nadolny, H.; Cardoso, G. B. X.; Pasini, A.; Bartz, M. L.; Sautter, K. D.; Thomazini, M. J.; Baretta, D.; Martinez, G. A.; Silva, E.; Antoniolli, Z. I.; Decaens, T.; Lavelle, P.; Sousa, J. P. F. A.; Carvalho, F. 2015. Biodiversidade da fauna do solo e sua contribuição para os serviços ambientais. In: Parron, L. M.; Garcia, J. R.; Oliveira, E. B. De; Brown, G. G.; Prado, R. B. (Eds.), *Serviços ambientais em sistemas agrícolas e florestais do Bioma Mata Atlântica*. 1ed. Embrapa, Brasília, pp. 121–154. Available for free download at: https://ainfo.cnptia.embrapa.br/digital/bitstream/item/131969/1/Livro-Servicos-Ambientais-Embrapa.pdf

http://dx.doi.org/10.19103/AS.2018.0033.11
Published by Burleigh Dodds Science Publishing Limited, 2018.

1 Introduction

There is a wealth of knowledge published on soil fauna, including many books as well as magazine articles and a very large number of scientific papers – using the keywords soil fauna and soil animals retrieves around 4000 and 15000 hits in the Web of Science (as of 9 March 2017). Over the last several decades the amount of available information on soil fauna has grown exponentially (Satchell, 1975, 1992), making it difficult even for professional soil zoologists to stay up to date on current research and latest results in the area.

Although Darwin opened the proverbial 'can of worms' of soil zoology by publishing a very popular book on earthworms at the end of the nineteenth century (Darwin, 1881; Feller et al., 2003), it was not until the mid-1950s that a Soil Zoology Commission of the International Society of Soil Science was officially recognized (Brown and Sautter, 2009), and the sub-discipline began to flourish, with the publishing of some of the first and 'classical' books on soil zoology (e.g. Drift, 1950, 1963; Kevan, 1955; Kunhelt, 1950; Murphy, 1962). Since then, many other very informative and synthetic books have reviewed the state of knowledge on overall soil fauna, or on various taxa, touching on aspects that range from specific issues such as biological functions (e.g. reproduction and feeding) up to general ecology and importance for ecosystem functions and services (e.g. Eisenbeis and Wichard, 1987; Lavelle and Spain, 2001; Wall et al., 2012). Of particular mention are two atlases on soil biodiversity that have been published recently, one on European soils (Jeffrey et al., 2010) and the other at the global scale (Origiazzi et al., 2016).

There is no doubt, today, that soil animals constitute one of the most important groups of organisms for the functioning of the world's terrestrial ecosystems, and that their presence, activity and biodiversity are in need of attention not only by the land user, but also by planning, management and conservation agencies (Nielsen et al., 2015; Wall et al., 2007). In this chapter, we provide ample evidence for these statements and synthesize the currently available information on soil fauna biodiversity and functional roles worldwide, with a particular emphasis on soil macroinvertebrates, providing also information on the main functions of soil animals, and especially their role in soil health and in the delivery of ecosystem services. We also briefly cover the importance of soil and ecosystem management strategies in determining soil fauna biodiversity and activity, and end the chapter by exploring some of the means by which contributions of soil fauna to ecosystem services can be measured and valued by society.

2 Soil fauna biodiversity

The soil, much more than a mere substrate for plant growth and food production, is a living entity, containing thousands of animals and millions of microorganisms. This biota forms complex trophic webs, at the base of which normally are plant roots, litter and soil organic matter. The soil invertebrates vary in size from a few micrometres (microfauna) or millimetres (mesofauna), up to many centimetres in length (macrofauna), as well as some vertebrates (megafauna), with life cycles that can vary from a few days up to more than ten years.

In Table 1 we present the spectrum of organisms in each of the four main size classes of soil fauna and their species richness worldwide. In total, the soil fauna may represent 1 million known species, although this number is slightly inflated because some groups have species that live in aquatic environments or above the soil surface, for example, in trees

Published by Burleigh Dodds Science Publishing Limited, 2018.

Table 1 Taxa representative of the soil micro-, meso-, macro- and megafauna; the number of described or estimated species richness worldwide; and the functional groups of the soil-associated species (expanded, corrected and updated from Brown et al., 2015, and compiled from many sources; for further information, see notes at end of table)

			Functional group			
Size class (width) Phylum, subphylum, class, order, family[a]	Common name	Species richness	Geophage/ bioturbator	Detritivore/ decomposer	Phytophage/ pest	Predator/ parasite
Microfauna (<0.1 mm)						
Nematoda	Nematodes, roundworms	24 773[b]		X	X	X
Rotifera	Rotifers, wheel animals	2030[b]		X		X
Tardigrada	Water bears, moss piglets	1157[b]		X		X
Subtotal microfauna		27 960				
Mesofauna (0.1–2 mm)						
Annelida: Enchytraeidae	Potworms	718[b]	X	X		
Chelicerata						
Acari	Mites, ticks	54 617[c]		X	X	X
Palpigradi	Microwhip scorpions	82				
Pseudoscorpionida	Pseudoscorpions	3454		X		X
Hexapoda						
Collembola	Springtails	8600		X	X	
Diplura	Two-pronged bristle tails	800		X		X
Protura	Coneheads	804		X		
Myriapoda						
Pauropoda		835		X		X
Symphyla	Garden centipedes	200		X	X	X
Subtotal mesofauna		70 110				
Macrofauna (invertebrates >2 mm)						
Annelida: Oligochaeta	Earthworms	3358[c,d]	X	X		
Chelicerata	Arachnids					

(*Continued*)

Table 1 (*Continued*)

Size class (width) Phylum, subphylum, class, order, family[a]	Common name	Species richness	Functional group: Geophage/ bioturbator	Detritivore/ decomposer	Phytophage/ pest	Predator/ parasite
Amblypygi	Whip spiders, tailless whip scorpions	170	X			X
Araneae	Spiders	43 579[c]	X		X	X
Opiliones	Harvestmen, daddy longlegs	6519[c]	X			X
Ricinulei	Hooded tick spiders	58	X			X
Schizomida and Thelyphonida	Whip scorpions, vinegaroons	383[c]	X			X
Scorpiones	Scorpions	1947[c]	X		X	X
Solifugae	Camel spiders, wind scorpions, sun spiders, solifuges	1113	X			X
Crustacea: Malacostracea						
Amphipoda: Talitridae	Sand fleas, sandhoppers	268[b]	X	X		
Isopoda	Pill-bugs, sowbugs, woodlice	4250[b]	X	X	X	
Hexapoda						
Archaeognatha and Zygentoma	Bristletails	1074[c]		X	X	
Blattodea[e]	Cockroaches	4622[c]	X	X	X	
Blattodea: Termitoidea[e]	Termites	2800[c]	X	X	X	
Coleoptera	Beetles	387 100[c,f]	X	X	X	X
Carabidae	Ground beetles	40 000[c]	X			X
Cerambycidae	Long-horn beetles	35 000[c]		X		
Chrysomelidae	Leaf beetles	36 500[c]	X	X	X	
Elateridae	Wireworms, click beetles	10 000[c]	X	X	X	X
Histeridae	Clown beetles	4300[c]	X	X		X
Scarabaeidae	White grubs, scarabs, Japanese beetles, dung beetles, June beetles, rose chafers, rhinoceros beetles	27 000[c]	X	X	X	

Staphylinidae	Rove beetles	56 000[c]	X	X		X
Tenebrionidae	Darkling beetles	20 000[c]	X		X	
Dermaptera	Earwigs	1978[c]	X	X	X	X
Diptera	Flies, maggots	159 294[c]	X	X	X	X
Hemiptera	Stink bugs, cicadas	103 590[c]	X	X	X	X
Hymenoptera: Formicidae	Ants	13 235[c]	X	X	X	X
Vespidae and Apidae	Wasps; solitary, stingless and ground-nesting bees	28 000[c]	X			X
Lepidoptera	Moths, butterflies	157 424[g]			X	X
Neuroptera: Myrmeleontidae	Ant lions	5868	X			X
Orthoptera: Grylloidea	Crickets	5200[c]	X	X	X	
Psocoptera	Booklice	5720[c]		X	X	
Mollusca						
Gastropoda	Snails	30 000	X	X	X	
Gastropoda	Slugs	5000	X	X	X	
Myriapoda						
Diplopoda	Millipedes	12 116[c]	X	X	X	
Chilopoda	Centipedes	3100[c]	X			X
Nematoda: Mermithidae[h]		599[b]				X
Nematomorpha[h]	Hairworms	351[b]				X
Platyhelminthes: 'Turbellaria'	Planarians	830[c]				X
Onychophora	Velvet worms	179				X
Subtotal macrofauna		988 895				
Megafauna (vertebrates >20 mm)						
Amphibia						
Caudata: Salamandridae	Salamanders	608[b,c]	X			X
Gymnophiona: Caeciliidae	Caecilians	189[c]	X			X

(Continued)

Table 1 (*Continued*)

Size class (width) Phylum, subphylum, class, order, family[a]	Common name	Species richness	Functional group: Geophage/ bioturbator	Detritivore/ decomposer	Phytophage/ pest	Predator/ parasite
Mammalia						
Rodentia	Pocket gophers, mole rats	250[c]	X		X	
Soricidae	Shrews	385[c]	X		X	X
Talpidae	Moles	46	X		X	X
Reptilia						
Amphisbenidae	Amphisbaenians, worm lizards	183[c]	X			X
Typhlopidae, Leptotyphlopidae, Anomalepididae	Blind snakes	378[c]	X			X
Subtotal megafauna		2039				
Total soil fauna		**1 089 000**				

Notes:
[a] Within each size class, animals are listed in alphabetical order, within each taxonomic classification level.
[b] The number of species indicated includes terrestrial and aquatic species.
[c] The number of species indicated includes those that live in the soil, litter and above ground.
[d] Although over 6000 species of earthworms have been described (Czusdi, 2012), the number presented is of the valid species according to DriloBASE (http://taxo.drilobase.org).
[e] This order now includes both cockroaches and termites, which were discovered to be eusocial cockroaches (Inward et al., 2007).
[f] The number of species includes all families; the main families with life cycle phases associated with the soil are detailed in the following lines.
[g] The number given is for all species of Lepidoptera; however, most are active only above-ground and only a small proportion of species have a life cycle phase associated with soil or litter. For instance, many species put their pupae in litter and butterflies of the Lycaenidae family, that includes >5000 species developed a symbiosis with ants (Pierce et al., 2002).
[h] These nematodes and nematode-like worms are animal (mostly insect) parasites and have a stage of their life cycle that is developed in the host and another that is free living, either aquatic and/or terrestrial.
Sources: for invertebrates, mainly from Zhang (2011a,b), but also Eisenbeis and Wichard (1987); Burakowski (1993); Schockaert et al. (2008), Brewer et al. (2012), Schmelz and Collado (2015), Bolton (2017), Orgiazzi et al. (2016), Beccaloni (2017); Bellinger et al. (2017); for vertebrates Pyron and Wiens (2011), Zhang (2011a), Pyron and Burbrink (2012), Pyron et al. (2013), Begal et al. (2007).

or bromeliads. Nevertheless, relatively few species of soil animals have had their biology or ecology thoroughly studied up to now, and this lack of knowledge is exacerbated by the lack of active taxonomists in many groups and taxa of soil animals (particularly the micro- and meso-invertebrates), creating what has been called the taxonomic impediment (Evenhuis, 2007; Hoagland, 1996).

Due to the high specific richness present in many taxa, performing an inventory of the soil fauna diversity in a given location can be an arduous and complex task. For instance, after many years of research, over 1000 invertebrate species m^{-2} were identified from a temperate forest in Germany, and this is probably the only place in the world for which such a wide taxonomic effort (many soil invertebrate taxa) has been performed (Beck et al., 1988). Although taxonomic richness may surpass 2200 species in some hectares of Amazonian rainforest (Mathieu, 2004), up to now a complete inventory of soil invertebrates has not been performed in any tropical site. In general, in a given site, only a few species (generally <20) will be earthworms (Fragoso and Lavelle, 1992) or pseudoscorpions (Franklin et al., 2008), while spiders and myriapods will be represented by several dozen species each (Adis, 2002). However, the vast majority of invertebrate species at a particular site will probably be nematodes (Huang and Cares, 2006; Nielsen et al., 2014), mites (Franklin et al., 2008) and insects (Barros et al., 2006), representing each over 100 species (being the Hexapoda mostly represented by beetle and ant species), since these groups normally have high local species richness (α diversity).

3 Soil fauna functional groups and classification

The reduced number of taxonomists specialized in soil invertebrates has many times been a hindrance to the identification of soil fauna species in biodiversity inventories and in ecological studies, particularly in the tropics and subtropics. To help overcome this deficit, ecologists frequently use the concept of functional groups in order to help describe soil fauna communities and interpret their importance for soil processes (Brussaard, 2012). The main functional groups of soil fauna are predators/parasites, detritivores/decomposers, geophages/bioturbators and phytophages/pests. In Table 1, we present the functional groups of all the main representatives of the soil micro-, meso-, macro- and megafauna.

The phytophages (including rhizophages) cause damage to plant shoots and roots and can be considered pests on many occasions. This group includes mainly the plant parasitic nematodes, cicadas, a few ants (e.g. leaf cutters, that despite not eating plants, cut them to feed fungal colonies), earwigs as well as some millipedes, beetles (especially scarabs, tenebrionids and crysomelids), mites, molluscs (snails and slugs), crickets, woodlice, springtails and symphylans. Other animals are considered mainly urban pests such as termites, cockroaches, scorpions, spiders, moths, ants, molluscs, fly larvae, some beetles, woodlice and millipedes.

The geophages and bioturbators ingest and/or transport soil, moving it within the profile. This group includes only a few smaller animals (e.g. enchytraeids) and soil macroinvertebrates that eat soil as their main sources of food (termites and earthworms, but also some beetles and millipedes; Brown et al., 2001b; Gassen, 2000; Toyota et al., 2006), and those that burrow into the soil creating galleries, tunnels and/or nests (ants, ant lions, many other insects and arachnids, and all the megafauna).

The predators and parasites negatively affect the life or populations of other animals and include a large gamut of soil and litter animals, especially those that hunt in the litter such as spiders (Chelicerata), velvet worms, planarians, beetles (especially staphylinids, carabids and elaterids), mites, ants, wasps, centipedes and earwigs. In particular, nematodes and nematode-like worms parasite a large number of invertebrate and vertebrate animals, including humans and insects (e.g. Nematomorpha: Mermithidae).

The detritivores and decomposers can be coprophagous (feeding on faeces), necrophagous (feeding on cadavers), or bacterial grazers, fungivorous and detritivorous (feeding on fungi or litter in various stages of decomposition). Among the most common coprophages are the millipedes, woodlice, dung beetles and earthworms. Necrophages include mainly some beetles (e.g. Histeridae) and fly larvae, while the detritivores include almost all soil and litter animals, excluding a few predators, parasites and pests (Table 1).

As mentioned earlier, soil fauna can also be classified into micro-, meso-, macro- and megafauna, according to their individual size (body length or diameter), as originally proposed by Dunger (1964), Wallwork (1970) and Swift et al. (1979). Width (diameter) seems to be a better parameter than length, which is much more variable within each category (Orgiazzi et al., 2016). Nevertheless, the size ranges of each category are not exact for all its representative members, often leading to confusion, so that some animals (e.g. beetles, ants and myriapods) are frequently placed both in the macro- and mesofauna, depending on their size. Hence, a group of specialists meeting under the auspices of the International Biodiversity Observation Year in Paris in 2000 proposed that the soil macrofauna be defined as 'an invertebrate group found within terrestrial soil samples which has more than 90% of its specimens (individuals) in such samples visible to the naked eye' (Ruiz et al., 2008).

The advantage of the size classification for soil fauna is that it roughly correlates with some of the main functional roles of these animals in the environment (Table 2), and because it is relatively easy to use. However, it still represents a coarse grouping of the soil fauna and does not consider the specificity of soil functions of different animals and their physiological variability, or their relative importance to soil processes over space and time (Briones, 2014). Nevertheless, this classification system remains one of the most widely used and found in the literature. Therefore, in the following paragraphs, we explore each of these main size groups of the soil fauna and discuss briefly their functional roles in the soil ecosystem.

3.1 Soil microfauna

The soil microfauna (Table 1) consists of microscopic invertebrates with body diameter <0.1 mm, and includes the nematodes, rotifers and water bears that live within soil water films. For many years the soil protozoa were considered as part of the soil microfauna (e.g. Coleman et al., 2004), but these eukaryotic microbes now belong to a separate kingdom (Protista) and are no longer part of the soil microfauna. However, they are still part of the soil microbiota, together with other soil microorganisms, such as bacteria, fungi and actinomycetes.

The soil microfauna have rapid life cycles and act mainly as micropredators or parasites (Table 2), feeding on other animals, plant roots and microorganisms (bacteria, protozoa, fungi, algae, actinomycetes). The rotifers and water bears are fascinating animals that are relatively little studied in the soil (mostly so in aquatic environments), and together with the free-living nematodes, act mainly as stimulators of nutrient cycling and mineralization, and in the control of soil microbial populations (Lavelle, 1997).

Published by Burleigh Dodds Science Publishing Limited, 2018.

Table 2 Effects of soil fauna size classes and functional groups on soil processes, biodiversity and plant production (modified and expanded from Lavelle, 1996)

Size class functional group	Effects on soil processes	Effects on biodiversity	Effects on plant/crop production
Macro and megafauna			
Ecosystem engineers	Bioturbation (regulation of soil physical properties and processes), changes in dynamics of soil organic matter, nutrient cycling and soil biological activity	Selection pressures on litter transformers and microbiota, mutualisms with microbiota, dissemination of organisms	Positive or negative, mainly indirect effects on root and shoot biomass and seed banks
Macro and mesofauna			
Litter transformers	Nutrient mineralization, organic matter protection and decomposition, some bioturbation through production of faecal pellets and small biopores	Selection pressures on litter microbiota	Positive (indirectly) through enhancement of nutrient availability and reduction in plant pathogens living in litter
Predators	Nutrient release and biological control	Selection pressures on microbiota and changes in abundance of beneficial and/or adverse organisms	Mostly indirect effects through changes in plant nutrient availability or populations of plant parasites or pathogens
Microfauna			
Micropredatory food web	Enhanced nutrient mineralization can affect soil structure through interaction with soil microbiota	Selection pressures on microbiota and changes in abundance of beneficial and/or adverse organisms	Mostly indirect effects through changes in plant nutrient availability or populations of plant parasites or pathogens
Mega, macro, meso and microfauna			
Phytophages and plant parasites	Some bioturbation (only macro- and megafauna)	Selection pressures on plants	Negative direct effects via herbivory (above- and below-ground plant parts), plant disease or parasitism

The nematodes, with 500 thousand to over one million estimated species worldwide (Groombridge, 1992), are the most abundant animals on the face of the Earth. Although various classification systems are available (Bongers and Bongers, 1998), nematodes can be roughly divided into five main functional groups, based mainly on feeding strategies (omnivorous, fungivorous, bacterivorous, plant parasites, animal parasites, including entomopathogenic species), that form an integral part of the soil food chain, and are very important for plant, soil and human health (Neher, 2001; Wall et al., 2015). The entomopathogenic mermithid nematodes as well as the Nematomorpha are often considered part of the soil macrofauna, due to their large size (up to several cm long). Historically, the plant and animal parasitic species have received priority attention due to their economic importance in the world's terrestrial and aquatic ecosystems (Yeates, 1979), but the free-living nematodes are also both ecologically and agronomically important in soils (Neher, 2001, 2010).

3.2 Soil mesofauna

The soil mesofauna (Table 1) are invertebrates with body width between 0.1 and 2 mm, and includes mainly the mites (Acari), springtails (Collembola), diplurans, proturans, enchytraeids, symphylans, little arachnids such as pseudoscorpions and other animals (e.g. some micro-coleopterans, very small ants, insect larvae and some animals that normally belong to the macrofauna are sometimes reported as mesofauna) that feed mainly on decaying organic matter or fungi or are predators of other smaller organisms (e.g. nematodes and protozoa). They participate actively in litter breakdown, increasing surface area contact for microbial attack, enhancing decomposition rates and releasing nutrients into the soil (Table 2).

Mites and springtails generally dominate in terms of abundance and diversity, the mites being more diverse, with >54 thousand species known worldwide (Table 1; Zhang, 2011a). The mites also have a large functional diversity (Walter and Proctor, 2013), although the most common soil/litter groups are detritivores, fungivores and omnivores, while predators tend to be less common. Predatory mites have a major role in controlling populations of other soil animals, mainly the microbiota, but also mesofaunal groups (e.g. enchytraeids and other mites). Springtails are less diverse than mites (approx. 8600 species worldwide; Bellinger et al., 2017), but exert significant detritivorous role in soils, contributing to organic matter decomposition and the control of fungal populations (Franklin and Moraes, 2006). They can be classified according to their distribution in the soil and litter into epiedaphic (generally large and pigmented surface litter dwellers), euedaphic (generally small and unpigmented topsoil and humic horizon dwellers) and hemiedaphic (intermediate between the two former types) species (Lavelle and Spain, 2001).

Of the 718 known species of enchytraeids (Schmelz and Collado, 2015) most are from Europe, where they have been relatively well studied compared to the rest of the world. These microbivores and saprovores (Didden, 1993) play an important role in organic matter decomposition and mineralization (Hendrix et al., 1986; Briones and Ineson, 2002), but also affect soil structure through burrowing, ingesting soil and producing faecal pellets (Didden, 1993; Van Vliet et al., 1993).

3.3 Soil macrofauna

The soil macrofauna are invertebrates with body width greater than 2 mm and includes over 20 taxonomic groups, especially the larger taxa commonly known as ants, termites, cockroaches, earwigs, crickets, cicada nymphs, beetles, fly larvae, millipedes, centipedes, snails, earthworms, woodlice, slugs, spiders, scorpions and other animals that in total amount to around 989 thousand species (Table 1; but total value includes many aquatic species as well). These animals can be omnivores, soil ingesters (geophages), or consume litter (detritivores), humified organic matter (humivores), plants (phytophages), wood (xylophages), roots (rhizophages), bacteria (bacterivores), fungi (fungivores) as well as animal tissues and fluids (predators, parasites, necrophages) (Brown et al., 2001a).

Among the soil macrofauna, ants, termites, earthworms and some scarab beetle larvae and millipede species have been called 'ecosystem engineers', because their activities lead to the production of biogenic structures (galleries, tunnels, nests, chambers, faecal pellets or casts) that modify soil physical properties and/or create habitats and change the availability of resources to other organisms (Brown et al., 2001a; Lavelle et al., 1997; Toyota et al., 2006; Table 2). In the following paragraphs, we present further details on these engineers, giving evidence of their contribution to various important soil processes and soil health.

Ants and termites are social insects that live in colonies that can number from a few up to more than one million individuals (Hölldobler and Wilson, 1990; Lee and Wood, 1971). Ants are highly diversified (13 235 valid species worldwide; Bolton, 2016), and have various ecological functions, acting not only as ecosystem engineers but also as fungus growers, leaf cutters, wood and soil organic matter decomposers and important predators of other organisms (Folgarait, 1998). However, unlike termites, ants do not ingest soil, they only transport it in their mandibles when constructing their nests. However, because they feed on other invertebrates, as well as on aphid secretions, they can affect the availability of organic materials and certain nutrients such as phosphorus in soils (Frouz and Jilková, 2008).

The termites include approximately 2800 species worldwide that feed on cellulosic materials, accelerating decomposition and recycling of the mineral nutrients from dead plants (Constantino, 2006). They build extensive nests including networks of chambers and tunnels in the soil, moving particles vertically and horizontally, creating aggregates and increasing soil porosity, aeration and water infiltration in the soil (Lavelle and Spain, 2001).

Both termites and some ants have developed a special symbiosis with protozoa and N_2 fixing bacteria in their guts (or fungus gardens), to compensate for the high C/N ratio of their diets (Bignell, 2000). Due to their importance as pests, their biology, behaviour and taxonomy have been relatively well studied, although important ecological questions still remain to be answered, particularly for tropical species.

Coleoptera is the largest insect order, including at least 100 families (and dozens of thousands of species; Table 1) that present a myriad of feeding habits such as rhizophagy, fungivory, frugivory, detritivory, necrophagy and predation (Costa et al., 1988; Burakowski, 1993). Because they use a variety of food sources and often burrow into the soil, they can also be important bioturbators, influencing soil porosity, organic matter redistribution in soil and the population of other ecosystem engineers such as earthworms, millipedes and ants (Brown et al., 2001b). Among the most important groups for soil processes are the dung beetles (Scarabaeidae), which by ingesting and burying faeces have major impacts on nutrient cycling, bioturbation, plant growth, seed dispersal and parasite control, as well as a smaller role on pollination and trophic food web regulation (Nichols et al., 2008).

The millipedes include about 12 000 species worldwide (Table 1; Brewer et al., 2012). Unfortunately, little is known about the biology and ecology of many species. Although most of the described species are considered to be litter transformers (sensu Lavelle, 1997), a few species are coprophagous and others are plant feeders, causing damage to plant seedlings and hence, considered as pests (e.g. soybean, manioc, sweet potato; Boock and Lordello, 1952; Ebregt et al., 2005; Hoffmann-Campo et al., 2013; Peracchi and Nunes, 1972). In addition, the Japanese train millipede (*Parafontaria laminata*) was shown to have major effects on bioturbation, affecting physical, chemical and biological properties of the soil (Toyota et al., 2006).

Earthworms are widely known, not only because of their widespread use as fish bait, but also because of their benefits to soil fertility. They mix soil and organic matter in their casts and create extensive networks of tunnels in the soil. Furthermore, some epigeic species can be used in composting fresh organic residues (including plant litter and animal manures) which results in a highly valuable organic fertilizer vernacularly called 'vermicompost', that is used extensively in seedling production (Lazcano and Domínguez, 2011). Around 3358 species are known worldwide, although cryptic diversity appears to be common in some of the more well-known and widely distributed species (Martinsson et al., 2015), while the biology and ecology of most native species, particularly in the tropics, remains mostly unknown (Brown and James, 2007), with the majority of the studies only addressing a few widespread or exotic species (Lowe and Butt, 2005).

3.4 Soil megafauna

Vertebrates living in the soil are considered megafauna by most authors (e.g. Gobat et al., 2004; Hole, 1981; Orgiazzi et al., 2016). This category includes around 2000 species of moles, shrews, fossorial rodents (mole rats, pocket gophers), amphibians (salamanders and caecilians) and reptiles (blind snakes and amphisbaenians) that burrow through the soil searching for food and habitation. Temporary burrowers such as chipmunks, burrowing owls, tortoises, rabbits, badgers, hedgehogs, foxes, marmots and armadillos, among others, are normally not considered part of the soil megafauna *sensu stricto*, although their importance for soil processes, particularly bioturbation or predation on soil animals is undeniable (Hole, 1981; Meadows and Meadows, 1992; Wilkinson et al., 2009).

All members of the soil megafauna are burrowers, and several of them are also ecosystem engineers, resulting in important physical changes to the soil and its plant, animal and microbial communities (Table 2). These soil vertebrates are major predators of earthworms, termites, ants, insect larvae (e.g. beetle grubs and fly larvae), snails, centipedes and isopods. Pocket gophers and mole rats, on the other hand, are herbivores that can have major impact on plant communities (Hagenah and Bennett, 2013; Huntly and Inouye, 1988; Lacey et al., 2000). The faeces and carcasses of soil vertebrates are a high-quality source of nutrients and energy for invertebrates and microorganisms in the soil and these animals play a very important role in the soil food web, generally acting as top predators (Orgiazzi et al., 2016). Furthermore, they are also an important food source for other, above-ground vertebrate predators, higher up in the food chain.

4 Role of soil fauna in delivering ecosystem services

Ecosystem services are the benefits that people obtain from ecosystems (MEA, 2005). They include four main types of services: provisioning (products obtained from ecosystems), cultural (non-material benefits obtained from ecosystems), regulating (benefits obtained from regulation of ecosystem processes) and supporting services (services necessary for the production of all other ecosystem services).

All the services identified by the Millennium Ecosystem Assessment (MEA, 2005) are affected by the soil fauna (Lavelle et al., 2006; Vandewalle et al., 2008) (Table 3). However, their effects are mainly indirect, resulting from feeding and burrowing activities (Table 2). Hence, soil animals can affect a wide range of soil processes that contribute directly and indirectly to soil health and essential provisioning services such as food, fuel and fibre production (through effects on plant growth). By changes in soil physical structure and hydraulic regime, they support services such as soil formation and biodiversity conservation (through changes in the soil as a habitat for other organisms). By processing soil and organic residues, they also support nutrient cycling and organic matter decomposition, affecting C sequestration and gas exchanges between the soil and the atmosphere (including greenhouse gas emissions), as well as residue and contaminant removal (through decomposition of residues or pesticide degradation), the latter being a regulating service. Through burrowing and soil mixing, they also regulate seed dispersal, pests and diseases, while changes in soil structure contribute to the regulation of soil water availability, infiltration and soil erosion. Finally, soil fauna also have important recreational (for hobbies such as collections or use of soil animals for fish bait) and educational values (cultural services; Table 3), and can be used for the development of pharmaceutical products (derived from the organisms or their sub-products) or directly consumed by human beings, pets or other animals (provisioning services). In the following sections we review the main contributions of soil fauna to each of the four categories of ecosystem services (provisioning, cultural, regulating and supporting).

4.1 Provisioning services

4.1.1 Direct use of soil animals

Several insects and other invertebrates can be used as human food or animal feed. Among these ants, termites, crickets, earthworms, molluscs, spiders, scorpions and beetle larvae are commonly consumed (Paoletti, 2004; Srivastava et al., 2009). Although despised by 'modern', mainly urban societies, these animals represent an important source of nutrients, proteins and amino acids and are currently eaten by various traditional cultures, especially in tropical and subtropical countries (Defoliart, 1997, 1999). Furthermore, some animals (for instance, earthworms, enchytraeids, maggots and beetle larvae) are used as fish bait or to feed fish in commercial fisheries or in aquaria. Moreover, some soil fauna (e.g. ants, earthworms and millipedes) have been used as sources of pharmaceutical or cosmetic products, or even explored for their potential use in biomedicine or in popular or traditional, indigenous medicine (Dossey, 2010; Ortiz et al., 1999; Zhenjun, 2003). Among these are anti-coagulants, anti-histamines, antibiotics, antiviral products, aphrodisiac substances or birth control products.

4.1.2 Food, fibre and fuel production

Soil animals can affect plant growth (primary production) positively and/or negatively both directly and/or indirectly. The direct positive effects of soil animals on plants are little known and could involve liberation of signal molecules (like indole-acetic acid (IAA), a plant growth-promoting hormone), produced mainly by rhizosphere microorganisms (Blouin et al., 2013). The most well-known example of positive effects of soil fauna on crop production is that of the earthworms (see reviews by Brown et al., 1999, 2004; Scheu, 2003;

Table 3 The main ecosystem services of the Millennium Ecosystem Assessment (MEA, 2005), the ecosystem processes associated with them, and the direct or indirect contribution of soil fauna to these processes. Modified and expanded from Lavelle et al. (2006), with data from Blouin et al. (2013), Decaëns et al. (2006, 2008), Del Toro et al. (2012) and Prather et al. (2013)

Category of service	Ecosystem service	Soil ecosystem-related process	Soil fauna contribution
Provisioning	Water availability	Soil water storage and infiltration into soil	Bioturbation, production of stable structures (casts, aggregates and galeries) that increase water infiltration and retention in soil
	Pharmaceutical products	Production of secondary molecules for medicinal use	Sources of biomolecules and pharmaceuticals of traditional and industrial interest
	Food	Animal biomass production	Food for human beings and animals
	Food, fibre and fuel production	Stimulation of the activity of symbionts and microorganisms involved in plant growth promotion	Selective stimulation of rhizosphere symbionts in the guts and biogenic aggregates (e.g. casts and burrows), changes in the activity of plant growth-promoting microbiota
		Production of signal molecules	Stimulation of the production of signal molecules by microorganisms
		Plant production	Consumption of living plant parts (rhizophagy, phytophagy), or increase in vigour, growth and/or productivity
Supporting	Biodiversity conservation	Maintenance of a biodiverse community	Change in the soil as a habitat for fauna, flora and microorganisms
	Soil formation	Pedogenesis	Bioturbation, particle selection, surface cast deposition, increase in soil formation rates
	Nutrient cycling	Decomposition and humification	Fragmentation, ingestion, stimulation of the microbial decomposer community
		Nutrient loss regulation	Mineralization, alteration of leaching rates, denitrification, N_2O emission, nitrification and enzyme activity in soils
Regulating	Pollination	Pollination	Pollinating insects that have a stage of their life cycle in the soil
	Seed dispersal	Seed dispersal	Transport, consumption and/or excretion of seeds with changed viability and germination potential

	Biological control	Protection against pests and diseases	Increase in the control of pests and diseases directly or indirectly (increasing plant fitness)
	Erosion and flood control	Run-off regulation	Formation of rugosity and deposition of surface casts that change water run-off and erosion rates
		Water storage and infiltration into the soil	Changes in soil structure, increasing porosity, and production of surface open biopores that enhance infiltration
	Climate regulation	Production/consumption of greenhouse gases	Changes in the rates of greenhouse gas emissions, especially N_2O (by denitrification) and CO_2 (by respiration/decomposition)
		C sequestration	Formation of stable aggregates and resistant humic substances
	Residue treatment	Detoxification and decomposition of organic residues	Enhancement of decomposition and stabilization of organic residues and absorption/desorption of pesticides and other organic compounds; alteration in the availability of heavy metals
Cultural	Recreation	NA	Food for fish, birds and other organisms, keeping as pets or collections as a hobby, themes for exhibits, art, literature, video games and movies
	Education	NA	Instrument for environmental education and for improving management of soils and crops

van Groenigen et al., 2016), although other soil animals can also have positive effects on plant growth (Edwards and Lofty, 1978; Setälä and Huhta, 1991). When soil fauna have a negative effect on plant production this is defined as environmental disservice. Some examples are phytophagy or rhizophagy that decrease plant biomass, although low levels of herbivore pressure or rhizophagy may result in compensatory growth, increasing plant production (Curry, 1987). A list of the phytophagous/rhizophagous soil animals can be seen in Table 1, and includes species belonging to a wide range of taxa.

Most of the effects of soil fauna on plants are indirect and involve changes in environmental conditions considered to be important for plant growth. Soil pH, aluminium toxicity, availability of essential plant nutrients (e.g. N, P, Ca and Mg), pest and disease pressure, competition, activity of plant growth-promoting microorganisms (including symbionts), soil aeration and the availability of water in the root zone, among others, can be greatly affected by soil fauna, altering plant nutrient uptake and growth rates (Lavelle et al., 1992). The activity of decomposers and mineralizers alters plant nutrient availability and soil C contents, affecting microbial activity as well as cation exchange capacity and aggregation, which are also important factors for plant growth. On the other hand, bioturbators mainly affect soil physical conditions for plant roots, but in some cases they can also affect nutrient and water availability.

4.2 Cultural services

Many soil animals can be and have been used as models for environmental education at various levels, from children up to adults. For instance, springtails, earthworms and enchytraeids (among many other soil animals) have been used by farmers, land managers or extension agents to assess soil quality (Bartz et al., 2013; Ernst, 1995; Jänsch et al., 2005; Pulleman et al., 2012; Rombke et al., 2005), or for practical exercises in schools, both in the laboratory and in the field, to showcase the importance of soil animals for soil structuring, or leaf litter and organic residue decomposition (e.g. household organic wastes: Appelhof et al., 1997; Hoffman, 1994; McLaughlin, 1986). Permanent or itinerant exhibitions focusing on the underground world and soil life have been successful in Europe and the United States, such as the 'Underground adventure' at the Field Museum of Natural History, Chicago (Field Museum, 2017), and 'The thin skin of earth – our soils' touring exhibit of the Senckenberg Museum in Germany (Senckenberg Museum, 2017). These events are very popular and show the potential for conscience building of the general public regarding soil, soil fauna and ecosystem services.

Finally, insect collections, or keeping invertebrates such as ants, earthworms, crickets, scorpions or spiders at home are very popular hobbies. Furthermore, soil invertebrates have been widely used as inspiration for artwork (e.g. the ancient Egyptians revered scarab beetles and scorpions; Kevan, 1985), crafts, educational literature (e.g. children's books on various insects, arachnids and worms), satires, comic books (e.g. Larson, 1998) and even movies (*Antz, The Ant Bully, Empire of the Ants, Marabunta, Arachnophobia, A Bug's Life, Miniscule, Spiderman,* among others), documentaries (e.g. *Creatures of the Undergrowth* and *Worm Hunters*) and video games!

4.3 Regulating services

4.3.1 Pollination

Only a few members of the soil fauna act as pollinators, and these are mainly arthropods like bees, flies, moths, butterflies, beetles and ants, that live part of their life cycle in close proximity to plants. Some bee species such as the bumblebees and stingless bees are important pollinators in agricultural and forestry systems (Freitas and Pereira, 2004). Although restricted to just some plant species, the scarab beetles can also be important and often obligatory pollinators of Araceae and Lowiacea plants (Nichols et al., 2008). Ants frequently visit flowers, but they pollinate only some plant species (Garcia et al., 1995; Gómez and Zamora, 1992; Peakall et al., 1991; Puterbaugh, 1998; Rico-Gray and Oliveira, 2007).

4.3.2 Seed dispersal

Several members of the soil fauna can be important primary or secondary seed dispersers. For instance, around 3000 plant species from more than 80 families from different ecosystems in the world produce seeds ready for ant transport (myrmecochory), and it has been estimated that at least 11 000 flowering plant species depend on ant dispersal (Lengyel et al., 2009). In the Brazilian dry Caatinga this dispersal is crucial, and more frequently in woody Euphorbiaceae species (Leal et al., 2007). Nonetheless, even plants that do not have elaiosomes (external seed structures, rich in nutrients) can often be transported by granivorous ants.

As mentioned earlier, coprophagous beetles have an important role in secondary seed dispersal by moving and burying animal faeces containing seeds. The relocation can be both vertical and horizontal. Hence, these seeds can have higher survival through less consumption and a more favourable soil microclimate (that enhances germination and emergence), and by reduction in seed agglutination and seedling competition (Nichols et al., 2008). Seeds buried at greater depths (e.g. 10 cm) by worms or insects will have lower germination and are less likely to be consumed than those deposited closer to (1–3 cm) or upon the soil surface (Sheperd and Chapman, 1998).

The passage of seeds through soil animal guts such as earthworm intestines can affect their germination, viability and seedling growth, favouring some species and inhibiting others (Eisenhauer et al., 2009). Therefore, seed dispersal by ants, beetles, earthworms and other soil invertebrates and vertebrates can deeply influence plant community composition in various ecosystems (Beattie and Hughes, 2002; Forey et al., 2011; Rissing, 1986), making this ecosystem service of extreme importance for the conservation of many plant species.

4.3.3 Biological control

Soil fauna feeding, burrowing, casting and nesting activities can modify activity of pests and pathogens as well as populations and activity of beneficial microorganisms (Brown et al., 2004; Lavelle et al., 2004). For instance, Blouin et al. (2005), Boyer et al. (1999) and Loranger-Merciris et al. (2012) found that earthworms could reduce the negative effects of nematodes on maize, rice and banana, respectively. Other authors observed reduced fungus or bacterial disease incidence and increases in root colonization by mycorrhizae or N_2 fixing bacteria, increasing crop production (see review in Brown et al., 2004). It has also been proposed that earthworms can affect plant gene expression (Blouin et al., 2005), affecting their ability to tolerate pests and increasing overall plant health.

Besides their effects on nutrient cycling, the activity of dung beetles can also help animal well-being in animal production systems, burying up to 12 tons of bovine faeces in one year. For instance, Bergstrom et al. (1976) observed a 24–90% reduction in the number of parasitic nematode eggs in sheep and cattle faeces when dung beetles were present, and Nichols et al. (2008) showed several examples of decreases in parasites and hematophagous flies through coprophagic activity. In Brazil, the introduced beetle species *Digitonthophagus gazella*, as well as native beetle species can significantly reduce horn fly (*Haematobia irritans*) populations (Bianchini et al., 1992; Horner and Gomes, 1990) in pastoral systems.

4.3.4 Erosion and flood control

Animal bioturbation or soil particle ingestion and egestion produces biogenic structures that can profoundly affect soil structure and architecture, affecting water run-off and infiltration rates and soil water holding capacity (Wilkinson et al., 2009), the latter also being a provisioning service. Various studies have shown the positive impact of bioturbation by termites (Leonard and Rajot, 2001; Mando and Miedema, 1997), ants (Lobry De Bruyn and Conacher, 1994) and earthworms (Ehlers, 1975; Van Schaik et al., 2014) on water infiltration in soils. Macropores open to the soil surface can act as preferential pathways for rainwater percolation (Van Schaik et al., 2014) and often serve as channels for root growth (Ehlers, 1975; Brown et al., 2004).

Integrated crop-livestock systems that include pasture grasses in the rotation favour the appearance and increase in numbers of soil macroinvertebrates that create these macropores that increase water infiltration and reduce surface run-off (Collof et al., 2010). Furthermore, stable biogenic structures on the soil surface increase rugosity, break surface tension and reduce crusting and water run-off. However, in the absence of plant cover and under intense rainfall events, these structures (particularly fresh worm casts) can rapidly be eroded and run-off more easily (Van Hoof, 1983), a phenomenon also documented by Darwin (1881). Nonetheless, bioturbation by soil ecosystem engineers is generally considered to be more beneficial than prejudicial to the soil (Lavelle et al., 1997), except in those cases when the engineer is an invasive organism in the ecosystem (Bohlen et al., 2004).

4.3.5 Climate regulation

The intense microbial activity in soil animal guts and faeces can lead to increased greenhouse gas emissions, namely CO_2, CH_4 and N_2O (Brümmer et al., 2009; Depkat-Jacob et al., 2012, 2013; Drake and Horn, 2007; Ngugi et al., 2011; Lubbers et al., 2013). Animal digestive tracts and biogenic structures such as earthworm casts are anaerobic or micro-aerophilic regions that have high concentrations of C, mineral N and microbial activity, ideal conditions for promoting the emissions of N_2O and CH_4. On the other hand, cast stabilization processes or construction of compact nests can reduce emission rates and lead to short-term (a few months or several years) increases in C and N sequestration (Brown et al., 2000; Fonte and Six, 2010; Zhang et al., 2013).

4.3.6 Treatment of residues and contaminants

Although the degradation of organic compounds is performed directly (and mainly) by microorganisms through biochemical processes, soil fauna can exert an important role in controlling the rate of these processes, due to their close interactions with the soil microbiota. For example, earthworms can accelerate the aerobic degradation of contaminants such as polycyclic aromatic hydrocarbons (PCBs) and of some pesticides through soil ingestion, reducing adsorption and increasing bioavailability to the microorganisms responsible for degradation processes (Andrea et al., 2004; Eijsackers et al., 2001; Papini and Andrea, 2001; Sanchez-Hernández et al., 2014). The ability of these organisms to modify the structure, biomass and functional profile of microbial communities (Sheehan et al., 2008) can also contribute to the degradation of various organic compounds (Natal da Luz et al., 2012; Shuttleworth and Cerniglia 1995; Weissenfels et al., 1992). Hence, earthworm inoculation at local scales may be an interesting bioremediation tool that requires further testing in order to enhance the biodegradation of persistent organic chemicals in soils and sediments (Hickman and Reid, 2008).

4.4 Supporting services

4.4.1 Biodiversity conservation

The physical activities of soil animals, including bioturbation, plant litter or residue processing and soil mixing can deeply affect soil properties and processes (Fig. 1) and in turn, profoundly affect the soil as a habitat for other soil animals, plants and microorganisms. This is the case of some ecosystem engineers such as earthworms and termites, that can, over time, lead to the formation of vegetated islands in flooded areas (Dangerfield et al., 1998; Zangerlé et al., 2016), creating patterns that can be observed even in satellite images.

4.4.2 Pedogenesis

Over 140 years ago, in his last book, Darwin (1881) showcased the bioturbating role of soil fauna, by presenting the role of earthworms in soil genesis (formation of subsuperficial soil horizons), and the burial over time of stones and archaeological remains (Roman Empire

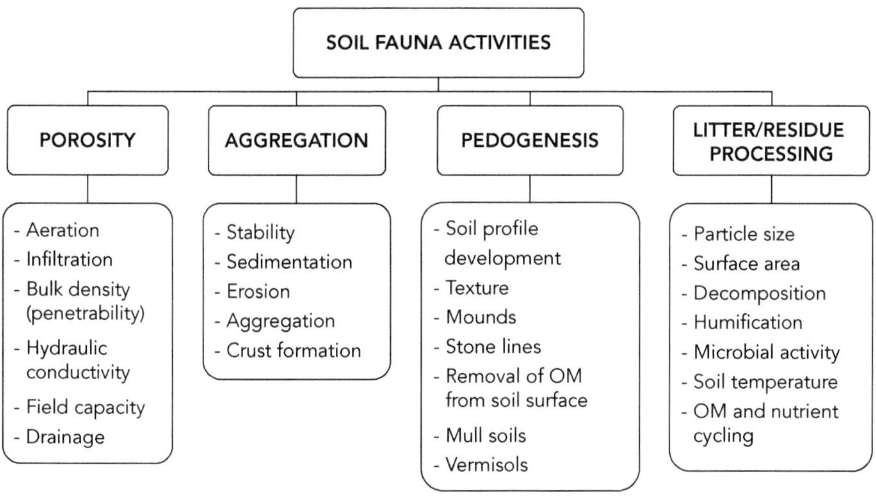

Figure 1 Relationship between soil fauna activities and the various soil properties and processes important for soil health and the delivery of ecosystem services (modified after Brown, 1995).

or Neolithic) in England (Meysman et al., 2006). He was the first to show that earthworms could affect pedogenesis through both biochemical and biomechanical processes (Johnson, 2002). The genesis of Brazilian Latossols (Oxisols) and the formation of stone lines by earthworms, ants and termites was emphasized by Miklós (2012), and the transport of finer soil particles by these animals can also affect the texture of the soil surface horizons, altering distribution of coarse sand and clay particles in the soil profile (Jouquet et al., 2011; Nooren et al., 1995).

In the long term, the bioturbating activity of worms can lead to the formation of 'mull'-like soil horizons or even Vermisols (a type of Mollisol), where castings dominate the soil profile down to almost 1 m depth (Buntley and Papendick, 1960). In countries such as Canada, New Zealand and the United States, the invasion of exotic Asian and European earthworms has deeply transformed forest and grassland soils over decades, creating a 'mull'-type A horizon (Nielsen and Hole, 1964; Langmaid, 1964; Stout, 1983). This phenomenon greatly increases pasture productivity (Stockdill, 1982), but can also negatively affect populations and diversity of other organisms (plants, animals and microorganisms) both above and within the soil (Bohlen et al., 2004; Eisenhauer et al., 2007; Yeates, 1981). On the other hand, soil fauna can be very important in regeneration processes of soils degraded by mining and other anthropogenic activities (Bal, 1982; Butt, 2008; Jouquet et al., 2014; Parker, 1989).

4.4.3 Nutrient cycling

Detritivores, geophages, necrophages, coprophages and predators are all important for the decomposition of organic matter, living and dead animals (Table 1). By ingesting and fragmenting litter, detritivores increase the surface area facilitating microbial attack and hence, accelerating residue decomposition (Fig. 1). The incorporation of this litter into the soil or into faeces also accelerates decomposition processes. Geophagous animals also affect organic matter decomposition rates by mixing residues within the soil profile. These actions contribute to nutrient mineralization and cycling, as well as having influence on the activity of other soil organisms (animals, plants and microorganisms).

4.5 Scale-dependent effects

The effective action of soil fauna in ecosystem processes varies greatly, depending on their abundance, community composition (especially in terms of functional diversity; De Bello et al., 2010), and environmental factors such as the type of ecosystem, climate and soil where they live. The abundance of certain fauna groups can reach millions of individuals m^{-2} for the microfauna, dozens to thousands of individuals m^{-2} for the mesofauna and macrofauna, and consequently, the ability of the different size classes of the soil fauna to affect ecosystem services is also distinct (Table 2). Soil micro- and mesofauna affect mainly plant health and nutrient cycling through their feeding activities and interactions with soil microbiota, while the soil macro- and megafauna have major effects on these and other soil processes through bioturbation, predation and litter fragmentation. Their biomass can reach up to several tons ha^{-1} (Brown et al., 2001a), surpassing many times the biomass of above-ground fauna. This is frequently the case for ecosystem engineers such as earthworms, termites and ants that generally represent around 80–90% of soil fauna abundance and/or biomass (Lavelle et al., 1994), and have major impacts on soil processes (Jouquet et al., 2011; Lavelle and Spain, 2001). Although the combined biomass of the soil fauna generally represent only a small fraction of overall soil biomass, particularly of its organic matter content (<2%), compared with microorganisms (Brussaard et al., 1990; Coleman et al., 2004), their interactions with plant roots and the soil microbiota, are highly important for plant and soil processes and health (Anderson, 1988; Brown et al., 2000; Coleman et al., 1988; Lavelle and Spain, 2001).

5 Soil management and soil fauna populations

The abundance and diversity of soil fauna in natural and agricultural ecosystems can be affected by various edaphic factors (soil type, predominant minerals, temperature, pH, organic matter content, moisture, texture, structure), plant communities (type of vegetation, plant cover), land use history (especially human impacts, but also geological), topography (orographic position, slope) and climate (amount and distribution of precipitation, ambient temperature, wind and relative humidity). Therefore, human interventions that affect these various factors both in natural and agricultural ecosystems can deeply affect the dynamics of soil communities and consequently, the ecological functions and services involved (Fig. 2).

At larger spatial scales (landscape), soil fauna is affected by climate and its interaction with soil type and vegetation cover, while at smaller scales (e.g. a farm), soil fauna are mainly affected by the land use type and soil management, especially plant identity and diversity (Brown et al., 2006; Fig. 2).

There is a strong link between biodiversity above- and below-ground, especially because plants and their diversity determine ecosystem function (Van Noordwijk and Swift, 1999) through:

- Effects on litter quality, quantity and time of deposition, and as potential energy source for the soil fauna;

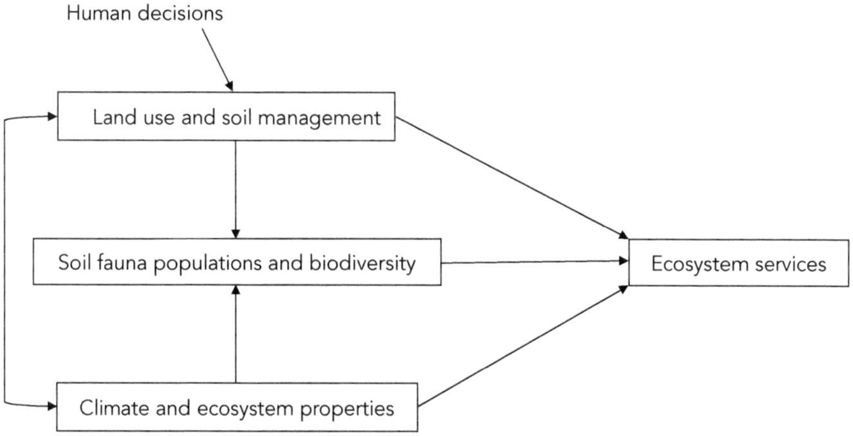

Figure 2 Schematic representation of the influence of environmental and human factors on soil animal biodiversity and the delivery of ecosystem services (modified and expanded from Cavigeli et al., 2012).

- Influence on the soil water balance and microclimate on the soil surface, via modification of evapotranspiration rates (through plant cover, type of vegetation and plant litter);
- Plant root growth that modulates the rhizosphere and the quantity and quality of root exudate C and organic materials for consumption (energy sources) and decomposition. Root exudates affect mostly microfauna in the rhizosphere, while root structural material serves as food for detritivores that fragment the residues, making them more accessible for microbial attack.

The type of ecosystem also determines the composition of soil fauna communities and their ability to affect soil properties and processes (Fig. 2). Soil management and changes in land use, such as urbanization, the conversion of tropical and subtropical rainforests to pastures or crops, and agricultural intensification (especially the adoption of high-input cropping systems such as monocultures and intensive tillage), cause profound impacts on biodiversity both above- and below-ground, although changes in the soil biota are often slower and more difficult to detect (Susilo et al., 2004). Nevertheless, as many soil animals are patchily distributed, displaying high endemism and limited dispersion rates (Lavelle and Lapied, 2003), many of these animals can become locally extinct (Barros et al., 2006), and recolonization could be much slower, with hindering the biological recovery of disturbed ecosystems.

Studies in Central Amazonia have clearly shown the importance of maintaining a diverse community of soil macrofauna for the soil to have a good structure and high fertility, since even organisms generally considered as beneficial (e.g. earthworms), when present in high abundance and biomass in the absence of other soil fauna, can cause soil dysfunction, such as compaction and surface sealing, impeding water and air circulation (Barros et al., 2001, 2004; Chauvel et al., 1999). Cultivation systems with different degrees of intensification, and systems with higher number of tree and crop species (e.g. agroforestry systems) also normally have higher abundance and diversity of soil fauna (Barros et al., 2006). Hence, maintenance of plant diversity, especially in agricultural systems, together with a diverse soil biota and permanent soil cover are important ingredients for maintaining a potentially active and dynamic (living) soil, resulting in a more sustainable production system.

6 Soil animals as indicators of soil management and soil health

Because soil organisms are sensitive and rapidly react to changes in soil and plant cover induced by human or natural processes, the populations and diversity of soil fauna can be used as bioindicators of soil use or its quality, providing insights into its current state and changes induced by internal or external (biotic and abiotic) forces over space and time (Brown, 1997a,b; Paoletti, 1999). These changes affect soil fauna distribution according to the availability of food resources, thus altering their inter- and intra-specific interactions. Generally speaking, species more sensitive to changes in the environment (e.g. epigeic species that live in the litter) can disappear with deforestation (Brown, 1997a) or with higher intensity of soil disturbance (e.g. tillage and pesticide applications) (Fig. 3).

On the other hand, conservation practices such as the use of green manures, and no-tillage and agroforestry systems, for instance, can positively affect soil fauna populations

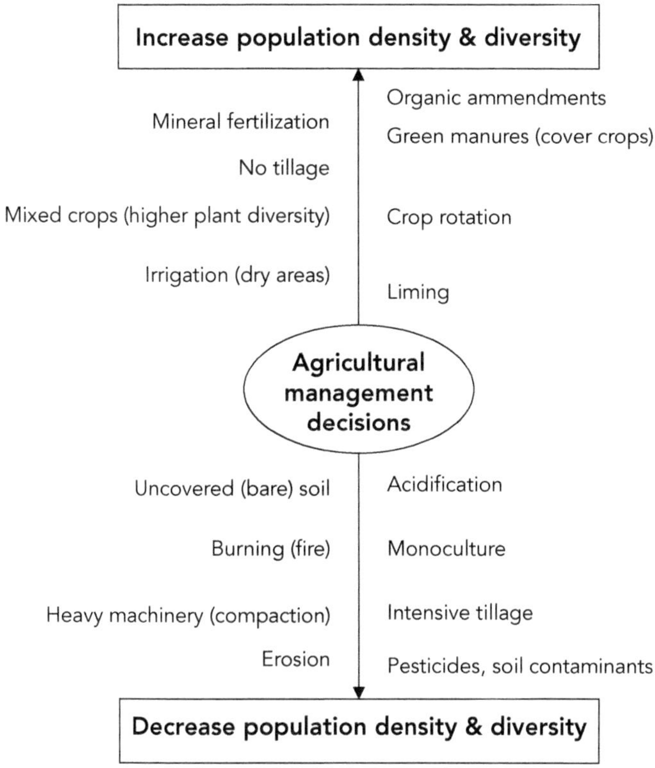

Figure 3 Diagram illustrating how different agricultural management practices at the local level (e.g. a farm) can affect soil fauna populations (density and diversity). The position of the various practices on the *y*-axis represents their hypothetical relative contribution (importance) to increasing or decreasing soil animal populations (modified after Brown et al., 2006).

and diversity (Fig. 3). In general, earthworm and collembola populations increase under no-tillage, since the absence of tillage and the maintenance of soil cover (especially use of green manures) increase organic matter and plant root growth (Brown et al., 2003; House and Parmelee, 1985), providing more food and shelter for these animals.

The use of soil fauna as soil quality bioindicators is a relatively recent phenomenon, and was only more widely adopted for monitoring soil quality in European countries in the early twenty-first century (Pulleman et al., 2012). Various characteristics of some faunal groups favour their use as indicators, including: convenient diversity (manageable number of species at a particular site; not too high); abundance in the soil or litter; ease of sampling, sorting and identification; short life cycle; sedentarism (non-migratory); wide distribution (in many habitats) but niche specificity; response to disturbance; and significant relationships between populations and environmental variables and soil chemical, physical and biological properties and processes, and with plant productivity (Freitas et al., 2006; Paoletti, 1999). Therefore, the abundance, biomass, richness and functional diversity of various soil invertebrates can be used as indicators to monitor qualitative and quantitative changes to the soil induced by land use and management practices (Lavelle et al., 1994; Paoletti, 1999). This also includes the use of several soil animal groups (e.g. mites, collembola, isopods, enchytraeids and earthworms) in ecotoxicological tests, many of which have already been standardized by ISO (Kuperman et al., 2009) and are widely adopted to verify the potential risks of soil contamination to the environment.

Among the main groups of soil macroinvertebrates, earthworms have received much more attention and are frequently used as soil quality bioindicators. This is despite the fact that their abundance and/or species composition do not always adequately indicate soil quality (Doube and Schmidt, 1997). On the other hand, the number of native or exotic species can be a valid indicator of the disturbance level of an ecosystem: higher number of native and lower number of exotic species are typically associated with lower disturbance, and vice-versa (Huerta et al., 2005). This can also be applied to termites (Lawton et al., 1998) and various other soil animals, although a general synthesis on the topic, including a wide range of the soil fauna, is still not available.

Earthworms also show other interesting characters for their use as bioindicators (Brown and Domínguez, 2010):

- they accumulate metals and other contaminants in their tissues and are sensitive to many pesticides and for this reason are commonly used in standardized ecotoxicological tests (ISO, 1993, 1998);
- their activity is intimately linked to physical soil properties and they also affect soil structure;
- they contribute to soil fertility through litter decomposition and stimulation of nutrient mineralization;
- they are easily recognized by farmers and are generally considered 'good for the soil'.

However, there are several problems associated with the use of soil fauna as soil quality indicators (Paoletti, 1999). Their distribution in the soil is not homogeneous and frequently aggregated, and their sampling involves many different methods, generally tailored for each main taxa, group or size class (Bignell, 2009). Therefore, the number and size of the soil or litter samples to be taken are determined by expertise knowledge on the ecology and biology of the organism in question and on the heterogeneity of the spatial distribution within the habitat or site to be studied. Hence, the methods are normally specific for each group or target organism pre-defined for evaluation. It is important to remember that social insects such as ants and termites have aggregated distribution and can forage large distances from their nests (Wood, 1988), while many smaller particularly non-winged animals have more restricted distributions (Giller, 1996).

7 Assessing the contribution of soil fauna to ecosystem services

The intrinsic difficulty of working with a highly diversified, cryptic and little visible soil fauna has greatly hindered estimates of its biodiversity and abundance and the evaluation of its functional importance in many of the world's terrestrial ecosystems. Furthermore, the relationships of biodiversity and abundance of soil fauna groups with ecosystem function are not normally direct. Obviously, the abundance of some key groups such as crop pests can be directly related to food production, but for the majority, the relationships between services and fauna are indirect and of complex mensuration. Therefore, the answer to the question - 'so what?' - in relation to the importance of soil fauna to ecosystem services continues to be a great challenge to soil ecologists (Barrios, 2007; Lavelle, 2000).

From a strictly agronomic perspective, some functions modulated by soil fauna can be replaced by chemicals (pesticides, fertilizers), and the use of fuels (soil tillage). However, the soil biota deserve special attention, since they are one of the main pillars in the search for a more holistic, sustainable agriculture. In this manner, much of the soil fauna can and must be considered allies of production systems.

But how is it that soil animals and their biodiversity affect ecosystem services? And how can human or environmental factors affect this relationship? And how can one achieve an economical valuation of their contributions? In the previous sections we dealt mainly with the first two questions. In the present and the following sections, we will address the last question, showing how to measure the contribution of soil fauna to ecosystem services and assess their potential value.

As the impact of soil fauna on ecosystem services is mainly indirect, and involves mostly supporting or regulating services, the evaluation of the contribution of soil fauna to these services is normally performed using indicators of fauna activity, although in a few cases evaluation can be direct (Table 3). In the following two sections we explore in more detail how two major ecosystem processes affected by soil fauna activities (bioturbation and decomposition) can be evaluated, and how they are related, indirectly, to soil ecosystem services.

7.1 Bioturbation

Soil bioturbation affects a wide range of ecosystem processes including pedogenesis, surface water run-off, soil architecture and structuration into pores and aggregates, and the ability of soils to hold water (Fig. 1). To assess the contribution of fauna to these processes, one must measure the proportional contribution of fauna by assessing: biological aggregates in soil (e.g. using the method proposed by Velásquez et al., 2007a,b, 2012); the number of biogenic structures (e.g. ant nests or earthworm castings) on the soil surface or within it; the stability of these aggregates/biostructures; water infiltration, hydraulic conductivity and soil water retention; soil porosity and the quantity of biopores by means of the cultural profile methodology (Silva et al., 2011; Tavares-Filho et al., 1999), or other methods of relatively easy application (e.g. the Visual Soil Assessment, VSA, Shepherd et al., 2009; and the Visual Evaluation of Soil Structure, VESS, Guimarães et al., 2011) (Table 3). These processes can be used as indicators of ecosystem services such as soil and water erosion and water availability in soils.

7.2 Organic matter decomposition

The importance of detritivores in organic matter decomposition has been assessed in many experiments around the world, since the 1960s and 70s, after the beginning of the International Biological Program of UNESCO (Huhta, 2007). Using nylon litter bags of different mesh sizes (Fig. 4), we can estimate the contribution of soil micro-, meso- and macrobiota to the disappearance (weight loss) of plant residues, indirectly indicating their contribution to decomposition (OECD, 2006). These litter bags can be placed horizontally on the soil surface or vertically in the topsoil layer to assess the activity of litter and soil feeding animals.

Published by Burleigh Dodds Science Publishing Limited, 2018.

Figure 4 Three mesh size classes of litter bags placed on the soil surface after wheat harvest in a no-tillage field at Embrapa Soybean (Londrina, Brazil) in October 2005. Photo George G. Brown (Embrapa).

The relative importance of dung beetle activity and that of other detritivores and predators can also be measured using pitfalls with baits (faeces, fruit, meat, etc.). This is an indirect method that measures attractiveness, and qualitatively assesses the populations of invertebrates attracted to the bait. But there are also methods that quantify the bait disappearance from the soil surface or the litter bags. For instance, the bait lamina (Fig. 5) evaluate feeding activity of euedaphic (endogeic) organisms by filling small circular holes in a plastic or PVC-like thin stick with bait and placing them in the topsoil (Törne, 1990). Food consumption rates allow a direct estimation of the intensity and vertical distribution of feeding activity by soil invertebrates, and at a smaller scale, of microbial activity (Kratz, 1998; Römbke, 2014). This method has been recommended in order to evaluate the functional value of the soil as a habitat for soil animals (ISO, 2014), but it also assesses the contribution of soil animals to decomposition and nutrient cycling.

Two recent studies in Amazonia (Grimaldi et al., 2014; Lavelle et al., 2016) and the Colombian *Llanos* (Lavelle et al., 2014) evaluated the relationship between the presence and activity of soil fauna and some ecosystem services such as agricultural production, climate regulation (via C sequestration and greenhouse gas emission), soil hydraulic functions (via water infiltration), soil stability (via macro-aggregation), nutrient provisioning to plants (via soil chemical quality assessment) and biodiversity conservation (number of species of various taxa). These services were measured using appropriate indicators (Table 3) and were correlated with soil macrofauna populations (overall community, including abundance of multiple taxa) using multivariate (co-inertia) analyses. These analyses compared matrices using several factors (e.g. socio-economic data, physico-chemical soil data, abiotic environmental data, and macrofauna populations and diversity) looking for significant correlations. The authors found significant relationships between soil biological quality (macrofauna) and soil physical and chemical quality, agricultural and primary productivity, C stocks in the soil and vegetation, biodiversity conservation and the socio-economic status of the landowners. Although this is an indirect assessment (correlation, not causation), the methods are robust and give a good idea of possible

Figure 5 On the left, individual bait-lamina sticks filled with bait (food), ready to be inserted into the topsoil, and on the right, bait lamina installed in a pasture at Embrapa Cerrados (Planaltina, Brazil). Photos Cintia C. Niva (Embrapa).

cause-effect relationships, particularly in those cases when there is some previous knowledge on the type of relationships between the parameters measured.

7.3 Economic value of soil fauna contributions to ecosystem services

The economic valuation of the impact of soil fauna on ecosystem services often encounters pitfalls, both methodological and practical, since many of the parameters or indicators to be measured are not directly correlated with ecosystem services, or because the measurements are in different units or scales (Anderson, 2009). Furthermore, many measurements depend on frequently subjective choices that need to be converted into monetary values. Disposition or willingness to pay for services derived from soil fauna can only be calculated in exceptional cases, when the animal or the service in question has a direct economic impact on the beneficiary (Table 4). For example, in the case of urban, crop, pasture and forestry pests, the willingness to pay for a programme or package of technologies of integrated pest management (IPM), involving a biocontrol agent, can easily be calculated.

However, in most cases, the presence or activity of soil fauna is associated with ecosystem processes (e.g. decomposition and bioturbation) that affect supporting, regulating or cultural services. In these cases, the economic importance must be calculated by assessing the contribution of these animals to these processes and the potential economic impact that the elimination of these animals would have on these processes and the associated ecosystem services. For example, the impact of macro and mesofauna on decomposition

Table 4 The main indicators and ecosystem processes to be measured in order to evaluate the potential economic value of soil fauna contributions to ecosystem services (see Table 2), and the value of some services as estimated by Pimentel et al. (1997) in US dollars

Ecosystem service	Ecosystem process or parameter to be measured	Calculation method to obtain economic value[1]	Estimated economic value (10^9 US\$ yr^{-1})[2]
Available water	Proportion and arrangement of biogenic structures, field capacity	Indirect: Replacement cost of water	?
Feed and food	Use as food for indigenous/local communities or for animal feed	Direct: Market value of purchase	<<180
Pharmaceutical products	Isolation and chemical characterization of the molecules, use in traditional or industrial medicine	Direct: Market cost of purchasing similar product	?
Nutrient cycling	Decomposition in litter bags and using bait lamina	Indirect: Cost of nutrient replacement	760
	Enzymatic activity, N_2O emission, denitrification/nitrification, OM fractions in aggregates	Indirect: Cost of nutrient replacement	?
Soil formation	Production of surface castings, soil architecture, soil dating, soil micromorphology	Indirect: Cost of soil replacement	25
Primary productivity	DNA of the castings, root colonization	Indirect: Cost avoided for inoculation	?
	Bio-essays with biogenic structures	Avoided and/or replacement cost	?
	Assessment of plant 'vigour' or 'health' associated with higher presence or aggregations of soil fauna	Indirect: Cost avoided for pest control	160
	Phytophagy, plant biomass associated with higher presence or aggregations of soil fauna	Indirect: Cost avoided for pest control	?
Biodiversity conservation	Effects on soil architecture and organic matter dynamics in the ecosystem	Indirect: Replacement cost and willingness to pay for conservation	?
Erosion and flood control	Production of surface biogenic structures	Indirect: Replacement cost of soil, cost avoided with floods	?
	Soil morphology (proportion of biological aggregates/pores), field capacity, infiltration rates, hydraulic conductivity	Indirect: Cost avoided with floods or for irrigation	?

Climate regulation	Greenhouse gas emissions from castings, nests and areas with higher presence or aggregations of soil fauna	Indirect: Replacement cost of N	?
	C present in biogenic aggregates/structures, NIRS of the biogenic structures	Indirect: Market value of C purchase	?
Pollination	Determination of the fauna and their presence/activity in plants dependent on pollination	Indirect: Avoided costs for pollination	<<200
Seed dispersal	Seeds associated with biogenic structures and functional domains in soil	Indirect: Avoided costs for seed dispersal	?
Residue treatment	Assessment of the contaminant residues in soil, assessment of activity and functional profile of the microbes stimulated by soil macro and mesofauna	Indirect: Avoided costs for cleaning contaminated area	121
Recreation	Use as fish bait, for photography or collections	Direct: Market cost of purchase; Indirect: Willingness to pay	>29
Education	Use in environmental education programmes and IPM, use as bioindicators of soil quality	Direct: Market cost; Indirect: Willingness to pay	?

[1] Based partially on Huguenin et al. (2006).
[2] The values calculated by Pimentel et al. (1997) considered all the soil biota (including microbiota), and not just the soil fauna.

estimated using litter bags can give an idea of the economic value of these animals, taking into account the nutrients released from the litter into the soil, particularly N and P that are of primary importance for agricultural production. Similarly, the disappearance of bait/food from the bait lamina tests serves as a measure of the contribution of soil biota to decomposition processes in soils. Hence, the substitution/replacement value in monetary units per hectare can be calculated, considering the cost saved from reducing the use of N and P-based fertilizers in an agro-ecosystem with a given population of macro and mesofauna. At some pastoral sites in New Zealand, for instance, the amount of N and P cycled annually by an earthworm community may reach as much as 8–42 and 2–9 kg ha^{-1}, respectively (Syers and Springett, 1984). Another way to calculate the biotic contribution to decomposition is via the cost of organic waste management or treatment, such as that associated with household or industrial organic wastes, and crop and animal residues and manures (Table 4). The cost of processing these residues worldwide amounted to 760 billion US dollar yr^{-1} (Table 4), as estimated by Pimentel et al. (1997).

Soil bioremediation and contaminant treatment are mainly performed by soil microorganisms. Nevertheless, soil fauna can contribute indirectly, by stimulating the detoxifying activity of microbial agents. The contribution of soil biota (including microbiota) to this process worldwide was estimated at around US$121 billion yr^{-1} (Pimentel et al., 1997; Table 4). The value of biological control estimated by the contribution of natural enemies to crop and forestry pest control worldwide reached US$160 billion yr^{-1}, and the value for pollination services, performed mainly by insects, birds and bats, was estimated at US$200 billion yr^{-1} (Table 4). The true contribution of soil fauna to this latter service is probably much lower, since only a few representatives of the soil fauna are actually pollinators.

The cost of soil erosion as affected positively or negatively by soil fauna activity can be calculated by the replacement/substitution cost of soil and of the soil nutrients in the same area (Pimentel et al., 1995). For soil biota, the contribution can be indirectly derived from calculating their contribution to soil formation *in situ*, and Pimentel et al. (1997) estimated this contribution in agricultural soils to be around US$25 billion yr^{-1} worldwide (Table 4). However, this value refers only to contribution in agricultural soils and does not include the value of soil formation under other land uses (e.g. forestry systems and regenerating ecosystems), or the importance of the burial and conservation of archaeological remains by bioturbation (also highlighted by Darwin, 1881). Another study in New Zealand calculated the mean contribution of earthworms to soil genesis in agricultural areas in the range of US$0.60–11.60 ha^{-1} yr^{-1}, considering soil formation rates of 1000 kg ha^{-1} yr^{-1} for an earthworm population weighing approximately 1000 kg ha^{-1} (Sandhu et al., 2010).

The direct benefits of the use of soil biota in ethnopharmacology and for the development of new medicinal products by the pharmaceutical industry were not assessed by Pimentel et al. (1997), but we know that various products used in both traditional and modern medicine are soil fauna-based (Anderson, 2009). This topic deserves much further attention and its economic value must still be quantified. Conversely, the direct use value of animal and plant products extracted worldwide from nature as food sources was estimated at US$180 billion yr^{-1} (Table 4), while recreational fishing in the USA (much of the fish bait used are soil animals) involves a market value of around US$26 billion yr^{-1} (Pimentel et al., 1997).

Unfortunately, there are no cases cited in the literature that include a wide range of ecosystem services, and there are still many hurdles to overcome concerning the economic valuation of soil fauna contributions. Hence, the economic value of most of the services shown in Table 3 continues to be unknown. For this purpose, data must still be gathered that relates ecosystem processes with the services and that quantifies faunal contribution to these processes. Furthermore, in the case of agricultural, pastoral and forestry production, data must be obtained on productivity and the costs and benefits involved in crop production, the market value of the inputs (e.g. pesticides, fertilizers and fuel), commodities and natural resources involved (e.g. 1 ton of soil and 1 m^3 of water), and waste and residue management. Finally, we must remember that there are important scale issues to be overcome (Anderson, 2009) and that some of these services are extremely difficult to measure or value. This is the case especially for cultural services (recreation, education), those of direct use that involve cultural aspects and human traditions, and the services of existential or strategic value, such as biodiversity conservation, that normally involve willingness of society to pay for these services (Pagiola et al., 2004).

8 Conclusion

Although soil fauna have been studied for decades by a wide range of specialists worldwide, there are still many countries, particularly in the tropics, for which very little data is available on soil fauna populations, their diversity and their potential role in soil processes and ecosystem services. Some economically more important groups or species (e.g. crop, pasture or forestry pests) have their taxonomy, biology and behaviour relatively well elucidated. However, many soil invertebrates and even many vertebrates still remain little studied, especially due to the limited number of taxonomists active in this field, and the relative difficulty of studying these cryptic and fossorial animals.

Significant advances in the knowledge of soil fauna can be gained by adopting a functional, taxonomic or size group approach, or by studying the ecological processes affected by soil fauna (e.g. decomposition, nutrient cycling and bioturbation), which also increases the possibility of using them as soil or environmental quality indicators (Barrios, 2007; Faber et al., 2013). However, regardless of the approach taken, it is important to always consult taxonomists in order to help identify the animals in question, thus validating the biological, behavioural and/or ecological data obtained.

There is a vast literature available on the impacts of land use and soil management on soil fauna populations. It is well known that the practices that reduce physical disturbance and that prioritize the maintenance of plant cover and more constant inputs of diverse organic materials are the most beneficial to soil fauna populations, as soils are their home and feeding grounds. Among these practices are the adoption of integrated crop-tree-pasture systems that include all three components, the use of no-tillage with crop rotations, organic production systems and agroforestry systems (Balbino et al., 2011; Brown et al., 2006; Cordeiro et al., 2004; Derpsch et al., 2010; Silva et al., 2006).

The living soil fauna make up a very small percentage of the soil living biomass, but their extensive activities and interactions with other organisms in soils (particularly microorganisms and plant roots) are highly relevant for our food security. It is the status and health of the living components in the soil that is 'loosely' called soil health. Knowing and recognizing soil fauna communities is a prerequisite in the search for ideal and sustainable soil management that not only conserves biodiversity, but also allows the important action of these animals in the ecosystem. Since each organism can have distinct effects on soil processes and plant production, their abundance and biomass may reach important thresholds, both negative and positive (Brown et al., 2001a). The soil may be one of the most diverse habitats on the face of the Earth (Giller, 1996), and the soil fauna are an important component of this diversity. Although they are mostly hidden (often 'invisible') to the naked eye due to their small size or because they are within the soil or the litter, this group delivers important ecosystem services that are unfortunately still very little recognized and valued. Slowly this important issue is getting political recognition (Römbke et al., 2016). It is mostly up to us, soil scientists, to provide effective and comparable means to assess and quantify these services (preferably *in situ*), and spread this (re)cognition to subsidize the sustainable management of terrestrial ecosystems, in benefit of both present and future human generations.

9 Where to look for further information

Much information on soil fauna can be viewed, gathered and downloaded from the internet using search engines and most soil animals are relatively well described in web-based encyclopaedia (e.g. Wikipedia). Some institutional, society or project-based sites also have web pages devoted to soil animals (e.g. AntWeb, DriloBase and Society of Nematologists), and many research and educational institutions worldwide have researchers and programmes devoted to the study of soil animals. For many years, the International Union of Soil Sciences (IUSS) sponsored the International Colloquium on Soil Zoology, which led to many excellent publications, including books and special issues of journals (see review in Brown and Sautter, 2009) devoted mainly to soil animals. These continue to be organized every four years, although the Soil Zoology Commission is no longer formally attached to the IUSS. There are several journals that regularly publish articles on the ecology and biology of soil animals. Among these are *Soil Organisms*, *Edaphologia*, *Pedobiologia*, *Applied Soil Ecology*, *Biology and Fertility of Soils*, *Soil Biology and Biochemistry*, and the *European Journal of Soil Biology*. We have provided a very long list of articles, book chapters and books in the list of references below, many of which are available for download online. However, much of the information on soil animals is spread out and at present there are few institutions and/or web resources that have regularly collated the vast amount of scattered information on soil animal biology, ecology and taxonomy.

Over the past decade, there have been several important projects on soil fauna, many of them funded by the EU (e.g. http://ec.europa.eu/environment/soil/biodiversity.htm; http://cordis.europa.eu/project/rcn/97538_en.html), as well as an international initiative led by the Food and Agriculture Organization of the United Nations (FAO) and the Convention on Biological Diversity (see http://www.fao.org/agriculture/crops/thematic-sitemap/theme/spi/soil-biodiversity/en/), resulting in important publications and syntheses on soil biodiversity (Brown et al., 2007; FAO, 2003; Gardi and Jeffery, 2009; Jeffery et al., 2010; Orgiazzi et al., 2015; Turbé et al., 2010). Furthermore, the scientist-based Global Soil Biodiversity Initiative (see http://globalsoilbiodiversity.org) has also been working towards global syntheses, conferences and widespread conscientization of the importance of soil biodiversity conservation, management and role in ecosystem function and the delivery of ecosystem services. This is a particularly hot topic of research at the moment, as there are still many questions to be answered. In the upcoming years, we will surely witness many exciting and interesting results showing applications of the concepts presented in this chapter, highlighting the crucial roles played by the denizen animals that live in the soil, and the consequences of their activities at various levels, from our own homes and backyards up to landscape and industrial scales.

Published by Burleigh Dodds Science Publishing Limited, 2018.

10 Acknowledgements

The authors thank Elizabeth Bach and an anonymous reviewer for their comments, Don Reickosky for the invitation to contribute this chapter, CNPq for scholarship and fellowship support, CAPES for scholarships to student authors, and Embrapa for allowing the reproduction of most of the text and many of the figures in this chapter. We also acknowledge the financial support of FAPDF, NSF, Fundação Araucária, Newton Fund, the EU and NERC.

11 References

Adis, J. (2002), *Amazonian Arachnida and Myriapoda*, Moscow, Pensoft.

Anderson, J. M. (1988), Spatiotemporal effects of invertebrates on soil processes, *Biology and Fertility of Soils*, 6, 216–27.

Anderson, J. M. (2009), Why should we care about soil fauna?, *Pesquisa Agropecuária Brasileira*, 44(8), 835–42.

Andréa, M. M., Papini S., Peres T. B., Bazarin, S., Savoy, V. L. T. and Matallo, M. B. (2004), Glyphosate: Influência na bioatividade do solo e ação de minhocas sobre sua dissipação em terra agrícola, *Planta Daninha*, 22, 95–100.

Appelhof, M., Fenton, M. F. and Harris, B. L. (1997), *Worms Eat Our Garbage. Classroom Activities for a Better Environment*, Kalamazoo, Flower Press.

Bal, L. (1982), *Zoological ripening of soils. I. The concept and impact in pedology, forestry and agriculture. II. The process in two Entisols under developing forest in a recently reclaimed Dutch polder*, Wageningen, Pudoc Center for Agricultural Publishing and Documentation (Agricultural Research Reports, 850).

Balbino, L. C., Barcelos, A. O. and Stone, L. F. (2011), *Marco referencial: integração lavoura-pecuária-floresta*, Brasilia, Embrapa Solos.

Barrios, E. (2007), Soil biota, ecosystem services and land productivity, *Ecological Economics*, 64, 269–85.

Barros, E., Curmi, P., Hallaire, V., Chauvel, A. and Lavelle, P. (2001), The role of macrofauna in the transformation and reversibility of soil structure of an oxisol in the process of forest to pasture conversion, *Geoderma*, 100, 1193–213.

Barros, E., Grimaldi, M., Sarrazin, M., Chauvel, A., Mitja, D., Desjardins, T. and Lavelle, P. (2004), Soil physical degradation and changes in macrofaunal communities in Central Amazon, *Applied Soil Ecology*, 26, 157–68.

Barros, E., Matthieu, J., Tapia-Coral, S., Nascimento, A. R. L. and Lavelle, P. (2006), Soil macrofauna communities in Brazilian Amazonia. In: Moreira F. M. S., Siqueira J. O. and Brussaard L. (Eds), *Soil Biodiversity in Amazonian and Other Brazilian Ecosystems*, Wallingford, CABI, pp. 43–55.

Bartz, M. L., Pasini, A. and Brown, G. G. (2013), Earthworms as soil quality indicators in Brazilian no-tillage systems, *Applied Soil Ecology*, 69, 39–48.

Beattie, A. J. and Hughes, L. (2002), Ant-plant interactions. In: Herrera, C. M. and Pellmyr, O. (Eds), *Plant-animal Interactions and Evolutionary Approach*, Oxford, Blackwell Science, pp. 211–35.

Beccaloni, G. W. (2017), *Cockroach Species File Online: Version 5.0/5.0*. Available at: http://Cockroach.SpeciesFile.org. Acessed on 28 February 2017.

Beck, L., Dumpert, K., Franke, U., Mittmann, H.-W., Römbke, J. and Schönborn, W. (1988), Vergleichende ökologische Untersuchungen in einem Buchenwald nach Einwirkung von Umweltchemikalien. In: Scheele, B. and Verfondern, M. (Eds), *Auffindung von Indikatoren zur prospektiven Bewertung der Belastbarkeit von Ökosystemen*, Jülich Spez, KFA, 439, pp. 548–702.

Begall, S., Burda, H. and Schleich, C. E. (2007), *Subterranean Rodents: News from Underground*, Berlin, Springer-Verlag.

Bellinger, P. F., Christiansen, K. A. and Janssens, F. 2017. *Checklist of the Collembola of the world*, Available at: http://www.collembola.org. Acessed on 28 February 2017.

Bergstrom, R. C., Maki, L. R. and Werner, B. A. (1976), Small dung beetles as biological control agents: Laboratory studies of beetle action on trichostrongylid eggs in sheep and cattle faeces, *Proceedings of the Helminthological Society of Washington*, 43(2), 171–4.

Bianchin, I., Honer, M. R. and Gomes, A. (1992), Controle integrado da mosca-dos-chifres na região Centro-Oeste, *Hora Veterinária*, 11(65), 43–6.

Bignell, D. E. (2000), Introduction to symbiosis. In: Abe, T., Bignell, D. E. and Higashi, M. (Eds), *Termites: Evolution, Sociality, Symbioses, Ecology*, Amsterdam, Springer, pp. 189–208.

Bignell, D. E. (2009), Towards a universal sampling protocol for soil biotas in the humid tropics, *Pesquisa Agropecuária Brasileira*, 44(8), 825–34.

Blouin, M., Zuily-Fodil, Y., Pham-Thi, A. T., Laffray, D., Reversat, G. and Pando, A. (2005), Belowground organism activities affect plant aboveground phenotype, inducing plant tolerance to parasites, *Ecology Letters*, 8, 202–8.

Blouin, M., Hodson, M. E., Delgado, E. A., Baker, G., Brussaard, L., Butt, K. R., Dai, J., Dendooven, L., Peres, G., Tondoh, J. E., Cluzeau, D. and Brun, J.-J. (2013), A review of earthworm impact on soil function and ecosystem services, *European Journal of Soil Science*, 64, 161–82.

Bohlen, P. J., Scheu, S., Hale, C., Mclean, M. A., Migge, S., Groffman, P. and Parkinson, D. (2004), Non-native invasive earthworms as agents of change in northern temperate forests, *Frontiers in Ecology and Environment*, 2, 427–35.

Bolton, B. (2017), An Online Catalog of the Ants of the World. Available from http://antcat.org. (Accessed 05 March 2017).

Bongers, T. and Bongers, M. (1998), Functional diversity of nematodes, *Applied Soil Ecology*, 10, 239–51.

Boock, O. J. and Loedello, L. G. E. (1952), Diplópoda depredador de tubérculos de batatinha, *Bragantia*, 12(10–12), 343–7.

Boyer, J., Michellon, R., Chabanne, A., Reversat, G. and Tibere, R. (1999), Effects of trefoil cover crop and earthworm inoculation on maize crop and soil organisms in Reunion Island, *Biology and Fertility of Soils*, 28, 364–70.

Brewer, M. S., Sierwald, P. and Bond, J. E. (2012), Millipede taxonomy after 250 years: Classification and taxonomic practices in a mega-diverse yet understudied arthropod group, *PLoS One*, 7(5), e37240.

Briones, M. J. (2014), Soil fauna and soil functions: A jigsaw puzzle, *Frontiers in Environmental Science*, 2(7), 1–22.

Briones, M. J. I. and Ineson, P. (2002), Use of ^{14}C carbon dating to determine feeding behaviour of enchytraeids, *Soil Biology & Biochemistry*, 34, 881–4.

Brown, G. G. and James, S. W. (2007), Ecologia, biodiversidade e biogeografia das minhocas no Brasil. In: Brown, G. G. and Fragoso, C. (Eds), *Minhocas na América Latina: biodiversidade e ecologia*, Londrina, Embrapa Soja, pp. 297–381.

Brown, G. G. and Sautter, K. D. (2009), Biodiversity, conservation and sustainable management of soil animals: The XV International Colloquium on Soil Zoology and XII International Colloquium on Apterygota, *Pesquisa Agropecuária Brasileira*, 44(8), i–ix.

Brown, G. G. and Domínguez, J. (2010), Uso das minhocas como bioindicadoras ambientais: princípios e práticas, *Acta Zoológica Mexicana: Nueva Série*, 26, 1–18.

Brown, G. G., Pashanasi, B., Villenave, C., Patrón, J. C., Senapati, B. K., Giri, S., Barois, I., Lavelle, P., Blanchart, E., Blakemore, R. J., Spain, A. V. and Boyer, J. (1999), Effects of earthworms on plant production in the tropics. In: Lavelle, P., Brussaard, L. and Hendrix, P. F. (Eds), *Earthworm Management in Tropical Agroecosystems*, Wallingford, CAB International, pp. 87–147.

Brown, G. G., Barois, I. and Lavelle, P. (2000), Regulation of soil organic matter dynamics and microbial activity in the drilosphere and the role of interactions with other edaphic functional domains, *European Journal of Soil Biology*, 36, 177–98.

Brown, G. G., Fragoso, C., Barois, I., Rojas, P., Patron, J. C., Bueno, J., Moreno, A. G., Lavelle, P., Ordaz, V. and Rodríguez, C. (2001a), Diversidad y rol funcional de la macrofauna edáfica en los ecosistemas tropicales mexicanos, *Acta Zoológica Mexicana: Nueva Série, n. especial*, 1, 79–110.

Brown, G. G., Pasini, A., Benito, N. P., Aquino, A. M. and Correia, M. E. F. (2001b), Diversity and functional role of soil macrofauna communities in Brazilian no-tillage agroecosystems. In: *Proceedings of the International Symposium on Managing Biodiversity in Agricultural Ecosystems*, Montreal, UNU/CBD, pp. 1–20.

Brown, G. G., Benito, N. P., Pasini, A., Sautter, K. D., Guimarães, M. F. and Torres, E. (2003), No-tillage greatly increases earthworm populations in Paraná State, Brazil, *Pedobiologia*, 47, 764–71.

Brown, G. G., Edwards, C. A. and Brussaard, L. (2004), How earthworms affect plant growth: burrowing into the mechanisms. In: Edwards, C. A. (Ed.), *Earthworm Ecology*, Boca Raton, CRC Press, pp. 13–49.

Brown, G. G., Römbke, J., Hofer, H., Verhaag, M., Sautter, K. and Santana, D. L. Q. (2006), Biodiversity and function of soil animals in Brazilian agroforestry systems. In: Gama-Rodrigues, A. C., Barros, N. F., Gama-Rodrigues, E. F., Freitas, M. S., Viana, A. P., Jasmin, J. M., Marciano, C. R. and Carneiro, J. G. A. (Eds), *Sistemas agroflorestais: bases científicas para o desenvolvimento sustentado*, Campos dos Goytacazes, Embrapa/UENF, pp. 217–42.

Brown, G. G., Swift, M. J., Bennack, D. E., Brussaard, L., Bunning, S. and Montáñez, A. (2007), Managing soil biodiversity in agricultural ecosystems. In: Jarvis, D., Padoch, C. and Cooper, D. (Eds.), *Managing Biodiversity in Agroecosystems*, New York, Columbia University Press, pp. 224–68.

Brown, G. G., Niva, C. C., Zagatto, M. R. G., Ferreira, S. A., Nadolny, H., Cardoso, G. B. X., Pasini, A., Bartz, M. L., Sautter, K. D., Thomazini, M. J., Baretta, D., Martinez, G. A., Silva, E., Antoniolli, Z. I., Decaens, T., Lavelle, P., Sousa, J. P. F. A. and Carvalho, F. (2015), Biodiversidade da fauna do solo e sua contribuição para os serviços ambientais. In: Parron, L. M., Garcia, J. R., Oliveira, E. B., Brown, G. G. and Prado, R. B. (Eds), *Serviços ambientais em sistemas agrícolas e florestais do Bioma Mata Atlântica*, 1st ed., Brasilia, Embrapa, pp. 121–54.

Brown Jr., K. S. (1997a), Insetos como rápidos e sensíveis indicadores de uso sustentável de recursos naturais. In: Martos, H. L., and Maia, N. B. (Eds), *Indicadores ambientais*, Sorocaba, PUCC/Shell Brasil, pp. 143–55.

Brown Jr., K. S. (1997b), Diversity, disturbance, and sustainable use of Neotropical forests: insects as indicators for conservation monitoring, *Journal of Insect Conservation*, 1, 25–42.

Brümmer, C., Papen, H., Wassmann, R. and Bruggemann, N. (2009), Fluxes of CH_4 and CO_2 from soil and termite mounds in south Sudanian savanna of Burkina Faso (West Africa), *Global Biogeochemical Cycles*, 23, GB1001.

Brussaard, L., Bouwman, L. A., Geurs, M., Hassink, J. and Zwart, K. B. (1990), Biomass, composition and temporal dynamics of soil organisms of a silt loam soil under conventional and integrated management, *Netherlands Journal of Agricultural Science*, 38, 283–302.

Brussaard, L., Ruiter, P. C. and Brown, G. G. (2007), Soil biodiversity for agricultural sustainability. *Agriculture, Ecosystems & Environment*, 121, 233–44.

Brussaard, L. (2012), Ecosystem services provided by the soil biota. In: Wall, D. H., Bardgett, R. D., Behan-Pelletier, V., Herrick, J. E., Jones, T. H., Ritz, K., Six, J., Strong, D. R. and Van Der Putten, W. H. (Eds), *Soil Ecology and Ecosystem Services*, Oxford, Oxford University Press, pp. 45–58.

Buntley, G. J. and Papendick, R. I. (1960), Worm-worked soils of eastern South Dakota, their morphology and classification, *Soil Science Society of America Proceedings*, 24, 128–32.

Burakoswki, B. (1993), *Laboratory Methods for Rearing Soil Beetles (Coleoptera).* Memorabilia Zoologica, 46, Polska Akademia Nauk, Muzeum I Instytut Zoologii, pp. 1–66.

Butt, K. R. (2008), Earthworms in soil restoration: Lessons learned from United Kingdom case studies of land reclamation, *Restoration Ecology*, 16, 637–41

Cavigeli, M. A., Maul, J. E. and Szlavecz, K. (2012), Managing soil biodiversity and ecosystem services. In: Wall, D. H., Bardgett, R. D., Behan-Pelletier, V., Herrick, J. E., Jones, T. H., Ritz, K., Six, J., Strong, D. R., and Van Der Putten, W. H. (Eds), *Soil Ecology and Ecosystem Services*, Oxford, Oxford University Press, pp. 337–56.

Chauvel, A., Grimaldi, M., Barros, E., Blanchart, E., Desjardins, T., Sarrazin, M. and Lavelle, P. (1999), Pasture damage by an Amazonian earthworm, *Nature*, 398, 32–3.

Coleman, D. C., Crossley Jr., D. A., Beare, M. H. and Hendrix, P. F. (1988), Interactions of organisms at root/soil and litter/soil interfaces in terrestrial ecosystems, *Agriculture, Ecosystems and Environment*, 24, 117–34.

Coleman, D. C., Crossley Jr., D. A. and Hendrix, P. F. (2004), *Fundamentals of Soil Ecology*, New York, Elsevier.

Colloff, M. J., Pullen, K. R. and Cunningham, S. A. (2010), Restoration of an ecosystem function to revegetation communities: The role of invertebrate macropores in enhancing soil water infiltration, *Restoration Ecology*, 18, 65–72.

Constantino, R. and Acioli, A. N. S. (2006), Termite diversity in Brazil (Insecta: Isoptera). In: Moreira, F. M. S., Siqueira, J. O. and Brussaard, L. (Eds), *Soil Biodiversity in Amazonian and other Brazilian Ecosystems*, Wallingford, CABI, pp. 117–28.

Cordeiro, F. C., Dias, F. C., Merlin, A. O., Correia, M. E. F., Aquino, A. M. and Brown, G. G. (2004), Diversidade da fauna invertebrada do solo como indicadora da qualidade do solo em sistemas orgânicos, *Revista Universidade Rural*, 24(2), 22–9.

Costa, C., Vanin, S. A. and Casari-Chen, S. A. (1988), *Larvas de Coleoptera do Brasil*, São Paulo, Museu de Zoologia, Universidade de São Paulo.

Csuzdi, C. (2012), Earthworm species, a searchable database, *Opuscula Zoologica*, 43(1), 97–9.

Curry, J. P. (1987), The invertebrate fauna of grassland and its influence on productivity. III. Effects on soil fertility and plant growth, *Grass and Forage Science*, 42, 325–41.

Dangerfield, J. M., Mccarthy, T. S. and Ellery, W. N. (1998), The mound-building termite *Macrotermes michaelseni* as an ecosystem engineer, *Journal of Tropical Ecology*, 14, 507–20.

Darwin, C. R. (1881), *The Formation of Vegetable Mould Through the Action of Worms with Observations on Their Habits*, London, John Murray.

De Bello, F., Lavorel, S., Díaz, S., Harrington, R., Cornelissen, J. H. C., Bardgett, R. D., Berg, M. P., Cipriotti, P., Feld, C. K., Hering, D., Silva, P. M., Potts, S. G., Sandin, L., Sousa, J. P., Storkey, J., Wardle, D. A. and Harrison, P. A. (2010), Towards an assessment of multiple ecosystem processes and services via functional traits, *Biodiversity and Conservation*, 19(10), 2873–93.

Decaëns, T., Jiménez, J. J., Gioia, C., Measey, G. J. and Lavelle, P. (2006), The values of soil animals for conservation biology, *European Journal of Soil Biology*, 42, S23–S38.

Decaëns, T., Lavelle, P. and Jiménez, J. J. (2008), Priorities for conservation of soil animals, *CAB Reviews: Perspectives in Agriculture, Veterinary Science, Nutrition and Natural Resources*, 3(14), 1–18.

Defoliart, G. R. (1997), An overview of the role of edible insects in preserving biodiversity, *Ecology of Food and Nutrition*, 36, 109–32.

Defoliart, G. R. (1999), Insects as food: Why the Western attitude is important, *Annual Review of Entomology*, 44, 21–50.

Del Toro, I., Ribbons, R. R. and Pelini, S. L. (2012), The little things that run the world revisited: A review of ant-mediated ecosystem services and disservices (Hymenoptera: Formicidae), *Myrmecological News*, 17, 133–46.

Depkat-Jacob, P. S., Hunger, S., Schulz, K., Brown, G. G., Tsai, S. M. and Drake, H. L. (2012), Emission of methane by *Eudrilus eugeniae* and other earthworms from Brazil, *Applied and Environmental Microbiology*, 78(8), 3014–19.

Depkat-Jacob, P. S., Brown, G. G., Tsai, S. M., Horn, M. A. and Drake, H. L. (2013), Emission of nitrous oxide and dinitrogen by diverse earthworm families from Brazil and resolution of associated denitrifying and nitrate-dissimilating taxa, *FEMS Microbiology Ecology*, 83, 375–81.

Derpsh, R., Friedrisch, T., Kassan, A. and Hongwe, L. (2010), Current status of no-till farming in the world an some of its main benefits, *International Journal of Agriculture and Biological Engineering*, 3, 1–26.

Didden, W. A. M. (1993), Ecology of terrestrial Enchytraeidae, *Pedobiologia*, 37, 2–29.

Dossey, A. T. (2010), Insects and their chemical weaponry: New potential for drug discovery, *Natural Product Reports*, 27, 1737–57.

Doube, B. M. and Schmidt, O. (1997), Can the abundance or activity of soil macrofauna be used to indicate the biological health of soils. In: Pankhurst, C. E., Doube, B. M. and Gupta, V. V. S. R. (Eds), *Biological Indicators of Soil Health*, Wallingford, CAB International, pp. 265–95.

Drake, H. L. and Horn, M. A. (2007), As the worm turns: the earthworm gut as a transient habitat for soil microbial biomes, *Annual Review of Microbiology*, 61, 169–89.

Drift, J. van der. (1950), *Analysis of the Animal Community in a Beech Forest Floor*, Wageningen, Ponsen en Looijen.

Drift, J. van der. (1963), *Soil organisms. Colloquium on Soil Fauna, Soil Microflora, and Their Relationships*, Amsterdam, North Holland Publishing Company.

Dunger, W. (1964), *Tiere im boden*, Wittenberg Lutherstadt, Ziemsen Verlag.

Ebregt, E., Struik, P. C., Odongo, B. and Abidin, P. E. (2005), Pest damage in sweet potato, groundnut and maize in north-eastern Uganda with special reference to damage by millipedes (Diplopoda), *NJAS – Wageningen Journal of Life Sciences*, 53(1), 49–69.

Edwards, C. A. and Lofty, J. R. (1978), The influence of arthropods and earthworms upon root growth of direct drilled cereals, *Journal of Applied Ecology*, 15(3), 789–95.

Ehlers, W. (1975), Observations on earthworm channels and infiltration on tilled and untilled Loess soil, *Soil Science*, 119(3), 242–49.

Eijsackers, H., Van Gestel, C. A. M., De Jonge, S., Muijs, B. and Slijkerman, D. (2001), Polycyclic aromatic hydrocarbon-polluted dredge peat sediments and earthworms: A mutual interference, *Ecotoxicology*, 10, 35–50.

Eisenbeis, G. and Wichard, W. (1987), *Atlas on the Biology of Soil Arthropods*, Berlin, Springer.

Eisenhauer, N., Partsch, S., Parkinson, D. and Scheu, S. (2007), Invasion of a deciduous forest by earthworms: Changes in soil chemistry, microflora, microarthropods and vegetation, *Soil Biology and Biochemistry*, 39, 1099–110.

Eisenhauer, N., Schuy, M., Butenschoen, O. and Scheu, S. (2009), Direct and indirect effects of endogeic earthworms on plant seeds, *Pedobiologia*, 52(3), 151–62.

Ernst, D. (1995), *The Farmer's Earthworm Handbook: Managing Your Underground Money-makers*, Brookfield, Lessiter Publications.

Evenhuis, N. L. (2007), Helping solve the 'other' taxonomic impediment: Completing the eight steps to total enlightenment and taxonomic Nirvana, *Zootaxa*, 1407, 3–12.

Faber, J. H., Creamer, R. E., Mulder, C., Römbke, J., Rutgers, M., Sousa, J. P., Stone, D. and Griffiths, B. S. (2013), The practicalities and pitfalls of establishing a policy-relevant and cost-effective soil biological monitoring scheme, *Integrated Environmental Assessment Management (IEAM)*, 9, 276–84.

FAO (2003), *Biological Management of Soil Ecosystems for Sustainable Agriculture*: Report of the International Technical Workshop, FAO, World Soil Resources Reports, 101.

Feller, C., Brown, G. G., Blanchart, E., Deleporte, P. and Chernyanskii, S. S. (2003), Charles Darwin, earthworms and the natural sciences: Various lessons from past to future, *Agriculture, Ecosystems and Environment*, 99, 29–49.

Field Museum. (2017), *Underground Adventure*. Available from: http://www.fieldmuseum.org/about/press/underground-adventure. Acessed 01 March 2017.

Folgarait, P. (1998), Ant biodiversity and its relationship to ecosystem functioning: A review, *Biodiversity and Conservation*, 7, 1221–44.

Fonte, S. and Six, J. (2010), Earthworms and litter management contributions to ecosystem services in a tropical agroforestry system, *Ecological Applications*, 20(4), 1061–73.

Forey, E., Barot, S., Decaens, T., Langlois, E., Laossi, K. R., Margerie, P., Scheu, S. and Eisenhauer, N. (2011), Importance of earthworm-seed interactions for the structure and composition of plant communities: A review, *Acta Oecologica*, 37(6), 594–603.

Fragoso, C. and Lavelle, P. (1992), Earthworm communities of tropical rain forests, *Soil Biology and Biochemistry*, 24, 1397–408.

Franklin, E. C. and Morais, J. W. (2006), Soil mesofauna in Central Amazon. In: Moreira, F. M. S., Siqueira, J. O. and Brussaard, L. (Eds), *Soil Biodiversity in Amazonian and Other Brazilian Ecosystems*, Wallingford, CABI, pp. 142–62.

Franklin, E., Aguiar, N. O. and Soares, E. D. L. (2008), Invertebrados do solo. In: Oliveira, M. L., Baccar, F. B., Bragoa-Neto, R. and Magnusson, W. E. (Eds), *Reserva Ducke: a biodiversidade amazônica através de uma grade*, Manaus, Áttema Design, pp. 109–22.

Freitas, A. V. L., Leal, I. R., Uehara-Prado, M. and Iannuzzi, L. (2006),. Insetos como indicadores de conservação da paisagem. In: Rocha, C. F. D., Bergalo, H. G., Sluys, M. V. and Alves, M. A. S. (Eds), *Biologia da conservação: Essências*, São Carlos, Editora Rima, pp. 1–28.

Freitas, B. M. and Pereira, J. O. P. (2004), *Solitary Bees: Conservation, Rearing and Management for Pollination*, Fortaleza, Imprensa Universitária, Universidade Federal do Ceará.

Frouz, J. and Jilková, V. (2008), The effect of ants on soil properties and processes (Hymenoptera: Formicidae), *Myrmecological News*, 11, 191–9.

Garcia, M. B., Antor, R. J. and Espadaler, X. (1995), Ant pollination of the palaeoendemic dioecious *Borderea pyrenaica* (Dioscoriaceae), *Plant Systematics and Evolution*, 198, 17–27.

Gardi, C. and Jeffery, S. (2009). *Soil biodiversity*. Ispra, JRC/EC.

Gassen, D. N. (2000), Os escarabeídeos na fertilidade de solo sob plantio direto. In: *Anais da Fertbio 2000, Biodinâmica do Solo*, Santa Maria, Universidade Federal de Santa Maria e SBCS., pp. 1–7.

Giller, P. E. (1996), The diversity of soil communities, the 'poor man's tropical rainforest', *Biodiversity and Conservation*, 5, 135–68.

Gobat, J.-M., Aragno, M. and Matthey, W. (2004), *The Living Soil. Fundamentals of Soil Science and Soil Biology*, Enfield, Science Publishers.

Gómez, J. M. and Zamora, R. (1992), Pollination by ants: Consequences of the quantitative effects on a mutualistic system, *Oecologia*, 91, 410–18.

Grimaldi, M., Oszwald, J., Dolédec, S., Hurtado, M. P., Miranda, I. S., Sartre, X. A., Assis, W. S., Castañeda, E., Desjardins, T., Dubs, F., Guevara, E., Gond, V., Lima, T. T. S., Marichal, R., Michelotti, F., Mitja, D., Noronha, N. C., Oliveira, M. N. D., Ramirez, B., Rodriguez, G., Sarrazin, M., Silva, M. L. S., Costa, L. G. S., Souza, S. L., Veiga, I., Velasquez, E. and Lavelle, P. (2014), Ecosystem services of regulation and support in Amazonian pioneer fronts: Searching for landscape drivers, *Landscape Ecology*, 29, 311–28.

Groombridge, B. (1992), *Global Biodiversity. Status of the Earth's Living Resources*, London, Chapman & Hall.

Guimarães, R. M. L., Ball, B. C. and Tormena, C. A. (2011), Improvements in the visual evaluation of soil structure, *Soil Use and Management*, 27(3), 395–403.

Hagenah, N. and Bennett, N. C. (2013), Mole rats act as ecosystem engineers within a biodiversity hotspot, the Cape Fynbos, *Journal of Zoology*, 289, 19–26.

Hendrix, P. F., Parmelee, R. W., Crossley, D. A., Coleman, D. C., Odum, E. P. and Groffman, P. M. (1986), Detritus food webs in conventional and no-tillage agroecosystems, *Bioscience*, 36, 374–80.

Hickman, Z. A. and Reid, B. J. (2008), Earthworm assisted bioremediation of organic contaminants, *Environment International*, 34, 1072–81.

Hoagland, K. E. (1996), The taxonomic impediment and the Convention of Biodiversity, *Association of Systematics Collections Newsletter*, 24(5), 61–2, 66–7.

Hoffman, J. (1994), *Backyard Scientist, Exploring Earthworms with Me*, Irvine, Backyard Scientist.

Hoffmann-Campo, C. B., Oliveira, L. J., Moscardi, F., Corrêa-Ferreira, B. S. and Corso, I. C. (2013), Pragas que atacam plântulas, hastes e pecíolos da soja. In: Hoffmann-Campo, C. B., Corrêa-Ferreira, B. S. and Moscardi, F. (Eds), *Soja: manejo integrado de insetos e outros Artrópodes-Praga*, Londrina, Embrapa Soja, pp. 145–212.

Hole, F. D. (1981), Effects of animals on soil, *Geoderma*, 25(1–2), 75–112.

Hölldobler, B. and Wilson, E. O. (1990), *The Ants*, Cambridge, Harvard University Press.

Honer, M. R. and Gomes, A. (1990), *O manejo integrado de mosca dos chifres, berne e carrapato em gado de corte*, Campo Grande, *EMBRAPA-CNPGC, Circular Técnica, 22*.

House, G. J. and Parmelee, R. W. (1985), Comparison of soil arthropods and earthworms from conventional and no-tillage agroecosystems, *Soil & Tillage Research*, 5, 351–60.

Huang, S. P. and Cares, J. E. (2006), Nematode communities in soils under different land-use systems in Brazilian Amazon and Savannah vegetation. In: Moreira, F. M. S., Siqueira, J. O. and Brussaard, L. (Eds), *Soil Biodiversity in Amazonian and Other Brazilian Ecosystems*, Wallingford: CABI, pp. 163–83.

Huerta, E., Rodríguez-Olán, J., Evia-Castillo, I., Montejo-Meneses, E. and García-Hernández, R. (2005), La diversidad de lombrices de tierra (Annelida, Oligochaeta) en el estado de Tabasco, México, *Universidad y Ciencia*, 21(42), 75–85.

Huguenin, M. T., Leggett, C. G. and Paterson, R. W. (2006), Economic valuation of soil fauna, *European Journal of Soil Biology*, 42, S16–S22.

Huhta, V. (2007), The role of soil fauna in ecosystems: A historical review, *Pedobiologia*, 50, 489–95.

Huntly, N. and Inouye, R. (1988), Pocket gophers in ecosystems: Patterns and mechanisms, *Bioscience*, 38, 786–93.

International Organization for Standardization. (1993), ISO 11268-1, *Soil quality – Effects of pollutants on earthworms (Eisenia fetida) – Part 1: Determination of acute toxicity using artificial soil substrate*, Geneva, ISO.

International Organization for Standardization. (1998), ISO 11268-2, *Soil quality – Efects of pollutants on earthworms (Eisenia fetida) – Part 2: Determination of effects on reproduction*, Geneva, ISO.

International Organization for Standardization. (2014), ISO/DIS 18311, *Soil quality – Method for testing effects of soil contaminants on the feeding activity of soil dwelling organisms: Bait lamina test*, Geneva, ISO.

Inward, D., Beccaloni, G. and Eggleton, P. (2007), Death of an order: A comprehensive molecular phylogenetic study confirms that termites are eusocial cockroaches, *Biology Letters*, 3(3), 331–5.

Jänsch, S., Römbke, J. and Didden, W. (2005), The use of enchytraeids in ecological soil classification and assessment concepts, *Ecotoxicology and Environmental Safety*, 62(2), 266–77.

Jeffery, S., Gardi, C., Jones, A., Montanarella, L., Marmo, L., Miko, L., Ritz, K., Peres, G., Römbke, J. and van der Putten, W. H. (2010), *European Atlas of Soil Biodiversity*, Luxembourg, European Union.

Johnson, D. L. (2002), Darwin would be proud: Bioturbation, dynamic denudation, and the power of theory in science, *Geoarchaeology*, 17(1), 7–40.

Jouquet, P., Traoré, S., Choosai, C., Hartmann, C. and Bignell, D. (2011), Influence of termites on ecosystem functioning. Ecosystem services provided by termites, *European Journal of Soil Biology*, 47, 215–22.

Jouquet, P., Blanchart, E. and Capowiez, I. (2014), Utilization of earthworms and termites for the restoration of ecosystem functioning, *Applied Soil Ecology*, 73, 34–40.

Kevan, D. K. Mc. E. (1955), *Soil zoology*. In: *Proceedings of the University of Nottingham, Second Easter School in Agricultural Science*, London, Butterworths.

Kevan, D. K. Mc. E. (1985), Soil zoology, then and now: Mostly then, *Quaestiones Entomologicae*, 21, 371.7–472.

Kratz, W. (1998), The bait-lamina test, *Environmental Science and Pollution Research*, 5(2), 94–6.

Kühnelt, W. (1950), *Bodenbiologie: mit besonderer Berücksichtigung der Tierwelt*, Vienna, Herold Verlag.

Kuperman, R. G., Checkai, R. T., Garcia, M. V. B., Römbke, J., Stephenson, G. L. and Sousa, J. P. (2009), State of the science and the way forward for the ecotoxicological assessment of contaminated land, *Pesquisa Agropecuária Brasileira*, 44(8), 811–24.

Lacey, E. A., Patton, J. L. and Cameron, G. N. (2000), *Life Underground: The Biology of Subterranean Rodents*, Chicago, Chicago University Press.

Langmaid, K. K. (1964), Some effects of earthworm invasion in virgin podzols, *Canadian Journal of Soil Science*, 44, 34–7.

Larson, G. (1998), *There's a Hair in My Dirt! A Worm's Story*, New York, Harper Collins.

Lavelle, P. (1996), Diversity of soil fauna and ecosystem function, *Biology International*, 33, 3–16.

Lavelle, P. (1997), Faunal activities and soil processes: Adaptive strategies that determine ecosystem function, *Advances in Ecological Research*, 27, 93–132.

Lavelle, P. (2000), Ecological challenges for soil science, *Soil Science*, 165, 73–86.

Lavelle, P. and Lapied, E. (2003), Endangered earthworms of Amazonia: A homage to Gilberto Righi, *Pedobiologia*, 47, 419–27.

Lavelle, P. and Spain, A. V. (2001), *Soil Ecology*, Amsterdam, Kluwer.

Lavelle, P., Blanchart, E., Martin, A., Spain, A. V. and Martin, S. (1992), Impact of soil fauna on the properties of soils in the humid tropics. In: Lal, R. and Sanchez, P. (Eds), *Myths and Science of Soils of the Tropics*, Madison, SSSA Special Publication, 29, pp. 157–85.

Lavelle, P., Dangerfield, M., Fragoso, C., Eschenbrenner, V., López-Hernández, D., Pashanasi, B. and Brussaard, L. (1994), The relationship between soil macrofauna and tropical soil fertility. In: Woomer, P. L. and Swift, M. J. (Eds), *The Biological Management of Tropical Soil Fertility*, Chichester, John Wiley and Sons, pp. 137–69.

Lavelle, P., Bignell, D., Lepage, M., Wolters, V., Roger, P., Ineson, P., Heal, O. W. and Ghillion, S. (1997), Soil function in a changing world: The role of invertebrate ecosystem engineers, *European Journal of Soil Biology*, 33, 159–93.

Lavelle, P., Blouin, M., Boyer, J., Cadet, P., Laffray, D., Pham-Thi, A.-T., Reversat, G., Settle, W. and Zuily, Y. (2004), Plant parasite control and soil fauna diversity, *Comptes Rendues de l'Académie des Sciences de Paris, Serie Biologie*, 327, 629–38.

Lavelle, P., Decaëns, T., Aubert, M., Barot, S., Blouin, M., Bureau, F., Margerie, P., Mora, P. and Rossi, J.-P. (2006), Soil invertebrates and ecosystem services, *European Journal of Soil Biology*, 42, S3–S15.

Lavelle, P., Rodríguez, N., Arguello, O., Bernal, J., Botero, C., Chaparro, P., Gómez, Y., Gutiérrez, A., Hurtado, M. P., Loaiza S., Pullido S., X., Rodríguez, E., Sanabria, C., Velásquez, E. and Fonte, S. J. (2014), Soil ecosystem services and land use in the rapidly changing Orinoco River Basin of Colombia, *Agriculture, Ecosystems and Environment*, 185, 106–17.

Lavelle, P., Dolédec, S., Sartre, X. A., Decaëns, T., Gond, V., Grimaldi, M., Oszwald, J., Hubert, B., Ramirez, B., Veiga, I., Souza, S., Assis, W. S., Michelotti, F., Martins, M., Feijoo, A., Bommelm, P., Castañeda, E., Chacon, P., Desjardins, T., Dubs, F., Gordillo, E., Guevara, E., Fonte, S., Hurtado, M. P., Lena, P., Lima, T., Marichal, R., Mitja, D., Miranda, I., Otero, T., Praxedes, C., Poccard, R., Roberto, P., Rodriguez, G., Sanabria, C., Teslouiko, S., Velasquez, A., Velasquez, E. and Velasquez, J. (2016), Unsustainable landscapes of deforested Amazonia: An analysis of the relationships among landscapes and the social, economic and environmental profiles of farms at different ages following deforestation, *Global Environmental Change*, 40, 137–55.

Lawton, J. H., Bignell, D. E., Bolton, B., Bloemers, G. F., Eggleton, P., Hammond P. M., Hodda, M., Holt, R. D., Larsen, T. B., Mawdsley, N. A., Stork, N. E., Srivastava, D. S. and Watt, A. D. (1998), Biodiversity indicators, indicator taxa and effects of habitat modification in tropical forest nature, *Nature*, 391, 72–6.

Lazcano, C. and Domínguez, J. (2011), The use of vermicompost in sustainable agriculture: Impact on plant growth and soil fertility. In: Miransari, M. (Ed.), *Soil Nutrients*, New York, Nova Science Publishers, pp. 230–54.

Leal, I. R., Wirth, R. and Tabarelli, M. (2007), Seed dispersal by ants in the semi-arid caatinga of north-east Brazil, *Annals of Botany*, 99, 885–94.

Lee, K. E. and Wood, T. G. (1971), *Termites and Soils*, London, Academic Press.

Lengyel, S., Gove, A. D., Latimer, A. M., Majer, J. D. and Dunn, R. R. (2009), Ants sow the seeds of global diversification in flowering plants, *PLoS One*, 4(5), e5480.

Leonard, J. and Rajot, J. L. (2001), Influence of termites on runoff and infiltration: Quantification and analysis, *Geoderma*, 104, 17–40.

Lobry De Bruyn, L. A. and Conacher, A. J. (1994), The effect of ant biopores on water infiltration in soils in undisturbed bushland and farmland in a semi-arid environment, *Pedobiologia*, 38, 193–207.

Loranger-Merciris, G., Cabidoche, Y.-M., Deloné, B., Quénéhervé, P. and Ozier-Lafontaine, H. (2012), How earthworm activities affect banana plant response to nematodes parasitism, *Applied Soil Ecology*, 52, 1–8.

Published by Burleigh Dodds Science Publishing Limited, 2018.

Lowe, C. N. and Butt, K. R. (2005), Culture techniques for soil dwelling earthworms: A review, *Pedobiologia*, 49, 401–13.

Lubbers, I. M., Van Groenigen, K. J., Fonte, S. J., Six, J., Brussaard L. and Van Groenigen, J. W. (2013), Greenhouse gas emissions from soils increased by earthworms, *Nature Climate Change*, 3, 187–94.

Mando, A. and Miedema, R. (1997), Termite-induced change in soil structure after mulching degraded (crusted) soil in the Sahel, *Applied Soil Ecology*, 6, 241–9.

Martinsson, S., Cui, Y., Martin, P. J., Pinder, A., Quinlan K., Wetzel, M. J. and Erséus, C. (2015), DNA-barcoding of invasive European earthworms (Clitellata: Lumbricidae) in south-western Australia, *Biological Invasions*, 17, 2527–32.

Mathieu J. (2004), *Étude de la macrofaune du sol dans une zone de déforestation en Amazonie du sud-est, dans le contexte de l'agriculture familiale*. PhD thesis in Ecology, Université de Paris VI.

McLaughlin, M. (1986), *Earthworms, Dirt and Rotten Leaves*, New York, Macmillan Publishing Company.

Meadows, P. S. and Meadows, A. (1992), The environmental impact of burrowing animals and animal burrows. In: *Proceedings of a Symposia of the Zoological Society of London*, Oxford, Oxford University Press.

Meysman, F. J. R., Middelburg, J. J. and Heip, C. H. R. (2006), Bioturbation: A fresh look at Darwin's last idea, *TRENDS in Ecology and Evolution*, 21(12), 688–95.

Millennium Ecosystem Assessment (2005), *Ecosystem and Human Well-being: Synthesis*, Washington D.C., Island Press.

Miklós, A. A. W. (2012), Biogênese do solo, Revista do Departamento de Geografia – USP, Volume Especial 30 Anos, pp. 190–229.

Murphy, P. W. (1962), *Progress in Soil Zoology*, London, Butterworths.

Natal-Da-Luz, T., Lee, I., Verweij, R., Morais, P. V., Van Velzen, M., Sousa, J. P. and Van Gestel, C. A. M. (2012), The influence of earthworm activity on microbial communities related with the degradation of persistent pollutants, *Environmental Toxicology and Chemistry*, 31(4), 794–803.

Neher, D. A. (2001), Role of nematodes in soil health and their use as indicators, *Journal of Nematology*, 33(4), 161–8.

Neher, D. A. (2010), Ecology of plant and free-living nematodes in natural and agricultural soil, *Annual Review of Phytopathology*, 48, 371–94.

Ngugi, D. K., Ji, R. and Brune, A. (2011), Nitrogen mineralization, denitrification, and nitrate ammonification by soil-feeding termites: A N-15-based approach, *Biogeochemistry*, 103, 355–69.

Nichols, E., Spector, S., Louzada, J., Larsen, T., Amezquita, S. and Favilla, M. E. (2008), Ecological functions and ecosystem services provided by Scarabaeinae dung beetles, *Biological Conservation*, 141, 1461–74.

Nielsen, G. A. and Hole, F. D. (1964), Earthworms and the development of coprogenous A_1 horizons in forest soils of Wisconsin, *Soil Science Society of America Proceedings*, 28, 426–30.

Nielsen, U. F., Wall, D. H. and Six, J. (2015), Soil biodiversity and the environment, *Annual Review of Environment and Resources*, 40, 4.1–4.28.

Nielsen, U. F., Ayres, E., Wall, D. H., Li, G., Bardgett, R. D., Wu, T. and Garey, J. R. (2014), Global-scale patterns of assemblage structure of soil nematodes in relation to climate and ecosystem properties, *Global Ecology and Biogeography*, 23, 968–78.

Nooren, C. A. M., Van Breemen, N., Stoorvogel J. J. and Jongmans, A. G. (1995), The role of earthworms in the formation of sandy surface soils in a tropical forest in Ivory Coast, *Geoderma*, 65, 135–48.

Organization for Economic Co-Operation and Development. (2006), *Guidance document on the breakdown of organic matter in litterbags*. Paris, OECD, Series on Testing and Assessment, 56.

Orgiazzi, A., Bardgett R. D., Barrios, E., Behan-Pelletier V., Briones M. J. I., Chotte, J-L., De Deyn, G. B., Eggleton, P., Fierer N., Fraser, T., Hedlund, K., Jeffery, S., Johnson, N. C., Jones, A., Kandeler, E., Kaneko, N., Lavelle, P., Lemanceau, P., Miko, L., Montanarella, L., Moreira, F. M. S.,

Ramirez, K. S., Scheu, S., Singh, B. K., Six, J., van der Putten W. H. and Wall, D. H. (2016), *Global Soil Biodiversity Atlas*, Luxembourg, European Union.

Ortiz, B., Fragoso, C., M'boukou, I., Pashanasi, B., Senapati, B. and Contreras, A. (1999), Perception and use of earthworms in tropical farming systems. In: Lavelle P., Brussaard, L. and Hendrix, P. (Eds), *Earthworm Management in Tropical Agroecosystems*, Wallingford, CABI, pp. 239–49.

Pagiola, S., von Ritter, K. and Bishop, J. (2004), *Assessing the Economic Value of Ecosystem Conservation*. Washington D.C., World Bank, Environment Development Paper No. 101.

Paoletti, M. G. (1999), *Invertebrate Biodiversity as Bioindicators of Sustainable Landscapes: Practical Use of Invertebrates to Assess Sustainable Land Use*, Amsterdam, Elsevier.

Paoletti, M. G. (2004), *Ecological Implications of Minilivestock: Role of Insects, Rodents, Frogs and Snails for Sustainable Development*, Enfield, Science Publishers.

Papini, S. and Andréa, M. M. (2001), Dissipação de simazina em solo por ação de minhocas (*Eisenia fetida), Revista Brasileira de Ciência do Solo*, 25, 593–9.

Parker, C. A. (1989), Soil biota and plants in the rehabilitation of degraded agricultural soils. In: Majer, J. D. (Ed.), *Animals and Primary Succession: The Role of Fauna in Reclaimed Lands*, Cambridge, Cambridge University Press, pp. 423–38.

Peakall, R., Handel, S. N. and Beattie, A. J. (1991), The evidence for, and importance of, ant pollination. In: Huxley, C. R. and Cutler, D. F. (Eds), *Ant-plant Interactions*, Oxford, Oxford University Press, pp. 421–8.

Peracchi, A. L. and Nunes, W. O. (1972), Sôbre um diplópodo prejudicial à cultura da mandioca (*Manihot esculenta), Pesquisa Agropecuária Brasileira*, 7, 181–3.

Pierce, N. E., Braby, M. F., Heath, A., Lohman D. J., Mathew, J., Rand, D. B. and Travassos, M. A. (2002), The ecology and evolution of ant association in the Lycaenidae (Lepidoptera), *Annual Review of Entomology*, 47, 733–71.

Pimentel, D., Harvey, C., Resosudarmo, P., Sinclair, K., Kurz, D., Mcnair, M., Crist, S., Shpritz, L., Fitton, L., Saffouri, R. and Blair, R. (1995), Environmental and economic costs of soil erosion and conservation benefits, *Science*, 267, 5201, 1117–23.

Pimentel, D., Wilson, C., Mccullum, C., Huang, R., Dwen, P., Flack, J., Tran, Q., Saltman, T. and Cliff, B. (1997), Economic and environmental benefits of biodiversity, *BioScience*, 47, 747–57.

Prather, C. S., Pelini, S. L., Laws, A., Rivest, E., Woltz, M., Bloch, C. P., Del Toro, I., Ho, C. K., Kominoski J., Newbold, T. A. S., Parsons S. and Joern, A. (2013), Invertebrates, ecosystem services and climate change, *Biological Reviews of the Cambridge Philosophical Society*, 88(2), 327–48.

Pulleman, M., Creamer, R., Hamer, U., Helder, J., Pelosi, C., Pérès, G. and Rutgers, M. (2012), Soil biodiversity, biological indicators and soil ecosystem services: an overview of European approaches, *Current Opinion in Environmental Sustainability*, 4(5), 529–38.

Puterbaugh, M. N. (1998), The roles of ants as flower visitors: experimental analysis in three alpine plant species, *Oikos*, 83, 36–46.

Pyron, R. A. and Wiens, J. J. (2011), A large-scale phylogeny of Amphibia including over 2800 species, and a revised classification of extant frogs, salamanders, and caecilians, *Molecular Phylogenetics and Evolution*, 61, 543–83.

Pyron R. A. and Burbrink, F. T. (2012), Extinction, ecological opportunity, and the origins of global snake diversity, *Evolution*, 66(1), 163–78.

Pyron, R. A., Burbrink, F. T. and Wiens, J. J. (2013), A phylogeny and revised classification of Squamata, including 4161 species of lizards and snakes, *BMC Evolutionary Biology*, 13, 93.

Rico-Gray, V. and Oliveira, P. S. (2007), *The Ecology and Evolution of Ant-plant Interactions*, Chicago, The University of Chicago Press.

Rissing, S. W. (1986), Indirect effects of granivory by harvester ants: plant species composition and reproductive increase near ant nest, *Oecologia*, 68, 231–4.

Römbke, J., Jänsch S. and Didden, W. (2005), The use of earthworms in ecological soil classification and assessment concepts, *Ecotoxicology and Environmental Safety*, 62(2), 249–65.

Römbke, J. (2014), The feeding activity of invertebrates as a functional indicator in soil, *Plant and Soil*, 383, 43–6.

Römbke, J., Gardi C., Creamer, R. and Miko, L. (2016), Soil biodiversity data: actual and potential use in European and national legislation, *Applied Soil Ecology*, 97, 125–33.

Ruiz, N., Lavelle P. and Jiménez, J. J. (2008), *Soil Macrofauna Field Manual. Technical Level*, Rome, FAO.

Sánchez-Hernández, J. C., Aira, M. and Domínguez J. (2014), Extracellular pesticide detoxification in the gastrointestinal tract of the earthworms *Aporrectodea caliginosa*, *Soil Biology and Biochemistry*, 79, 1–4.

Sandhu, H. S., Wratten, S. D. and Cullen, R. (2010), The role of supporting ecosystem services in conventional and organic arable farmland, *Ecological Complexity*, 7, 302–10.

Satchell, J. E. (1975), Soil biology and the doctoral syndrome. In: Vanek, J. (Ed.), *Progress in soil zoology. Proceedings of the V International Colloquium on Soil Zoology*, Prague, Akademia, pp. 603–10.

Satchell, J. E. (1992), Take the money – call the tune, *Soil Biology and Biochemistry*, 24(12), 1193–6.

Scheu, S. (2003), Effects of earthworms on plant growth: Patterns and perspectives, *Pedobiologia*, 47, 846–56.

Schmelz, R. M. and Collado, R. (2015), Checklist of taxa of Enchytraeidae (Oligochaeta, Annelida), *Soil Organisms*, 87, 149–52.

Schockaert, E. R., Hooge, M., Sluys R., Schilling, S., Tyler, S. and Artois T. (2008), Global diversity of free living flatworms (Platyhelminthes, 'Turbellaria') in freshwater, *Hydrobiologia*, 595, 41–8.

Senckenberg Museum. (2017), *Touring exhibition: The thin skin of earth – our soils*. Available at: http://www.senckenberg.de/root/index.php?page_id=17871&preview=true. Accessed on 1 March 2017.

Setälä, H. and Huhta, V. (1991), Soil fauna increase *Betula pendula* growth: laboratory experiments with coniferous forest floor, *Ecology*, 72(2), 665–71.

Sheehan, C., Kirwan, L., Connolly, J. and Bolger, T. (2008), The effects of earthworm functional diversity on microbial biomass and the microbial community level physiological profile of soils, *European Journal of Soil Biology*, 44, 65–70.

Sheperd, V. E. and Chapman, C. A. (1998), Dung beetles as secondary seed dispersers: Impact on seed predation and germination, *Journal of Tropical Ecology*, 14(2), 199–215.

Shepherd, G., Stagnari, F., Pisante, M. and Benítes, J. (2009), *Visual Soil Assessment: Field Guides*, Rome, FAO.

Shuttleworth, K. L. and Cerniglia, C. A. (1995), Environmental aspects of PAH biodegradation, *Applied Biochemistry and Biotechnology*, 54, 291–301.

Silva, R. F., Aquino, A. M., Mercante, F. M. and Guimarães, M. F. (2006), Macrofauna invertebrada do solo sob diferentes sistemas de produção em Latossolo da região do Cerrado, *Pesquisa Agropecuária Brasileira*, 41(4), 697–704.

Silva R. F., Guimarães, M. F., Aquino, A. M. and Mercante, F. M. (2011), Análise conjunta de atributos físicos e biológicos do solo sob sistema de integração lavourapecuária, *Pesquisa Agropecuária Brasileira*, 46(10), 1277–83.

Srivastava S. K., Babu, N. and Pandey, H. (2009), Traditional insect bioprospecting: as human food and medicine, *Indian Journal of Traditional Knowledge*, 8, 485–94.

Stockdill S. M. J. (1982), Effects of introduced earthworms on the productivity of New Zealand pastures, *Pedobiologia*, 24, 29–35.

Stout, J. D. (1983), Organic matter turnover by earthworms, In: Satchell, J. E. (Ed.), *Earthworm Ecology: From Darwin to Vermiculture*, New York, Chapman and Hall, pp. 35–48.

Susilo, F. X., Neutal, A. M., Van Noordwijk, M., Hairiah K., Brown, G. and Swift, M. J. (2004), Soil biodiversity and food webs. In: Van Noordwijk, M., Cadisch, G. and Ong, C. K. (Eds), *Below-ground Interactions in Tropical Agroecosystems: Concept and Models with Multiple Plant Components*, Wallingford, CAB International Publishing, pp. 285–308.

Swift, M. J., Heal, O. W. and Anderson, J. M. (1979), *Decomposition in Terrestrial Ecosystems*, Berkeley, University of California Press.

Syers, J. K. and Springett, J. A. (1984), Earthworms and soil fertility, *Plant and Soil*, 76, 93–104.

Tavares Filho, J., Ralisch, R., Guimarães, M. F., Medina, C. C., Balbino, L. C. and Neves, C. S. V. J. (1999), Método do perfil cultural para avaliações do estado físico do solos em condições tropicais, *Revista Brasileira de Ciência do Solo*, 23, 393–9.
Törne, E. V. (1990), Assessing feeding activities of soil-living animals. I. Bait-lamina-tests, *Pedobiologia*, 342, 89–101.
Toyota, A., Kaneko, N. and Ito, M. T. (2006), Soil ecosystem engineering by the train millipede *Parafontaria laminata* in a Japanese larch forest, *Soil Biology and Biochemistry*, 38, 1840–50.
Turbé, A., Toni, A., Benito, P., Lavelle, P., Lavelle, P., Ruiz, N., Van der Putten, W. H., Labouze, E. and Mudgal, S. (2010), *Soil biodiversity: functions, threats and tools for policy makers*. Parie, Bio Intelligence Service, IRD, and NIOO, Report for European Commission (DG Environment).
Vandewalle, M., Sykes, M. T., Harrison, P. A., Luck, G. W., Berry, P., Bugter R., Dawson, T. P., Feld, C. K., Harrington, R., Haslett J. R., Hering, D., Jones, K. B., Jongman R., Lavorel, S., Martins Da Silva, P., Moora, M., Paterson, J., Rounsvell, M. D. A., Sandin, L., Sattele, J., Sousa, J. P. and Zobel, M. (2008), *Review paper on concepts of dynamic ecosystems and their services. The Rubicode Project*, Report to the European Union Sixth Framework Programme.
Van Groenigen J. W., Lubbers, I. M., Vos, H. M. J., Brown, G. G., De Deyn, G. B. and Van Groenigen, K. J. (2016), Earthworms increase plant production: A meta-analysis, *Nature Scientific Reports*, 4(6365), 1–7.
Van Hoof, P. (1983), Earthworm activity as a cause of splash erosion in a Luxembourg forest, *Geoderma*, 31, 195–204.
Van Noordwijk, M. and Swift, M. J. (1999), Belowground biodiversity and sustainability of complex agroecosystems. In: Gafur, A., Susilo, F. X., Utomo, M. and Van Noordwijk, M. (Eds), *Management of Agrobiodiversity for Sustainable Land Use and Global Environmental Benefits*, Bogor, ICRAF, pp. 8–28.
Van Schaik, L., Palm, J., Klaus, J., Zehe, E. and Schröder, B. (2014), Linking spatial earthworm distribution to macropore numbers and hydrological effectiveness, *Ecohydrology*, 7(2), 401–8.
Van Vliet, P. C. J., West, L. T. and Hendrix, P. F. and Coleman, D. C. (1995), The influence of Enchytraeidae (Oligochaeta) on the soil porosity of small microcosms, *Geoderma*, 56, 1–4, 287–99.
Velásquez, E., Lavelle, P. and Andrade, M. (2007a), GISQ, a multifunctional indicator of soil quality, *Soil Biology and Biochemistry*, 39, 3066–80.
Velásquez, E., Pelosi, C., Brunet, D., Grimaldi, M., Martins, M., Rendeiro, A. C., Barrios, E. and Lavelle, P. (2007b), This ped is my ped: Visual separation and near infrared spectra allow determination of the origins of soil macroaggregates, *Pedobiologia*, 51, 75–87.
Velásquez, E., Fonte, S. J., Barot, S., Grimaldi, M., Desjardins, T. and Lavelle, P. (2012), Soil macrofauna-mediated impacts of plant species composition on soil functioning in Amazonian pasture, *Applied Soil Ecology*, 56, 43–50.
Wall, D. H., Behan-Pelletier, V., Covich, A. P. and Snelgrove, P. (2007), *Hidden assets: Biodiversity below-surface*, Paris, UNESCO-SCOPE Policy Briefs No. 5.
Wall, D. H., Bardgett, R. D., Behan-Pelletier, V., Herrick, J. E., Jones, T. H., Ritz, K., Six, J., Strong, D. R. and Van Der Putten, W. H. (2012), *Soil Ecology and Ecosystem Services*, Oxford, Oxford University Press.
Wall, D. H., Nielsen, U. F. and Six, J. (2015), Soil biodiversity and human health, *Nature*, 528, 69–76.
Wallwork, J. A. (1970), *Ecology of Soil Animals*, London, McGraw-Hill.
Walter, D. E. and Proctor, H. C. (2013), *Mites – Ecology, Evolution and Behaviour: Life at a Microscale*, 2nd ed., New York, Springer.
Weissenfels, W. D., Klewer, H.-J. and Langhoff, J. (1992), Adsorption of polycyclic aromatic hydrocarbons (PAHs) by soil particules: influence on biodegradation and biotoxicity, *Applied Microbiology and Biotechnology*, 36, 689–96.
Wilkinson, M. T., Richards, P. J. and Humphreys, G. S. (2009), Breaking ground: pedological, geological, and ecological implications of soil bioturbation, *Earth-Science Reviews*, 97, 257–72.
Wood, T. G. (1988), Termites and soil environment, *Biology and Fertility of Soils*, 6, 228–36.

Yeates, G. W. (1979), Soil nematodes in terrestrial ecosystems, *Journal of Nematology*, 11(3), 213–29.
Yeates, G. W. (1981), Soil nematode populations depressed in the presence of earthworms, *Pedobiologia*, 22, 191–5.
Zangerlé, A., Renard, D., Iriarte, J., Suarez Jimenez, L. E., Adame Montoya, K. L., Juilleret, J. and McKey, D. (2016), The surales, self-organized earth-mound landscapes made by earthworms in a seasonal tropical wetland, *PLoS ONE*, 11(5), e0154269.
Zhang, Z.-Q. (2011a), Animal biodiversity: An outline of higher-level classification and survey of taxonomic richness, *Zootaxa*, 3148, 212–21.
Zhang, Z.-Q. (2011b), Phylum Arthropoda von Siebold, 1848. *Zootaxa*, 3148, 99–103.
Zhang, W., Hendrix, P. F., Dame, L. E., Burke, R. A., Wu, J., Neher, D. A., Li, J., Shao, Y. and Fu, S. (2013), Earthworms facilitate carbon sequestration through unequal amplification of carbon stabilization compared with mineralization, *Nature Communications*, 4, 2576.
Zhenjun, S. (2003), *Vermiculture and vermiprotein*, Beijing, China Agricultural University Press.

Part 3

Soil dynamics

Chapter 10

The role of soil hydrology in soil health

Melissa Miller and Henry Lin, Pennsylvania State University, USA

1 Introduction

The interactive relationship between soil and water is fundamental to plant growth and many other ecosystem services provided by soils, from hosting large microbial biodiversity to providing habitat to soil fauna and from filtering water to regulating trace gas emissions. These interactions are complex and dynamic, relying on the capacity of soils to retain and filter water as well as transport and transform nutrients and other chemicals. Soil health, as defined by Doran et al. (1996), is '*the capacity of a living soil to function, within natural or managed ecosystem boundaries, to sustain plant and animal productivity, maintain or enhance water and air quality, and promote plant and animal health*'. This indicates that the definition of soil health must go beyond the traditional view of soil as a medium for crop growth alone. In sustainable agriculture, soils must not only provide sufficient water for crop growth, but also prevent rapid transport of fertilizers, pesticides and sediment to water sources and reduce the potential for soil loss and degradation. These require a balance of storage and fluxes in soils to allow for necessary soil functions, including maintaining an appropriate balance between water-filled and air-filled pore spaces for plant and microbial health, adequate infiltration to prevent surface ponding and runoff, and suitable drainage to prevent saturated subsurface conditions. Integral to all of these soil functions is the storage and movement of water in soils.

The study of soil health has thus far focused on a number of metrics regarding water flow and retention in soils: soil texture, available water-holding capacity, surface and subsurface penetration resistance, wet aggregate stability and organic matter content, to name only a few (Moebius-Clune et al., 2016). To fully examine how soils function with respect to water in soils, we must look beyond these metrics in order to understand related processes and

http://dx.doi.org/10.19103/AS.2017.0033.12

the landscape setting in which these functions occur in different soil types (Lin et al., 2006). Soil water dynamics depend not only on surface soil conditions, but also on the deeper subsurface and various interfaces in a soil profile (Lin, 2012). The classical understanding of water flow in soils often assumes homogeneity, with uniform wetting fronts and relatively simple flow pathways, which leads to simplified models to predict water flow in soils. Recent understanding and advancements in soil hydrology and hydropedology have led to the recognition of diverse flow scenarios in soils accounting for heterogeneity and complex flow networks (Lin, 2012).

This chapter will discuss various soil hydrologic processes and properties that influence soil health along with crop growth, ecosystem services and environmental impacts. Soil physical, chemical and biological processes are often mediated by soil hydrologic conditions. These processes and related properties will be discussed in this chapter in relation to soil hydrologic functions in the context of soil health and various management practices.

2 Soil health indicators related to soil hydrology

The concept of soil health is derived from the need to continue to produce enough food for a growing population while also recognizing that established agricultural practices have been degrading soils, and reducing their capacity to function as a medium for food production and a source of environmental protection. The origin of the concept of soil health came somewhat out of the recognition that soil management was contributing to water quality degradation as well as the increased rate of soil degradation through water and wind erosion (Doran and Parkin, 1994). Hydrology is at the nexus of soil health, as water

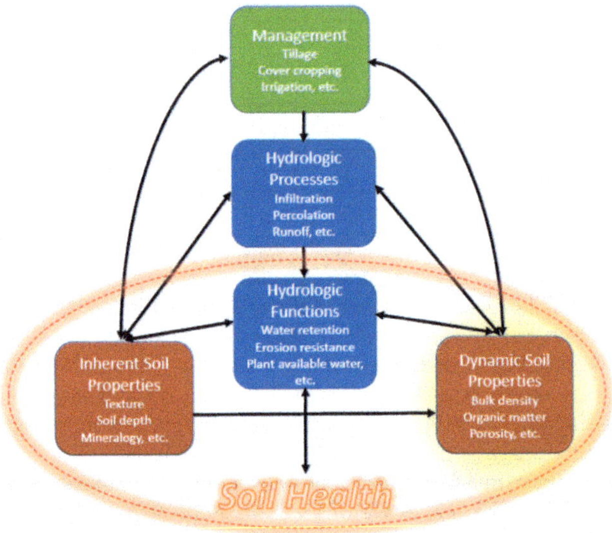

Figure 1 A conceptual framework to illustrate the links among soil hydrologic processes, properties, management and soil health, as discussed in this chapter.

acts as the primary transport agent for sediment, nutrients and contaminants, allowing the management of soil water to either promote or degrade soil and environmental quality.

Before assessing the relative importance of various soil health indicators, it is first important to understand the role they play in contributing to soil functions. Physical, chemical and biological properties all contribute to hydrologic soil functions, from water infiltration to retention and leaching. The key in understanding soil hydrology within the context of soil health is in maintaining these fundamental functions in soils while preventing degradation of the surrounding land, air and water. Each of the soil health indicators, comprising both dynamic and static soil properties, provides a context in which to understand hydrologic processes in soils (Fig. 1 and Table 1). Dynamic soil properties are those that change within limited time scales and can be relatively easily altered through management, while static properties are those properties that remain largely fixed over human time scales and cannot be easily altered. These soil properties are integrally tied to soil functions and there is a complex set of interactions that create both positive and negative feedback loops.

An important point about soil health is that soils are managed for beneficial properties, such as organic matter content, which in turn affect soil processes, which feeds into soil

Table 1 Soil hydrologic functions and related soil health indicators (compiled from Doran and Parkin, 1996; Karlen et al., 1997; Karlen and Stott 1994).

Soil hydrologic functions	Related soil health indicators
Infiltration	Texture Structure Aggregate stability Organic matter content Pore size and distribution Bulk density Microbial biomass
Resistance to erosion	Aggregate stability Microbial biomass Texture Structure Surface crusting and sealing Organic matter
Water filtration	Cation exchange capacity Texture Bulk density Surface crusting and sealing Porosity Organic matter
Water retention	Texture Structure Pore size and distribution Bulk density Organic matter
Subsurface drainage	Bulk density Texture Structure Pore size and distribution Organic matter

functions (Fig. 1). This increases the importance of understanding the interactions among soil properties and processes and, especially, the role of dynamic soil properties in soil hydrology and the overall soil health, as these are the properties altered through soil management.

2.1 Organic matter content

Organic matter plays a key role in maintaining soil moisture, increasing water-holding capacity and promoting the formation of water-stable aggregates, which, in turn, increase infiltration and water storage (Franzluebbers, 2002). The addition of organic matter in soils has long been shown to increase the structural stability and water-holding capacity for a wide range of soils with all textural classes (Hudson, 1994; Rawls et al., 2003).

The stability of organic carbon in soils along with its distribution with depth has important implications for soil structure as well as soil hydrology (Franzluebbers, 2002; Six et al., 2004; Dexter et al., 2008). Additional studies have shown the influence of soil organic carbon stratification within the soil profile as a contributing factor to maintain effective infiltration and water storage. Stratification refers to the gradient of organic matter observed in soils, where organic matter concentrations decrease with depth. Franzluebbers (2002) showed that increased stratification of organic matter increased aggregation and permanent pore development, leading to greater infiltration rates. The difference in infiltration rates between contrasting management strategies (e.g., no-till versus conventional tillage) was significant (three times greater in no till than conventional tillage) due to the impacts of both disturbance in the tillage system and organic matter addition in the no-till system. Similar effects were noted by Bruce and Langdale (1997) with respect to increasing aggregate stability with greater organic matter stratification.

2.2 Aggregate size and stability

Soil aggregates are a function of physical, chemical and biological properties and are vital to a multitude of soil hydrologic functions (Bronick and Lal, 2005). Water-stable aggregates play an important role in maintaining soil's ability to intercept water without complete reorganization of particles (which over time restricts infiltration) (Fig. 2). Soil aggregates generally form via interactions among soil mineral fractions, organic matter and soil biota (Bronick and Lal, 2005). As organic matter binds to clay particles with cations held at their exchange sites, aggregates are formed, often with aid from roots and fungal hyphae (Bronick and Lal, 2005; Tisdall and Oades, 1982). Glomalin, a specific glycoprotein produced by fungal hyphae, has been shown to be particularly beneficial in increasing aggregate stability (Wright and Upadhyaya, 1998; Wright et al., 1999; Wright and Upadhyaya, 1996). Root and microbial exudates act as glue to support aggregates and increase aggregate stability, while certain cations including Si^{4+}, Fe^{3+}, Al^{3+} and Ca^{2+} can play a role in acting as a bridge between clays and organic matter that can withstand rainsplash (Bronick and Lal, 2005). However, disaggregation can also occur via natural processes, including the decomposition of organic matter during periods of increased microbial activity and mobilization of structurally important clays or cations through dispersion by water (Bronick and Lal, 2005). This provides the context for the importance of maintaining water-stable aggregates in soils, which tend to have greater organic and biologic binders. Understanding the processes that lead to increases and decreases in soil aggregation can aid in managing soils to increase water retention and transmission.

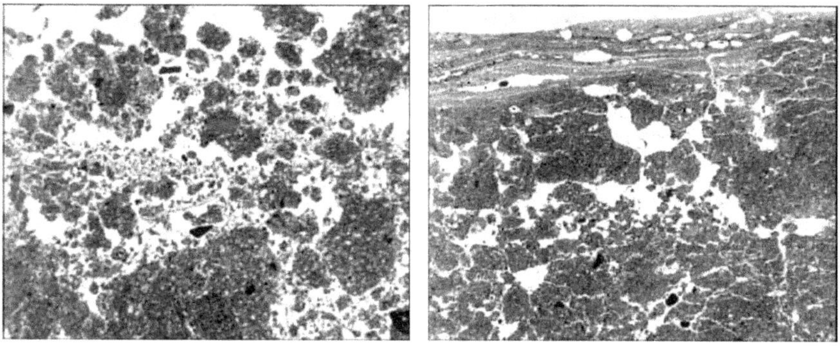

Figure 2 Photograph of the upper 100 mm of an undisturbed soil sample under deep ploughing (left) and after rainfall (right), illustrating the effects of poor aggregate stability and particle rearrangement on soil structure and infiltration capacity. White spaces represent pores, while the grey areas represent soil particles and organic matter (adapted from Pagliai et al., 2004).

The development of micro- and macro-aggregates follows different patterns of formation that depend on different combinations of organic inputs and soil conditions. More labile carbon has been associated with stronger aggregate stability, as labile carbon plays a stronger role in binding micro-aggregates to form macro-aggregates, while more stable forms of organic carbon are those responsible for aiding in the formation of micro-aggregates (Garcia-Orenes et al., 2012; Tisdall and Oades, 1982; Oades, 1993).

2.3 Bulk density

Bulk density, a measure of the mass per unit volume of dry soil, is a strong indicator of soil architecture, including total porosity and compaction (da Silva et al., 1997). As the volume of pore spaces declines, the mass of the soil per unit volume increases, resulting in greater bulk density. Soil compaction is widely recognized as one of the major factors affecting crop yields in cultivated soils due to the decrease in penetrability by roots and water/nutrient mobility (Pagliai et al., 2000). Management practices and soil properties such as texture and organic matter content all influence bulk density (Rawls, 1983; Huntington et al., 1989). Different soils and plants have different ideal values for bulk density, creating a challenge when using bulk density as a measure of compaction within the context of soil health and crop yields (Hakansson and Lipiec, 2000).

Soil bulk density is largely controlled by soil organic matter and clay content, where organic matter is commonly negatively related to bulk density and clay content is generally positively linked to bulk density (Lin et al., 2014). Soil bulk density impacts soil's saturated hydraulic conductivity in an inverse (linear or exponential) manner (Lin et al., 2014).

2.4 Pore space

Soil porosity can be measured by several metrics, all of which have important implications in water storage and movement in soils. Pore size, size distribution, length and connectivity are all key attributes that determine the pathways water moves through in soils. Pores are important not only for water storage, but also for root growth. Both roots and water follow paths of least resistance, and roots will also tend to grow in areas of the soil with access to water and may even take up water preferentially from areas with higher moisture (Green and Clothier, 1998). Thus, pore space and connectivity are critical in allowing water, gases and roots to move through the soil matrix effectively.

Macropores often function as water highways in soils, capable of allowing water to be transported downward at rates faster than what would occur in the soil matrix. Micropores are more effective at storing water and allowing for more diffuse flow through the soil matrix, thus filtering out contaminants more effectively. Water infiltration into and within soils depends heavily on pore size distribution in soils (Boyle et al., 1989). A balance of macropores and micropores in soils is required to hold enough water for crop growth while maintaining aeration for roots and to prevent ponding (Tisdall and Oades, 1982).

3 Soil health indicators and soil water dynamics

Soil health indicators selected for a particular study are based on soil functions of interest. Therefore, it is helpful to understand the processes involved in the soil hydrologic cycle and their effects on a full range of possible soil health indicators. This section provides a brief overview of the soil hydrologic cycle and its significance in the overall soil and environmental health, along with their relationships to the soil health indicators discussed above.

3.1 The soil water cycle

The use of soil water balance is an effective way to understand the linkage among various components involved in maintaining soil health and productivity and preventing degradation. The soil water cycle is generally described in terms of a water mass balance, whereby inputs minus outputs is equal to the change in soil water storage (ΔS) over a given time frame for a given volume of soil (Eq. [1]). When soil health is maintained, the water balance functions properly to supply water to plants, without degrading the surrounding environment over time via the transport of chemical fertilizers, pesticides and sediments. The water balance is constantly shifting both spatially and temporally and can be altered through management of soil hydrologic properties (Fig. 3):

$$\Delta S = (P + I - R) - (ET + D + SR), \quad [1]$$

where P = precipitation, I = irrigation, R = surface runoff, ET = evapotranspiration, D = deep percolation and SR = subsurface runoff.

While Eq. [1] is a heavily simplified version of complex processes occurring in soils, it does provide a useful framework for identifying specific management goals. Soil water storage can be influenced by managing for reductions in soil water losses through processes such as surface runoff and evapotranspiration (ET) or by managing for improved efficiency in capturing applied water through improving infiltration and soil water storage. Each of these components plays a role in determining the amount of plant-available water. Highlighted below are key fluxes that contribute to changes in soil water storage once inputs, including precipitation and irrigation water, have been applied.

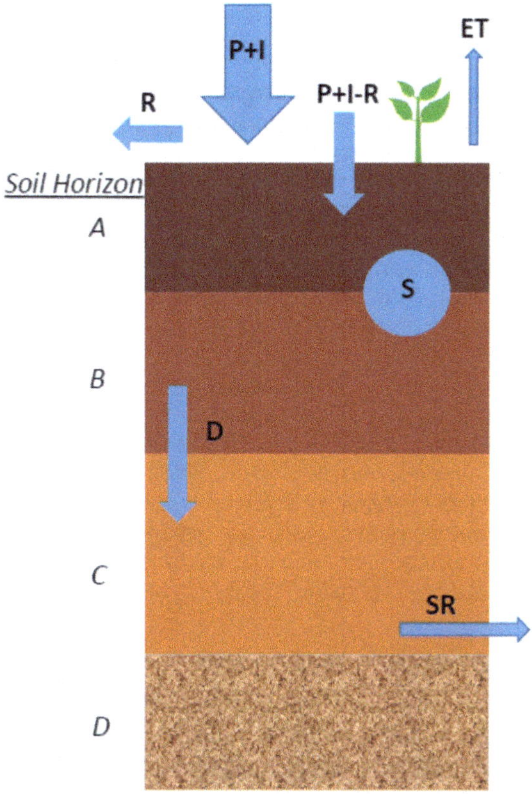

Figure 3 A simplified schematic of soil hydrologic processes, depicting water storage (S) and soil water fluxes into and out of a soil profile with multiple layers. P = precipitation, I = irrigation, R = surface runoff, ET = evapotranspiration, D = deep percolation and SR = subsurface runoff.

3.2 Infiltration

Soil surface infiltration rates play a key role in the accessibility of water to crops and the overall health of soil. Infiltration rates control the amount of applied water that makes its way into the soil and is highly variable across both time and space. Various factors influence soil water infiltration. Some are relatively static, including soil texture, rock fragments and the presence of a water-restricting layer within a soil profile, while others are dynamic and can either be easy or difficult to manage, including soil temperature, antecedent moisture condition, surface crusting, aggregate stability, organic matter content, hydrophobicity and macropores. Dominant macropores in agricultural systems consist of root channels and earthworm boreholes, which can extend below the root zone, creating pathways for deeper drainage (Edwards et al., 1990; Stroud et al., 2017). If precipitation rates exceed infiltration rates, excess water is lost as surface runoff. Because the amount of water received as precipitation is not controllable, managing soils for increased infiltration capacity is key. Good soil health requires good infiltration.

3.3 Runoff

Surface runoff, or overland flow, occurs when incoming rainfall or irrigation rate exceeds the infiltration capacity of the soil. This is often exacerbated by surface crusting or compaction, even at depths below the soil surface (such as plough pans). In more humid regions, saturation excess overland flow is more common (resulting from soils reaching satiation and no longer allowing additional water into the soil), while in arid regions, infiltration excess overland flow is more widespread (i.e., rainfall or irrigation rate exceeds soil infiltration rate).

Surface runoff often carries sediment, contaminants from pesticides and nutrients from fertilizers to surface water bodies, increasing the environmental impact of agriculture. Surface runoff is also a primary form of soil loss and degradation, where topsoil is eroded. However, surface runoff is one of the primary means of water delivery to streams in more arid regions. In this context, the quantity and quality of surface runoff must be considered.

Subsurface runoff can also occur under certain conditions. Most commonly, this is found in sloped landscapes with soils that contain a subsurface-restricting layer. The restricting layer can be found in the form of a plough pan, fragipan, argillic horizon or a soil–bedrock interface, among others. The occurrence of subsurface runoff can be important for two reasons. The first is that this represents additional loss of water from the desired area, which can be especially detrimental to overall soil health in near-surface horizons due to the limited depth of the wetting front. Secondly, this acts as a source of excess water to downslope areas, which recharges streams.

3.4 Evapotranspiration (ET)

ET is the sum of all evaporation and transpiration from a given soil. The rate of ET depends on the amount of available water for evaporation, the time of year, the amount of direct solar radiation, humidity, vapour pressure deficit and plant species (Campbell and Norman, 1998). Transpiration is essentially a measure of the amount of water taken up by plants and transpired back to the atmosphere. In the context of the soil water cycle, this loss is considered a productive use of soil water, meaning that the loss of water directly contributes to the desired use, in this case, plant growth. Evaporation is considered unproductive due to the net loss of water from the soil with no overall gain for production. Therefore, when considering management decisions regarding minimizing ET losses, it is evaporation, rather than transpiration, that needs to be minimized so as not to reduce yields. Transpiration rates vary significantly throughout the year and throughout the growing season, as they are affected by plant species, the growth stage, the amount of available soil water and the amount of direct solar radiation, all of which vary based on local climate and landscape setting (Campbell and Norman, 1998).

Direct measurements of ET are difficult to obtain. Most are based on potential ET, which assume an infinite supply of water, meaning that the measurements are regulated solely by the plant species, vapour pressure deficit of the atmosphere, wind and solar radiation. While these measurements are useful, they do not take into account actual soil conditions nor do they mimic the complexity of real-world soils.

3.5 Deep percolation

Deep percolation, often described as groundwater recharge, refers to the downward movement of water through the soil profile following the initial entry of water through infiltration. Percolation is an important process in supplying water to the deeper root zone and preventing surface and subsurface ponding. Percolation is heavily influenced by pedogenic processes and the resulting soil properties as well as antecedent moisture conditions. Soil structure and texture both control the movement of water through the soil profile, as do macropores and horizon boundaries (Lin, 2012). Macropores consist of earthworm boreholes, root channels (active or decayed), interpedal pores (especially in well-structured soils) and fissures created during shrinking processes in shrink–swell clays. Matrix flow occurs within smaller pores and transports water at much slower rates than water flow in macropores. This process is important in regulating water content within the upper portions of the profile that encompass the root zone and primary habitat zones for soil biota. It can be equally important to promote drainage in this zone as it is to encourage water retention, as roots and other soil organisms require a balance of air and water, especially with respect to aerobic microbial processes (Doran and Parkin, 1996).

3.6 Water storage

Soil water storage is the primary factor in determining whether the soil is suitable for crop growth and is dependent on all of the other processes in the soil water cycle. Smaller pores are known to hold water more tightly than larger macropores. However, water stored in smaller pores is also less easily accessible to plants because of the increased energy required to remove the water from those pores. As a result, soil water storage is dependent not only on all the other fluxes in the soil water cycle, but also on the internal physical, chemical and biological properties of soils, including soil texture, structure, pore size distribution, minerology, rooting patterns, horizonation and organic matter content (Rawls et al., 2003; Hudson, 1994; Franzluebbers, 2002). Additionally, water storage is affected by the drainage properties of the soil, including amount and distribution of biopores (Edwards et al., 1990). Plant-available water is generally used as a measure of the amount of water accessible to plants, which encompasses the moisture between the permanent wilting point and the field capacity.

4 Soil health management and soil hydrology

Soil health depends largely on management decisions made by producers and land owners, whether that relates to tillage practices or cropping patterns or other land use/management measures (Tugel et al., 2005). Changes in soil dynamic properties, defined as soil properties that change with land use/management (Tugel et al., 2005), can provide useful indicators for maintaining healthy soils. Soil management often has a much faster and greater impact on dynamic soil properties than naturally occurring processes or so-called inherent properties (Pagliai et al., 2000). This ranges from compaction as a result of heavy machinery to disruption of surface structure and microbial habitat due to tillage. On the contrary, effective management strategies with appropriate understanding of soil hydrology and related processes can improve plant growth conditions, while increasing the soil's capacity to sustain environmental quality. Below are some common management practices that significantly impact soil health as well as soil hydrology.

4.1 Tillage

One of the major influences of tillage on soils is the disruption of natural soil architecture/structure (Strudley et al., 2008). This affects both the ratio of solids to pore spaces and the distribution of those pore spaces, which have a direct influence on the passage of water through those soils. Tillage has been shown to have long-term negative influences on soil surface structure and accumulations of surface soil organic matter, which decrease infiltration rates and water storage overall (Franzluebbers, 2002). Pagliai et al. (2000) reported decreased pore length and an overall reduction in larger pores within the upper 10 cm of the soil profile with mouldboard ploughing as compared with less destructive practices, including the use of a ripper subsoiler. The observed effects of mouldboard ploughing on pore structure limit the soil's hydraulic conductivity and infiltration capacity. Similar patterns were also observed at greater soil depths. In many cases, the effects of tillage on porosity simply result in a sharp decrease in the diversity of pore size and length, often resulting in the creation of large pores and the destruction of smaller or intermediate pores directly following tillage (Fig. 4). The loss of pore length and soil structural stability

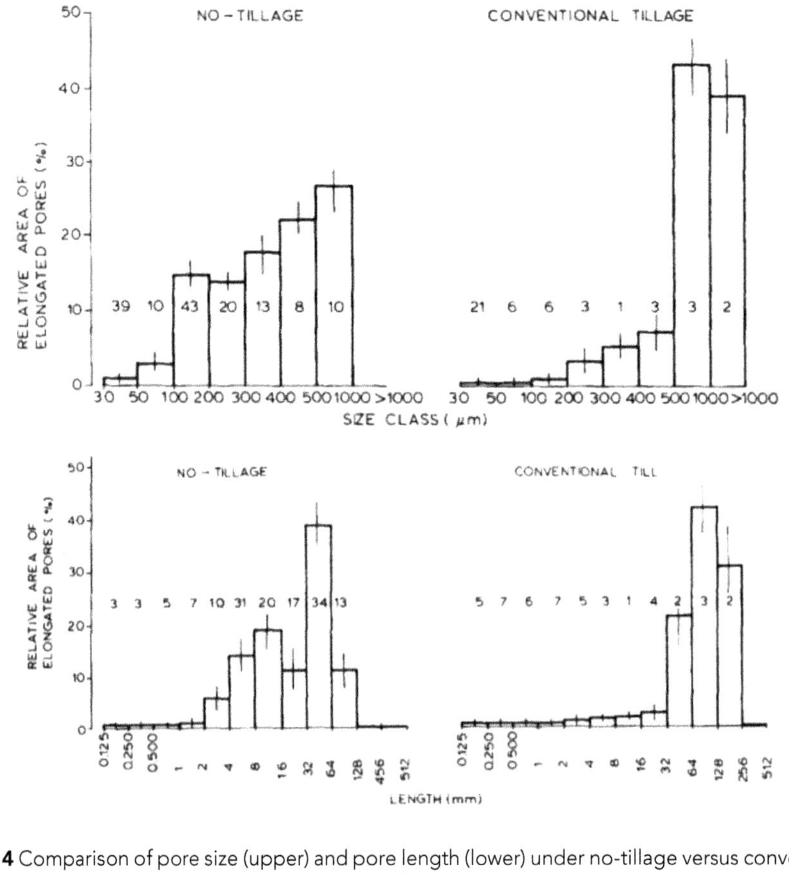

Figure 4 Comparison of pore size (upper) and pore length (lower) under no-tillage versus conventional tillage in a clay loam soil (adapted from Pagliai et al., 1984).

has been tied to increased occurrence of surface crusting and compaction, leading to decreased infiltration, increased surface runoff and increased soil loss (Pagliai et al., 2004).

Additional impacts include the destruction of biopores, including earthworm boreholes (Edwards et al., 1990). These biological conduits act as important features in transporting water through the soil profile more rapidly and can allow for increased movement beyond otherwise restrictive soil layers. Destruction of these pores further limits infiltration capacity beyond surface horizons (Edwards et al., 1990).

The short- and long-term effects of management practices can differ significantly. Studies assessing the short-term impacts of no-till management often show results contrary to the longer term studies. In several short-term studies, recently converted no-till and minimum-till systems showed greater degrees of surface compaction and increased bulk density, as compared to conventional tillage (Salem et al., 2014; Afzalinia and Zabihi, 2014). As soils are converted, the longer term effects of tillage, including surface crusting, are exacerbated for a time before the soil is able to return to normal function under minimal tillage. In contrast, results of studies showing the short-term effects of tillage indicate positive influences on porosity due to the mixing of soils, which creates larger temporary pores (Franzluebbers et al., 2002).

A multitude of studies have documented greater infiltration over the long term in no-till systems than in conventional systems (Franzluebbers, 2002; Bruce and Langdale, 1997; Dao, 1993, Arshad et al., 1999). Results showing the effects of tillage on bulk density are somewhat less consistent. A number of studies showed increases in bulk density with increasing tillage intensity and depth (Salem et al., 2014; Afzalinia and Zabihi, 2014), while other studies showed no change or decreases in bulk density (Afzalinia et al., 2011; Taser and Metinoglu, 2005). This is likely due to the fact that not all soils react to tillage in the same way as a result of differences in soil texture and structure. The degree to which compaction is detrimental, for instance, relies heavily on the ratio of sand to clay-sized particles.

While soil physical properties are often assessed to determine the possible effects of tillage on soil water processes, several studies have addressed changes in soil water dynamics and storage (Afzalinia et al., 2011; Bescansa et al., 2006). Water-holding capacity and plant-available water have been found to be greater under more conservative tillage systems due to the increased ratio of small to large pores than under conventional tillage because of the large openings created near the soil surface during tillage (Bescansa et al., 2006). These trends were observed in the upper portion of the soil profile and at depth throughout the growing season. These trends, usually associated with the increase in organic matter, were independent of increased crop residues, indicating that the structural and pore size distribution differences between conventional and no-till systems are great enough to impact soil water content alone (Bescansa et al., 2006). However, the addition of organic matter combined with limited tillage is beneficial in promoting soil water storage.

4.2 Organic matter additions

Soil organic matter content has become somewhat ubiquitous with no-till agriculture, largely because of the lack of litter and stubble removal following harvesting (Franzluebbers, 2002). With the maintenance of organic materials on the soil surface associated with minimal tillage or no-till agriculture, aggregation is increased and the prevalence of surface crusting is decreased (Pagliai et al., 2004). Many studies have shown increased organic matter in soils as a means of increasing water-holding capacity (Fig. 5) or seasonal

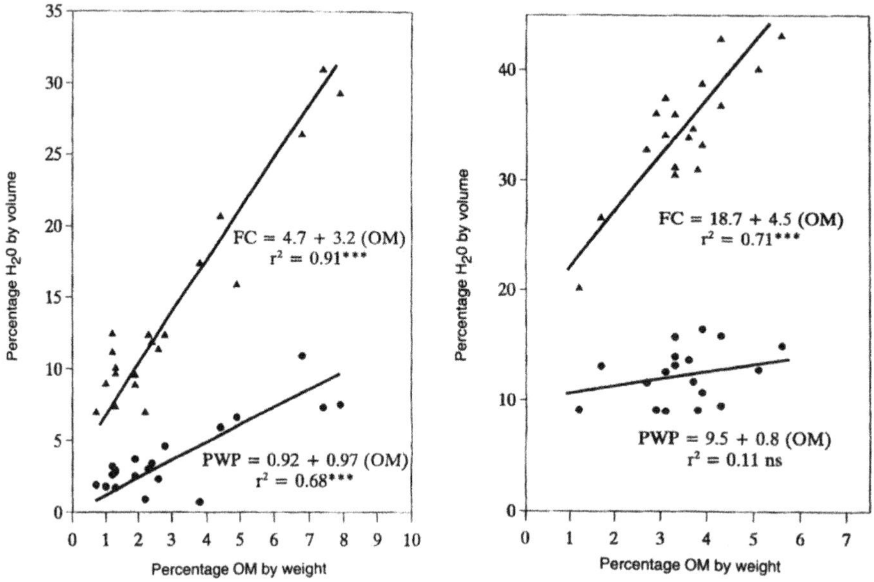

Figure 5 Organic matter (OM) content versus volumetric water content for surface sandy (left) and silt loam (right) soils, demonstrating the influence of organic matter on soil water storage (FC = field capacity, PWP = permanent wilting point) for different soil types (Hudson, 1994). Note the difference in the scale of the X and Y axes between the two plots.

water storage (Hudson, 1994; Rawls et al., 2003). However, that is not always the case. Bescansa et al. (2006) observed no difference between water storage in soils with a 50% difference in residue maintenance over the course of one growing season. This may be the result of textural differences in soils examined in different studies. Saxton and Rawls (2006) attributed the role of organic matter in increasing soil water-holding capacity to its binding properties in forming soil aggregates in a manner similar to that of clay bridging. Thus, organic matter inputs are more effective for coarser soils than for finer textured soils when used to increase water storage capacity.

No-till and reduced tillage have been used as a means of managing soils for hydrologic benefits by reducing structural degradation to promote infiltration and limit surface runoff, increasing organic matter inputs when reduced tillage or no-till systems are combined with plant residue applications and maintaining micro, and macropores for deeper subsurface percolation and water storage.

4.3 Cover cropping

The use of cover crops in soil health management has gained immense support over the years based on improvements in year-round soil cover, organic matter additions, reduction in sediment and topsoil loss from erosion and reductions in nutrient leaching from fields (Kaspar et al., 2001; Unger and Vigil, 1998; Dabney et al., 2001). They have been historically recognized for their capacity to increase nutrient concentrations through atmospheric fixation, in the case of leguminous cover crops, and reduce soil erosion through the increase in surface cover and roughness. More recent studies have confirmed that cover crops provide benefits to physical, chemical and biological soil properties, which influence soil hydrologic processes and promote ecosystem services beyond that of simply providing more water and nutrients to crops (Snapp et al., 2005).

Soils in different regions and climates react differently to cover crops because of the resulting changes in the soil water cycle that occur when planting during seasons that tend to be fallow (Unger and Vigil, 1998). On the hydrologic side, cover crops alter the soil water balance in a way that may be detrimental to crop growth during the growing season due to decreases in available soil moisture for the following crop (Dabney et al., 2001). However, this may be counteracted by terminating cover crops at the proper interval before planting, depending on the soil and the species of crop and cover crops planted (Unger and Vigil, 1998). However, these alterations to the water cycle may be overcome by improvements to soil physical quality, which improve soil water use efficiency and storage and reduce initial soil water losses to surface runoff (Table 2). Where cover crops are used specifically with the goal of increasing infiltration and organic matter content, soil health improvements outweigh losses in soil moisture in terms of long-term soil functionality and fertility (Unger and Vigil, 1998).

Dabney (1998) provided a brief summary of the effects of cover crops on hydrologic processes and properties, with a three-fold conclusion:

1 Cover crops reduce the likelihood of surface crusting and sealing by providing physical cover that reduces the physical impact of rainsplash and adds organic matter to increase aggregate stability. Dabney (1998) and Brill and Neil (1950) noted that these are beneficial in promoting infiltration and reducing surface runoff (Fig. 6). Following the termination of cover crop, the plant material is either incorporated into the topsoil or is left on the surface, adding additional organic matter to the soil. The manner in which the plant residue is incorporated depends on the type of cover crop (e.g., winter kill and winter hardy) and on the tillage regime (e.g., no-till, conventional tillage and reduced tillage) (Dabney et al., 2001).
2 In humid regions, cover crops can increase available water storage through the uptake of excess water that would otherwise be leached or saturate the soil. In poorly drained soils, cover crops can take up water during the traditionally fallow season, reducing the amount of water stored in the soil when the soil is not in production. This allows more available storage for rain events later in the season, closer to planting.

Table 2 Advantages and disadvantages of cover crops related to hydrologic processes in soils (modified from Dabney et al., 2001). Note that these advantages and disadvantages must be considered within the context of a tillage system applied to any given field.

Advantages	Disadvantages
Reduce soil erosion	Reduce soil moisture*
Increase residue cover	Increase seasonal evapotranspiration**
Increase water infiltration into soil	
Increase soil organic carbon	
Improve soil physical properties	

*May only occur during cover crop growth stage for soils in humid and temperate regions
** Only applies to systems where cover crop residue is removed following the cover cropping season (Klocke et al., 2009; Mitchell et al., 2012)

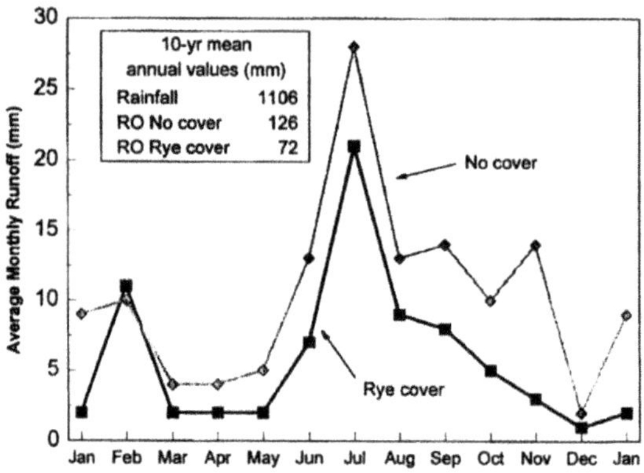

Figure 6 Average monthly runoff averaged over 10 years for plots planted with and without cover crops (adapted from Brill and Neil, 1950).

However, in semi-arid and arid climates, this may have a net negative impact on soil water storage during the growing season due to the increased losses to transpiration during the cover crop growth period (Reeves, 1994; Dabney et al 2001).

3 Cover crops increase macroporosity both directly and indirectly. Cover crops with deep rooting depths can aid in increasing the hydraulic conductivity of underlying soil layers by creating large channels for more rapid water flow, thus increasing deep percolation and enhancing overall drainage. They can also increase macroporosity indirectly by providing more suitable habitat than fallow soils for soil macrofauna, including earthworms, whose channels provide effective pathways for rapid vertical water movement (Kemper et al., 2011).

5 A broader perspective on soil health

Soil health thus far has focused on specific indicators that provide information about the soil's ability to function within ecosystem boundaries (Doran and Parkin, 1994). These indicators are useful because of the accessibility of their measurements in a variety of situations. However, with such a heavy emphasis on individual measurements for representing the overall soil physical, chemical and biological health, the broader landscape context and integrated assessment of soil health can be lost. In the case of soil hydrology and its ties to soil health as well as the sustainable long-term management of soil health, these two aspects cannot be ignored. They are further discussed in the following sections.

5.1 Landscape setting and soil architecture

Soil hydrology is primarily dictated by three categories of factors: 1) environmental factors such as climate and vegetation, 2) soil properties (such as soil depth, texture and structure), and 3) soil's position in a landscape (such as slope, aspect and landform). Within the existing framework of soil health, the first two categories of factors are generally addressed (Moebius-Clune et al., 2016; Doran and Parkin, 1994). The local landscape setting, however, is often missing and thus should also be included in assessing and managing soil health, at least with respect to hydrology. It is the interaction between soil and water at all positions in the landscape and throughout the soil profile that controls the true dynamics and distribution of water and nutrients in soils. A big picture view of soils attained simply by looking farther down and beyond the pedon or plot scale can shed better light on the complexity involved in soil health and its links to soil water dynamics.

A common approach to assessing soil structure in the context of promoting water retention and root growth focuses on soil aggregates and their stability. A broader concept of soil architecture, as emphasized in hydropedology (Lin, 2012), provides a means of addressing soil structure at various scales as well as considering the soil as a system. Soil architecture refers to the entirety of how the soil is organized across scales from the pore scale to the landscape scale and includes a 3D system from soil surface to deeper soil layers and encompasses three interlinked components at each scale – solid components, pore space and their interfaces (Lin, 2012). This concept of soil architecture is much broader and comprehensive than the classical concept of soil structure (i.e., soil aggregation). Sustainable management of soils has recognized more and more the necessity of addressing a soil profile as a whole, not just at the surface, and the importance of depth-integrated *in situ* evaluation (Lin, 2006; Koch et al, 2015). However, *in situ* measurements to assess soil hydrologic function beyond surface horizons remain limited, in part due to the difficulty of incorporating measurements at greater depths (Salem et al., 2015). The importance of understanding soils beyond surface horizons stems from the larger scale interactions among soils, crops and management. More specifically, crop roots interact with soils at greater depths and the influence of management including tillage is often found beyond soil surface (Koch et al., 2015; Salem et al., 2015). The effects of layering on soil moisture have been documented for decades, showing the influence of soil properties in lower horizons on soil moisture in upper horizons and vice versa (Miller, 1973; Richards, 1955). The influence of layering on soil hydrology is dependent on a number of factors, including the texture and structure of each horizon and the depth to horizon boundaries (Miller, 1973).

5.2 Integrated soil health and fundamental principles for caring for soils

Lin and Lin (2015) introduced a new framework for addressing soil health in the context of global sustainability and holistic management. They proposed the following four fundamental principles for caring about soils:

1 Recognizing that humans do not grow plants; rather, soils grow plants, and thus humans need to care and nurture soils;
2 Promoting better living through biology rather than pure chemistry (e.g., fertilizers) or pure physics (e.g., irrigation);
3 Nurturing soil's internal quality to maintain its internal health and balancing soil use for external productivity; and
4 Adopting a mosaic and more organic approach to farming and gardening instead of monocultures and mechanical approach.

While these four principles are novel in their integrated and holistic view of soil management, they also coincide directly with the goals of soil health management as well as practices surrounding the links among soils, the broader ecosystems and humans. The idea behind a holistic approach to addressing soil health and its link to soil hydrology is to move away from the concept that soil hydrology is only a physical science. Instead, the idea of integrated soil fertility (Lin, 2014) frames a web of processes that encompass physical, chemical and biological aspects in soils. In essence, these principles encourage soil health management strategies that integrate both biotic and abiotic components (Doran and Zeiss, 2000; Laishram et al., 2012). Soils must be considered an integral component and base of any functioning ecosystem and be managed as an ecosystem in and of itself to sustain vital soil functions.

6 Conclusion

Water is the quintessential link among all of the factors considered central to soil health, from crop growth, to preventing soil loss and degradation, to providing habitat for soil biota and reducing air and water pollution. In essence, the relationship between soil, water and the living ecosystem is at the nexus of the soil health movement.

Soil water dynamics involve a complex set of interactions among physical, chemical and biological factors that affect soil health. These properties, in conjunction with amendments made through human management, affect the functionality of soils. Soil hydrologic functions are among the key factors that influence the overall health of soil, including water filtration, soil water storage, plant-available water and drainage condition. Soils are generally managed for measurable properties, such as organic matter content, pH, bulk density, aggregate stability, moisture content and aeration, among others. The reason that these metrics have taken hold is because they provide a good indication of soil functionality, and the processes upon which those functions are based are dependent on the soil water cycle and water fluxes and storages in soils across diverse ecosystems.

A broader perspective on soil health is suggested in this chapter, including the landscape setting and the integrated assessment of soil health, which are essential in terms of soil hydrologic functions and their impacts on the overall health of soils. In addition, the time scales of soil health indicators and their measurements with respect to soil water dynamics are also vital to be considered.

7 Where to look for further information

Future research on the topic requires knowledge about soil–plant–water interactions and the influence of management on those interactions. Below are several resources related to the nexus between soil health and soil hydrology:

The '*Comprehensive Assessment of Soil Health: The Cornell Framework Manual*' is an excellent reference for advancing one's knowledge about soil health in general. This manual provides information about general soil health and up-to-date information about relevant soil tests to evaluate soil health, with comprehensive explanations relating to why certain indicators are of value. This resource is particularly useful in the northeastern United States.

The USDA Natural Resource Conservation Service's *Soil Health Initiative* also provides excellent information regarding soil health practices related to water and water quality. Several resources are available at their website, with access to literature reviews and citations related to soil health under the 'Resources and Publications' tab. Smith (2016), available via the aforementioned web page, is an excellent reference for a more in-depth analysis relating hydraulic properties and management, specifically related to soil health management practices.

8 References

Afzalinia, S. and Zabihi, J. (2014), 'Soil compaction variation during corn growing season under conservation tillage', *Soil and Tillage Research*, 137, 1–6.

Afzalinia, S., Behaeen, M. A., Karami, A., Dezfuli, A. and Ghasari, A. (2011), 'Effect of conservation tillage on the soil properties and cotton yield', 11th International Congress on Mechanization and Energy in Agriculture, 21–23 September, Istanbul, Turkey.

Arshad, M. A., Franzluebbers, A. J. and Azooz, R. H. (1999), 'Components of surface soil structure under conventional and no-tillage in northwestern Canada', *Soil and Tillage Research*, 53, 41–7.

Bescansa, P., Imaz, M. J., Virto, I., Enrique, A. and Hoogmoed, W. B. (2006), 'Soil water retention as affected by tillage and residue management in semiarid Spain', *Soil Tillage Research*, 87, 19–27.

Boyle, M., Frankenberger, W. T. and Stolzy, L. H. (1989), 'Influence of organic matter on soil aggregation and water infiltration', *Journal of Production Agriculture*, 2(4), 290–9.

Brill, G. D. and Neal, O. R. (1950), 'Seasonal occurrence of runoff and erosion from a sandy soil in vegetable production', *Agronomy Journal*, 42, 192–5.

Bronick, C. J. and Lal, R. (2005), 'Soil structure and management: A review', *Geoderma*, 124(1–2), 3–22.

Bruce, R. R. and Langdale, G. W. (1997), 'Soil carbon level dependence upon crop culture variables in a thermic–udic region'. *Soil Organic Matter in Temperate Agroecosystems: Long-term Experiments in North America*. CRC Press, Boca Raton: 247–61.

Campbell, G. S. and Norman, J. M. (1998), *An Introduction to Environmental Biophysics: Second Edition*. Springer-Verlag, New York.

Dabney, S. M. (1998), Cover crop impacts on watershed hydrology, *Journal of Soil and Water Conservation*, 53(3), 207–13.

Dabney, S. M., Delgado, J. A. and Reeves, D. W. (2001), 'Using winter cover crops to improve soil and water quality', *Communications in Soil Science and Plant Analysis*, 32(7–8), 1221–50.

Dao, T. H. (1993), 'Tillage and winter wheat residue management effects on water infiltration and storage'. *Soil Science Society of America Journal*, 57(6), 1586–95.

Da Silva, A. P., Kay, B. D. and Perfect, E. (1997), 'Management versus inherent soil properties effects on bulk density and relative compaction', *Soil and Tillage Research*, 44(1–2), 81–93.

Dexter, A. D., Richard, G., Arrouays, D., Czyz, E. A., Jolivet, C. and Duval, O. (2008), 'Complexedorganic matter controls soil physical properties', *Geoderma*, 144, 620–7.

Doran, J. W. and Parkin, T. B. (1994), 'Defining Soil Quality for a Sustainable Environment', Soil Science Society of America, SSSA Special Publication no. 35.

Doran, J. W. and Parkin, T. B. (1996), 'Quantitative Indicators of Soil Quality: A Minimum Data Set', *Methods for Assessing Soil Quality*. Soil Science Society of America.

Doran J. W., Sarrantonio, M. and Liebig, M. (1996), 'Soil health and sustainability'. *Advances in Agronomy*, Elsevier, 56, 1–54.

Doran, J. W. and Zeiss, M. R. (2000), 'Soil health and sustainability: Managing the biotic component of soil quality'. Agronomy and Horticulture-Faculty Publications, Paper 15.

Edwards, W. M., Shipitalo, M. J., Owens, L. B. and Norton, L. D. (1990), 'Effect of *Lumbricus terrestris* L. burrows on hydrology of continuous no-till corn fields', *Geoderma*, 46(1–3), 73–84.

Franzluebbers, A. J. (2002), 'Water infiltration and soil structure related to organic matter and its stratification with depth', *Soil and Tillage Research*, 66(2), 197–205.

García-Orenes, F., Roldán, A., Mataix-Solera, J., Cerdà, A., Campoy, M., Arcenegui, V. and Caravaca, F. (2012), 'Soil structural stability and erosion rates influenced by agricultural management practices in a semiarid Mediterranean agroecosystem', *Soil Use Management*, 28, 571–9.

Green, S. and Clothier, B. (1999), 'The root zone dynamics of water uptake by a mature apple tree', *Plant and Soil*, 206(1), 61–77.

Håkansson, I. and Lipiec, J. (2000), 'A review of the usefulness of relative bulk density values in studies of soil structure and compaction', *Soil and Tillage Research*, 53(2), 71–85.

Hudson, B. D. (1994), 'Soil organic matter and available water capacity', *Journal of Soil and Water Conservation*, 49(2), 189–94.

Huntington, T. G., Johnson, C. E., Johnson, A. H., Siccana, T. G. and Ryan, D. F. (1989), 'Carbon, organic matter, and bulk density relationships in a forested Spodosol', *Soil Science*, 148, 380–6.

Karlen, D. L., Mausbach, M. J., Doran, J. W., Cline, R. G., Harris, R. F. and Schuman, G. E. (1997), 'Soil quality: A concept, definition, and framework for evaluation', *Soil Science Society of America Journal*, 61, 4–10.

Karlen, D. L. and Stott, D. E. (1994), 'A framework for evaluating physical and chemical indicators of soil quality', In: Doran, J. W., Coleman, D. C., Bezdicek, D. F. and Stewart, B. A. (Eds.). *Defining Soil Quality for a Sustainable Environment. SSSA Special Publication no. 35.* SSSA and ASA, Madison, WI, 53–72.

Kaspar, T. C., Radke, J. K. and Laflen, J. M. (2001), 'Small grain cover crops and wheel traffic effects on infiltration, runoff, and erosion', *Journal of Soil and Water Conservation*, 56(2), 160–4.

Kemper, W. D., Schneider, N. N. and Sinclair, T. R. (2011), 'No-till can increase earthworm populations and rooting depths', *Journal of Soil and Water Conservation*, 66(1), 13A–17A.

Klocke, N. L., Currie, R. S. and Aiken, R. M. (2009), 'Soil water evaporation and crop residues', *Transactions of the ASABE*, 52(1), 103–10.

Koch, A., Chappell, A., Eyres, M. and Scott, E. (2015), 'Monitor soil degradation or triage for soil security? An Australian challenge', *Sustainability*, 7(5), 4870–92.

Laishram, J., Saxena, K. G., Maikhuri, R. K. and Rao, K. S. (2012), 'Soil quality and soil health: A review', *International Journal of Ecology and Environmental Sciences*, 38(1), 19–37.

Lin, H.S. (editor) 2012. *Hydropedology: Synergistic Integration of Soil Science and Hydrology.* Waltham, MA: Elsevier.

Lin, H. S. (2014), 'A new worldview of soils', *Soil Science Society of America Journal*, 78, 1831–44.

Lin, H. S., Bouma, J., Pachepsky, Y., Western, A., Thompson, J., Van Genuchten, R. and Lilly, A. (2006), 'Hydropedology: Synergistic integration of pedology and hydrology', *Water Resources Research*, 42(5), 1–13.

Lin, H. S. and Lin, A. J. (2015), 'Caring for soils for a sustainable world' *Soil Horizons*, 56.

Lin, H. S., W. Zhang and Yu, H. (2014), 'Hydropedology: Linking dynamic soil properties with soil survey data'. *Application of Soil Physics in Environmental Analyses*. Springer, New York, 23–50.

Miller, D. E. (1973), 'Water retention and flow in layered soil profiles', *Field Soil Water Regime*, Soil Science Society of America, Inc., 107–17.

Mitchell, J. P., Singh, P. N., Wallender, W. W., Munk, D. S., Wroble, J. F., Horwath, W. R. and Hanson, B. R. (2012), 'No-tillage and high-residue practices reduce soil water evaporation', *Calif Agriculture*, 66, 55–61.

Moebius-Clune, B. N., Moebius-Clune, D. J., Gugino, B. K., Idowu, O. J., Schindelbeck, R. R., Ristow, A. J., van Es, H. M., Thies, J. E., Shayler, H. A., McBride, M. B., Wolfe, D. W. and Abawi, G.S. (2016), *Comprehensive Assessment of Soil Health: The Cornell FrameworkManual*, Cornell University, Ithaca.

Oades, J. M. (1993), 'The role of biology in the formation, stabilization and degradation of soil Structure', *Geoderma*, 56, 377–400.

Pagliai, M., La Marca, M., Lucamante, G. and Genovese, L. (1984), 'Effects of zero and conventional tillage on the length and irregularity of elongated pores in a clay loam soil under viticulture'. *Soil and Tillage Research*, 4, 433–44.

Pagliai, M., Pellegrini, S., Vignozzi, N., Rousseva, S. and Grasselli, O. (2000), 'The quantification of the effect of subsoil compaction on soil porosity and related physical properties under conventional to reduced management practices', *Advances in Geoecology*, 32, 305–13.

Pagliai, M., Vignozzi, N. and Pellegrini, S. (2004), 'Soil structure and the effect of management Practices'. *Soil and Tillage Research*, 79(2 SPEC.ISS), 131–43.

Rawls, W. J. (1983), 'Estimating soil bulk density from particle size analysis and organic matter Content', *Soil Science*, 136, 48–55.

Rawls, W. J., Pachepsky, Y. A., Ritchie, J. C., Sobecki, T. M. and Bloodworth, H. (2003), 'Effect of soil organic carbon on soil water retention', *Geoderma*, 116, 61–76.

Reeves, D. W. (1994), 'Cover crops and rotations', In: *Crops Residue Management: Advances in Soil Science*. Lewis Publishers, 125–72.

Richards, L. A. (1955), 'Retention and transmission of water in soil', *Yearbook of Agriculture*, U.S. Department of Agriculture, 144–51.

Salem, H. M., Valero, C., Muñoz, M. Á., Rodríguez, M. G. and Silva, L. L. (2015), 'Short-term effects of four tillage practices on soil physical properties, soil water potential, and maize yield', *Geoderma*, 237–8, 60–70.

Saxton, K. E. and Rawls, W. J. (2006), 'Soil water characteristic estimates by texture and organic matter for hydrologic solutions', *Soil Science Society of America Journal*, 70, 569–1578.

Six, J., Bossuyt, H., Degryze, S. and Denef, K. (2004), 'A history of research on the link between (micro) aggregates, soil biota, and soil organic matter dynamics'. *Soil and Tillage Research*, 79, 7–31.

Smith, C. W. (2016), *Effects of Implementation of Soil Health Management Practices on Infiltration, Saturated Hydraulic Conductivity (K_{sat}), and Runoff*. USDA and NRCS, Lincoln, NE.

Snapp, S. S., Swinton, S. M., Labarta, R., Mutch, D., Black, J. R., Leep, R., Nyiraneza, J. and O'Neil, K. (2005), 'Evaluating cover crops for benefits, costs and performance within cropping system niche', *Agronomy Journal*, 97, 322–32.

Stroud, J. L., Irons, D. E., Watts, C. W., Storkey, J., Morris, N. L., Stobart, R. M., Fielding, H. A. and Whitmore, A. P. (2017), 'Cover cropping with oilseed radish (*Raphanus sativus*) alone does not enhance deep burrowing earthworm (*Lumbricus terrestris*) midden counts', *Soil and Tillage Research*, 165, 11–15.

Strudley, M. W., Green, T. R. and Ascough II J. C. (2008), 'Tillage effects on soil hydraulic properties in space and time: State of the science', *Soil and Tillage Research*, 99, 4–48.

Taser, O. and Metinoglu, F. (2005), 'Physical and mechanical properties of a clay soil as affected by tillage systems for wheat growth', *Acta Agriculturae Scandinavica. Section B. Soil and Plant Science*, 55, 186–91.

Tisdall, J. M. and Oades, J. M. (1982), 'Organic matter and water-stable aggregates in soils', *Journal of Soil Science*, 33(2), 141–63.

Tugel, A. J., Herrick, J. E., Brown, J. R., Mausbach, M. J., Puckett, W. and Hipple, K. (2005), 'Soil change, soil survey, and natural resource decision making: A blueprint for action', *Soil Science Society of America Journal*, 69, 73–747.

Unger, P. W. and Vigil, M. F. (1998), 'Cover crop effects on soil water relationships', *Journal of Soil and Water Conservation*, 53(3), 200–7.

Wright, S. F., Starr, J. L. and Paltineanu, I. C. (1999), 'Changes in aggregate stability and concentration of Glomalin during Tillage management transition', *Soil Science Society of America Journal*, 63(6), 1825–9.

Wright, S. F. and Upadhyaya, A. (1998), 'A survey of soils for aggregate stability and glomalin, a glycoprotein produced by hyphae of arbuscular mycorrhizal fungi'. *Plant Soil*, 198, 97–107.

Wright, S. F. and Upadhyaya, A. (1996), 'Extraction of an abundant and unusual protein from soil and comparison with hyphal protein of arbuscular mycorrhizal fungi', *Soil Science*, 161, 575–86.

Chapter 11

Nutrient cycling in soils

E. A. Stockdale, Head of Farming Systems Research, NIAB, UK

1 Introduction

The processes of soil formation and the ongoing processes which provide the functions at the heart of the 'ecological factory' of goods and services on which humanity depends (Wall et al. 2012) result from interactions between biology, geology, climate and hydrology. Of the global challenges that humanity faces, the need to feed an increasing population from finite resources is probably the longest recognised and has resulted in the rapid intensification of agricultural systems globally from the mid-twentieth century onwards. Technological developments in tillage, agrochemicals and plant breeding have been key drivers supporting the increase in agricultural productivity; as a consequence the biological interactions at the heart of soil processes including, but not only, nutrient cycling were increasingly marginalised and replaced by direct inputs during the intensification of agricultural systems (Giller et al. 1997). In this period, the focus was mainly on crop health through external inputs, fertilisers, pesticides, herbicides *inter alia*. However, more recently the key role of soil organisms has been re-recognised and maintaining soil health has become an explicit target of farm management strategies. By understanding and working with the natural biological cycles and implementing nutrient management practices that support soil health whilst supplying crop nutrient requirements, farmers can increase the

http://dx.doi.org/10.19103/AS.2017.0033.13

resilience of the system to extreme weather and increase the sustainability of agricultural production systems without any loss of productivity or profitability.

In all terrestrial ecosystems, carbon (C) accumulated from the atmosphere, largely by plants through photosynthesis, is shaped, transformed and intimately mixed with mineral material through a series of quantitatively small, but continually recurrent biological decomposition processes, whose first, and still most famous, student was Charles Darwin (1881). The key defining characteristic of soil is the dynamic cycling of C, nitrogen (N), phosphorus (P) and many other major and trace elements in the plant–soil system. In natural ecosystems at equilibrium, C and nutrient dynamics are known to be tightly coupled and nutrient availability to support plant growth often constrains ecosystem C gain and net primary productivity (NPP). These dynamics are interrupted and disrupted by the human management of ecosystems through agriculture. Vitousek et al. (1997) showed clearly for the N cycle that, during the development of agricultural systems, humans have at least doubled the transfer of N from the vast and unreactive pool of N in the stable gaseous form (N_2) in the atmosphere to the biologically available pools on land. Changes in both the amount and the forms of the nutrients cycling in the soil–plant–water–atmosphere system have had a range of concomitant impacts on: 1) the composition and function of terrestrial, aquatic and marine ecosystems; 2) greenhouse gas concentrations in the atmosphere; and 3) acidification of soils and water bodies.

As the twenty-first century advances, awareness of these, and other locally important environmental impacts, has led to an increased focus on the efficiency of nutrient use within agricultural systems (Cassman et al. 2002). At the same time, alongside the changes to the climate itself, it seems almost certain that fossil-fuel-based resources, including N fertiliser and pesticides, will become increasingly expensive and scarce. Increasing scarcity of other mineral resources, for example, phosphate rock, also seems likely. Balancing environmental and economic outputs in a socially appropriate context will therefore require the adoption of practices and policies which enable benefits from applied organic and inorganic fertilisers to be maximised while off-site impacts, including nutrient flow to waterways and greenhouse gas emissions, are minimised (Foley et al. 2011). The focus has moved from system productivity to crop, soil and ecosystem health.

In this chapter, sequential sections will draw on current literature to build understanding of nutrient cycling in soils beginning with some key concepts and then exploring the underpinning process of decomposition (C cycling), which can also be considered to be the keystone process of all soil biological function and overall soil health. Following sections will explore the close links between the C and N cycles and then consider the links between organic and geochemical cycling processes in the P cycle. Each section will also explore the implications for management of the particular soil nutrient cycle within sustainable agricultural systems. Finally the chapter will consider the approaches to integrated nutrient management developed in the tropics in response to constrained availability of fertiliser and increasingly adopted in the developed world where environmental concerns about nutrient loss are driving approaches to reduce fertiliser use.

2 The underlying concepts of nutrient cycling

From the beginning of agricultural science, the role of the soil in the supply of nutrients for plant growth has been a central question for research (Nortcliff and Gregory 2013).

Understanding the essential plant nutrients, their amounts, availability to crop plants, chemical reactions they undergo in soil, loss mechanisms, the processes making them unavailable or less available to crop plants and means of replenishing them has been an ongoing focus of agronomic research. Much is, therefore, understood about how chemical, physical and biological processes within the soil matrix moderate the availability of additions of soluble nutrients to plants (e.g. Cooke 1981). Plants also manipulate the local soil environment around the roots, as well as responding to it (e.g. Hinsinger 2001). The importance of soil moisture in mediating crop nutrient uptake and controlling loss processes means that in practice, consideration should be given to soil structure and its role in the control of both water and nutrient dynamics.

Almost all nutrients taken up by plants are absorbed by the root system (both passively and actively) as simple (usually ionic) forms from soil solution (Barber 1995). However, the nutrients held in the soil solution form only a very small proportion of the total soil reserve: typically, <1% of soil N and <0.01% of soil potassium (K) (Jarvis et al. 1996; Kofoed 1978). Measurements of the total amount of any nutrient within the soil rarely give useful information about nutrient cycling or plant availability. Nutrients are found in a range of non-soluble organic and mineral forms in soil; while some of these pools are considered to be immediately plant-available, others will release nutrients into soil solution only through slow weathering processes. In some cases, the pools are distinct and directly measurable through physical or chemical extractions; however, care is needed in selection of methods and interpretation as the methods and associated interpretation frameworks have often been developed for particular soil/climate combinations (Renneson et al. 2016). For a consideration of nutrient cycling, the transfer of nutrients between pools in the soil is critically important, particularly those processes which control the concentration in the soil solution of the forms available for absorption by plant roots. Such processes may be the result of either biological or chemical transformations, depending in part on whether the nutrient was originally present in an organic or mineral form. Though simple chemical equilibria play a part, nutrient cycling processes in soil are largely the outcome of a network of biochemical process resulting from the function of a huge diversity of soil organisms and their interactions (Stockdale and Brookes 2006; Bardgett and Wardle 2010). Hence discussions of soil health and practices which seek to maintain or improve soil health in agricultural systems necessarily impact on nutrient cycling.

Nutrient cycling in soil is often measured by quantification of available pools of soil nutrients (and crop uptake) through time; however, it is well known that such data give an indication only of the net outcome of a complex set of interacting biochemical processes (Jarvis et al. 1996). At regional/national scales, in the late twentieth century there was a focus on development of broadly effective approaches for the assessment of soil nutrient status to guide the development of recommendations for lime and fertiliser application to crops (Cooke 1981; Havlin et al. 2014). Thousands of pot and field experiments have been carried out to evaluate soil fertility (given conditions x, how well does the plant grow, take up nutrients and water and produce grain/fruit) and to establish management recommendations (given conditions x, with an addition of y, then a yield increase of z can be expected) for example, for inputs of fertiliser and lime. Hence in many countries/regions robust general advisory guidelines to support fertiliser management for crops are available.

In the soil, spatial and temporal heterogeneity in both the soil's physico-chemical characteristics and external drivers are very high and both the size and extent of soil populations and plant rooting systems are also locally very variable. Small changes in any one of the properties can influence the direction and magnitude of nutrient flows. At farm scale, management factors (tillage, crop choice, fertilisation, etc.) also play a key role in determining nutrient flows; hence, adaptive approaches are needed to develop effective site-specific nutrient management approaches (Gregory et al. 2006). Technological development in global positioning and precision application/cultivation means that in industrialised agriculture there is increasing capacity to develop site-specific (and even within field) approaches to nutrient management.

3 Carbon (C) cycling

Very rapid and dynamic transfers to the soil of the C fixed through photosynthesis can be mediated by roots into the rhizosphere. Studies in upland grassland in the UK where rooting is very dense showed very rapid transfer of pulse-labelled photosynthetically fixed C with about 5–8% of C fixed by plants appearing in arbuscular mycorrhizal fungi and 2% in rhizosphere bacteria within hours (Ostle et al. 2003). It has been estimated that 10–44% of the total C fixed by the plant is released to the soil via the root (Bais et al. 2006). More complex inputs of organic C in roots, litter and other plant and animal residues are processed by decomposition within the soil. C cycling and decomposition of organic matter in soils are driven mainly by biological mechanisms, with abiotic chemical oxidation likely to account for less than 5% of organic matter decomposition (Lavelle et al. 1993).

The flux of carbon dioxide (CO_2) from soils is the net effect of all heterotrophic aerobic decomposition processes (Fig. 1). Hence, where the total amount of C input to soil is increased, whether through NPP or in managed inputs such as manures/composts, both the total CO_2 efflux and the C stored in the soil are likely to increase (e.g. Jacinthe et al. 2002). For soils in C-balance, the decomposition flux to the atmosphere is equal to the deposition of plant-derived C to soil, but it is increasingly recognised that this balance may be disrupted by environmental and land-use changes, potentially resulting in significant (positive or negative) feedbacks to rising CO_2 concentrations in the atmosphere (Houghton 2007; Langley et al. 2009). However, uncertainties in measuring soil C-balance (Saby et al. 2008) and a lack of knowledge of the drivers and controls of soil organic matter mineralisation (Bradford et al. 2008) mean that the role of soils in global C-cycling budgets is still poorly explored (Trumbore 2006).

The cycling of C within the soil is largely controlled by the intermeshing vital processes of many soil organisms. The fact that soil-derived organisms could decompose protein, cellulose and a wide range of organic substrates was well known by the end of the nineteenth century with distinct metabolic pathways shown under aerobic and anaerobic conditions. Jenkinson's (1977) eloquent description of the microbial biomass as 'the eye of the needle through which all the organic materials must pass' still holds; microorganisms in soil are now known to degrade sooner or later virtually all organic compounds including persistent xenobiotics and natural polyphenolic compounds (Nannipieri et al. 2003). Across a range of biomes, plant growth (measured as NPP) is linked to soil C status and soil microbial biomass has been directly correlated with plant productivity (Fierer et al. 2009). The microbial biomass is far from being a static component of the total soil organic matter pool. Jenkinson and Rayner (1977) estimated a turnover time of 2.5 years for microbial biomass C measured in the Broadbalk Continuous Wheat Experiment under UK field conditions, and a turnover time for N of 1.5 years was proposed for the soil

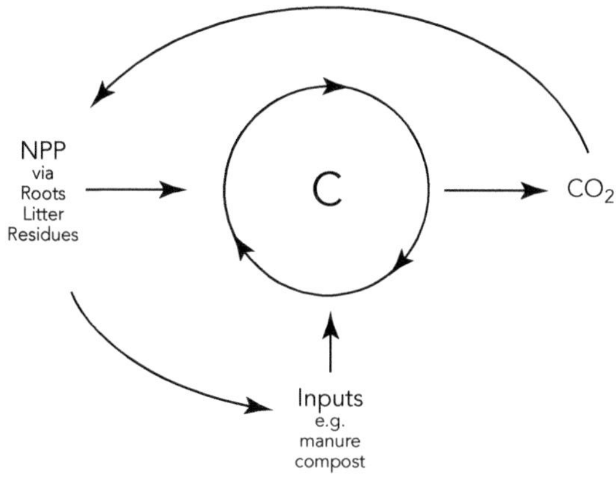

Figure 1 The simple C balance model. The soil carbon cycle returns the CO_2 fixed during photosynthesis to the atmosphere, but the C might cycle within the soil for many decades first.

microbial biomass in soils of the same experiment, again measured under field conditions (Jenkinson and Parry 1989). The population dynamics of the soil microbial community, as well as its size, are therefore important in regulating the cycling of C in soil; it has therefore sometimes been suggested as a key measure of soil health (Gonzalez-Quiñones et al. 2011).

Decomposition in soils takes place through a complex enzymatic cascade in which a wide range of bacterial and fungal species and their associated predatory food webs interact (Adl 2003). The size, diversity and activity of these microbial and faunal populations are therefore key determinants of the C cycling capacity of a soil, but the rate of decomposition and extent to which C is assimilated or lost as CO_2 will also depend on a range of fixed and manageable site and soil factors (Fig. 2). In the past, soil organic matter was considered to be composed of inherently stable large-molecular-size and chemically unique compounds known as 'humic substances'; however, as can be seen from the discussion of C cycling above, soil organic matter is better considered to be a continuum of progressively decomposing organic compounds (Lehmann and Kleber 2015).

The supply of energy to the soil ecosystem (directly from photosynthesis via root exudation and through organic C inputs, root decay and plant residues) is clearly a major metabolic constraint to the cycling of C within the soil ecosystem; but this is also mediated by the capacity of the microbial community to access and decompose organic materials and other nutrients, which is controlled by soil structure and water content (Carson et al. 2010). The spatial location of soil organisms within the soil's three-dimensional matrix is also of importance as non-preferred soil spaces/pores (i.e. where biological function is low or absent) are more likely to stabilise soil organic matter (Ekschmitt et al. 2008). Physical protection through association between organic materials and soil mineral surfaces is then critical in determining the stabilisation of organic compounds resulting from decomposition. Sollins et al. (1996) presented a conceptual model of decomposition and stabilisation of organic matter in soil showing how a range of stabilisation mechanisms interact in space and time. A range of soil physico-chemical properties directly affect soil organisms and their activity; for example, pH has been shown to have a range of direct and indirect impacts on C dynamics (Kidd et al. 2017). Consequently, at a local scale

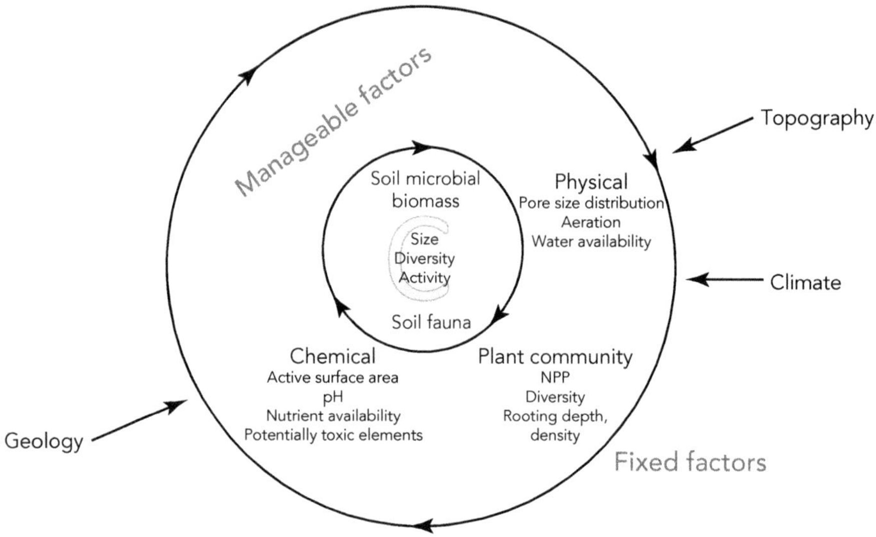

Figure 2 A more complex view of the proximal controls (soil community, physical, chemical, plant community) affecting the soil C cycle and the interaction of fixed and manageable factors that directly and indirectly control C cycling and stabilisation.

(within an agro-ecological zone), site factors such as slope, soil depth, mineralogy and soil texture have a major influence on soil organic matter (Ingram and Fernandes 2001) and the size of the soil microbial biomass (Gonzalez-Quiñones et al. 2011).

The quality of plant material has a large influence on the 'burst' of initial decomposition (Waksman and Tenney 1927); in this phase of decomposition, the size of the soil microbial biomass was not a key limiting factor (Jenkinson and Powlson 1976a,b). Plant material has a variable C:N ratio depending on species, growth stage and the environmental conditions during growth (Ågren and Weih 2012). In comparison, the soil microbial biomass has a narrower C:N ratio; globally soil microbial biomass has been found to have an average C:N:P ratio of 60:7:1 (Hartman and Richardson 2013). A range of plant, soil physical and chemical properties and fixed environmental factors together with residue management practices (e.g. incorporation method) interact to determine the decomposition of added organic materials (see review by Kumar and Goh 2000). Kaiser et al. (2014) showed that resource stoichiometry (C:N ratio), especially of the soluble bioavailable organic matter from inputs, controls competition between microbial groups with different life-history traits (cellular C:N ratios and species-specific turnover rates). They also showed that this competition in turn governs changes in community dynamics and it is these dynamics which control the rate of litter decomposition. Kaiser et al. (2015) further explored the dynamics within microbial communities showing the interacting roles of microorganisms producing extracellular enzymes ('decomposers') and microorganisms exploiting the catalytic activities of others ('cheaters') in the regulation of organic matter turnover. In practice, crop selection and the mixing of crop residues from separate sources before incorporation have been used to manage nutrient release during microbial decomposition (Palm et al. 1997). However, the deliberate manipulation of these processes within agricultural systems is tricky. Wardle et al. (1997) observed that increasing species richness of plant litter inputs (beyond two species) led to unpredictable responses in ecosystem processes and properties.

Decomposition is a central process that also underpins the delivery of many of the ecosystem functions essential for soil health in sustainable agricultural systems; its role in other nutrient cycles is discussed further in the following sections. In addition, polysaccharides and glycoproteins exuded in mucilages and synthesised during decomposition promote better soil structure through their ability to bind inorganic soil particles into stable aggregates (Oades 1993). Research indicates that longer chain polysaccharide molecules are important in promoting aggregate stability and hence improving water infiltration (Elliot and Lynch 1984). In addition, regular additions of fresh organic materials stimulate the activity of ecological engineers (earthworms, termites, collembola, mites, etc.), which mix organic matter with mineral particles and modify pore size and continuity as they move through soil (Lavelle et al. 2016). It has also long been known that some soils are naturally suppressive of soil-borne disease; a wide range of mechanisms (including parasitism, direct and indirect antagonism) are considered to have a role (Mazzola 2002). Ebrahimi et al. (2016) also showed that organic matter amendments reduce the viability and reproduction of potato cyst nematode as a result of direct effects on pore water chemistry but also on indirect effects through impacts on soil structure and soil microbial communities.

Soil organic matter contents reduce markedly where land is converted from primary forest to cropping, but there is a mixed picture of the impacts of more recent changes resulting from agricultural intensification (drainage, cropping intensification, tillage, fertilisation) and climate change (van Wesemael et al. 2010; Reynolds et al. 2013). There is limited evidence of a clear threshold/critical value for total soil organic matter content (e.g. Loveland and Webb 2003). Nonetheless, reviews of literature and practice all highlight the importance of improved organic matter management for sustainable agriculture and there is a growing body of evidence that 'fresh' or 'active' plant residues are more important than the total size of the soil organic matter pool (Loveland and Webb 2003). Stockdale and Watson (2012) proposed the adoption of a number of principles that should guide the development of locally adapted system-oriented approaches to improve soil biological fertility and overall soil health, ideally in combination:

- increase organic matter inputs to soil
- increase diversity of above-ground plant species
- reduce tillage intensity.

On-farm practices that promote C cycling in soil by increasing the amount and managing the quality of OM inputs generally have positive benefits for soil biota and overall soil health, with expected enhancement of, or low risk to, other soil functions and the wider environment.

4 Nitrogen (N) cycling

N is usually the major yield-limiting nutrient in agricultural systems (Berry et al. 2002) but its cycling also has major environmental/ecological consequences, since both nitrate (NO_3^-)

leaching and nitrous oxide (N_2O) emissions can result from N cycling in soil. Plants take up N from the soil solution as ammonium (NH_4^+) and nitrate; in some cases, simple organic N compounds are also taken up by plants (Jones et al. 2004). However, more than 90% of the total N found in soils is found within high-molecular weight organic polymers; N is an essential component of amino acids, peptides and proteins and is incorporated into a wide range of other compounds essential for the functioning of living cells such as nucleic acids and chitin. Consequently, interlocking C and N cycles in soil have been a focus of soil research from the earliest studies (e.g. Jenkinson 1966, 1971) as the release of N through mineralisation is critical for N supply to crops. It has been shown that the gross rates of N mineralisation are driven by C mineralisation (decomposition) rates (Murphy et al. 2007). Understanding and management of the seasonality of the soil microbial biomass and the microbial processes controlling nutrient supply is as important as determining the annual flux of nutrients through the soil microbial biomass for management of the N cycle in practice. Climatic factors, in particular, the interaction of temperature and moisture, largely control the rate of biological processes and hence organic matter decomposition and N cycling (Nishio and Fujimoto 1989; Recous et al. 1999). Seasonal variation in plant growth also leads to temporal variability in organic matter inputs to soil (Murphy et al. 2007). Temporal variability in the biomass and activity of soil organisms can therefore be as significant as spatial variability in determining N availability at field scale. Spatial patterns and activity of soil organisms often show greater fluctuations than underlying patterns of soil resources (Ettema and Wardle 2002).

The cycling of N in soil, although influenced by chemical and physical factors, occurs primarily as a result of a series of connected and interacting biological processes (Fig. 3). The direct and indirect effects of trophic (food webs) and non-trophic interactions (manipulation of habitat) between soil organisms and growing plants at a range of spatial scales within the soil result, at an ecosystem level, in the processes we measure as the soil N cycle. However, at each level the contest for each individual is one of survival and maximisation of resources over evolutionary time (Brussaard 1998). The pool of ammonium released through mineralisation in soil is usually small but cycles very rapidly, as both general microbial assimilation of N to meet growth demands (immobilisation) and specialist (largely autotrophic) nitrification processes compete for ammonium as it is released. The increase in both mineral N pools (nitrate and ammonium) as a result of mineralisation is the result of the difference between the gross mineralisation and immobilisation processes and is often measured as an indicator of N supply (net mineralisation). The role of each group or species in determining the amount of N cycling in soil and its fate has been studied in a number of agricultural systems (see Moore 1994) and there is some indication that more intensive agricultural systems impact on the diversity of soil organisms compared with conservation techniques such as minimum tillage and integrated farming practices (Stockdale and Cookson 2003). However, there is little unequivocal evidence that directly links soil biodiversity and nutrient cycling in soils (Swift et al. 1998; Bradford et al. 2014). There is also a need to consider more explicitly the critical biological feedbacks that influence the abiotic factors at an ecosystem level (Brussaard 1998), and to determine keystone species amongst the bewildering diversity of organisms contributing to N cycling processes in soils, whose inclusion in models may be critical, particularly if global change differentially affects such species (Andrén et al. 2001) and models are increasingly used at a wide range of temporal and spatial scales.

In many farming systems the amounts of N cycling through soils during a year are more than enough, even where no fertiliser is applied, to satisfy crop N demand (Fillery 2001; Table 1). However, asynchrony is a feature of many agricultural systems and the use of

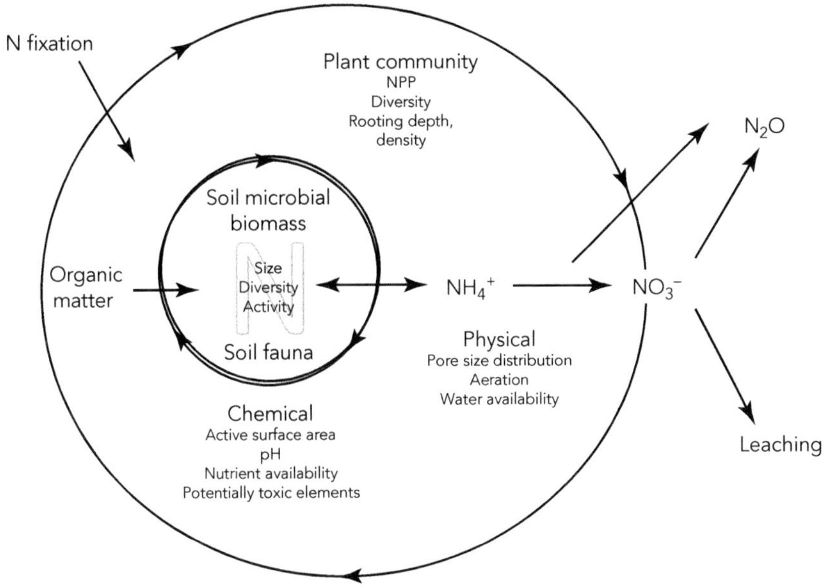

Figure 3 The soil N cycle showing the internal N cycle driven by decomposition and pools of soluble mineral N. The main proximal controls (soil community, physical, chemical, plant community) affecting the N cycle within the soil and the key processes leading to N loss from the nitrate pool are also indicated.

N fertilisers is commonly employed to overcome the difficulties of matching biological supply with plant N demand. On a global scale the application of N fertilisers is now a greater source of N supply to agricultural crops than that via biological processes (Jenkinson 2001). In the UK, about half of crop N uptake is supplied directly by fertilisers applied in that season (Jenkinson 2001), whilst in southern Australia because of the lower fertiliser N applications and lower fertiliser efficiency, due to water limitation, on average 80% of crop uptake is supplied via biological processes (Angus 2001). It is clear, however, that in both systems soil N supply is an important component of crop N uptake.

Asynchrony is characterised by (i) the accumulation of plant-available N in soil in excess of current crop and microbial demand, at a time when those N pools are subject to N loss processes or (ii) characterised by insufficient availability when crop demand is high. Abiotic factors, such as climate and soil parent material, are widely recognised as the major determinants of ecosystem composition and function; their effects on N fluxes often overshadow even large differences between management treatments (Andrén et al. 1999). Whilst these abiotic factors affecting asynchrony cannot be controlled, it is important that if we are to better understand the timing and location of both N supply and demand, their effects on N fluxes should be understood. The development of ^{15}N isotopic dilution techniques has provided an opportunity to obtain a much clearer understanding of soil N cycling by enabling the process rates that form specific pathways within the soil N cycle to be quantified (see review by Murphy et al. 2003). Studies of the processes controlling soil N supply confirm that there is a strong seasonality in gross N fluxes in soil, which have been related to the complex interacting seasonal patterns of soil temperature, moisture content

Table 1 Nitrogen balance sheets (kg N ha^{-1} for plough layer) for a wheat crop following rotations in experimental trials from Australia (East Beverley, 0–10 cm) and the UK (Woburn, 0–23 cm)

	Australia[a]		United Kingdom[b]	
	No legume in system	Legume-based system	Legume-based system	Grass–ley rotation
	Wheat:Wheat	Lupin:Wheat	Beans:Wheat	Grass:Wheat
N pools in soil				
Soil total N	1008	1002	3027	3376
Microbial biomass-N	64	68	84	163
Inorganic N in soil profile (0–90 cm at sowing)	28	32	48	13
Nitrogen supply (per year)				
Gross mineralisation	100	120	440	689
Gross immobilisation	57	61	357	511
Net mineralisation	43	59	83	178
Fertiliser applied	46	0	0	0
Crop N demand				
Plant N uptake (total above-ground at harvest)	13	44	36	54

[a] Data from Murphy et al (1998).
[b] Data from D. V. Murphy and E. A. Stockdale; some published in Murphy et al. (2007).

and residue quality (Murphy et al. 2007). Climatic conditions also strongly influence crop growth and development (and hence N demand) in any season, which in turn determines the quality of residue inputs. Given the inherent variability in temperature and rainfall patterns, it is therefore clear that, even where the same treatments and management practices are applied, synchrony between soil N supply and crop demand may be achieved more effectively in some cropping seasons than others (Ma et al. 1999).

Snapshot measurements of soil mineral N can only give a partial prediction of soil N supply for crop growth since the bulk of soil N is found in organic forms; as part of a long-term monitoring scheme, the average amount of mineral N measured in topsoil during autumn in the UK was only 76 compared to 7000 kg N ha^{-1} in organic matter (Shepherd et al. 1996). Nonetheless, there has been increased emphasis on the measurement of soil mineral N (usually nitrate) in the soil before planting or at a specific time during the crop growing season to give an indication of likely soil N supply (e.g. Meisinger et al. 1992). The availability of biologically derived N to plants is difficult to predict and manage since it is regulated by a number of variables including (i) the quantity/quality of the residue material, (ii) rate of fragmentation and soil incorporation of residues by fauna, (iii) amount and timing of microbial N release, (iv) microbial requirements for N and (v) N loss from the rooting zone. These processes are influenced by both environmental factors (e.g. temperature, wet-dry cycles), which can be difficult to predict, and agricultural management practices (e.g. tillage, green manures, residue management), which can be manipulated to enhance supply and demand.

Few processes directly lead to the loss of NH_4^+ produced in soil (rather than fertiliser applied at the surface); in contrast NO_3^- is readily lost from soils in drainage (by leaching) and/or by denitrification at anaerobic microsites. Nitrification is therefore a key process in regulating the potential for loss of plant-available N from soil (Stockdale et al. 2002). Bowles et al. (2015) found substantial variability in soil inorganic N concentrations and potential for N losses across cropping fields under a variety of management systems. Fields showing evidence of tightly coupled plant-soil N cycling, a desirable scenario in which high crop yields are supported by adequate N availability but low potential for N loss, had the highest total and labile soil C and N and received organic matter inputs with a range of N availability. In these fields, there was also elevated expression of a key gene involved in root N assimilation, cytosolic glutamine synthetase GS1, which confirmed that plant N assimilation was high even when inorganic N pools were low. More work is needed to investigate how crop/variety choice, rotation design and management of organic inputs can be combined most effectively in practice to optimise N efficiency at rotational level and reduce nitrate losses.

In most dynamic models, soil N dynamics are driven by the C cycle (see Shaffer et al. 2001). The models encode our current level of understanding of the soil–crop system and have a similar but non-identical structure. N and C are returned to the soil by growing crops, not only at harvest as stubble and straw, but also throughout the growing season as root exudates, dead leaves and fragments of roots. The model first simulates the decomposition of organic C according to the soil type as it moves through the various compartments (released as CO_2, incorporated into microbial biomass or into humus) and then calculates the N content of these compartments from their C:N ratios (for humus and biomass, these are assumed to remain constant). If the C:N ratio rises, due to a higher C:N composition of the crop debris, N is immobilised first from NH_4^+ and then from NO_3^- in the soil. If the C:N ratio falls, N is mineralised to NH_4^+. Ammonium may then be nitrified to NO_3^-, and NO_3^- may be lost by denitrification or leaching. Using models to underpin fertiliser decisions in-season is difficult because of the influence of the variable and largely unpredictable weather conditions. However, models have been validated successfully using retrospective yield and weather data (e.g. Weiss et al. 2003; Stone and Meinke 2005) as this demonstrates that the model can be used reliably to study the impacts of previous decisions and models may also be used to assist in designing rotations to optimise the synchrony of N supply from the soil and crop demand and identify the most promising management options for in-field testing.

5 Phosphorus (P) cycling

Nearly all the P in terrestrial ecosystems is ultimately derived from the weathering of rocks. The total amount of P found in the upper continental crust is around 700 mg kg^{-1} with marked differences between rock types; the average total P in the relatively young temperate soils of the UK soils is 900 mg kg^{-1} (Rawlins et al. 2012). However, only a very small fraction (<0.1%) of the total P is immediately available to plants as HPO_4^{2-} or $H_2PO_4^-$ in soil solution. Given these low solution concentrations of soluble P and the slow diffusion rates of soluble P in soil, effective crop establishment and rooting is critically important in determining crop P availability.

Studies of soil chronosequences have allowed the main changes in the P pools during soil formation to be established (Walker and Syers 1976). During early stages of soil development, weathering of soluble phosphates from primary minerals provides much of plant demand for P; P is transformed into organic forms as a result of uptake and bio-transformation by soil organisms and plants. Over a quarter of the total P in soil is often found in organic forms. Over time, the stock of P in weatherable minerals becomes exhausted and P is also gradually lost from the ecosystem through erosion and leaching. Over longer timescales, P can become associated with secondary minerals such as aluminium (Al) and iron (Fe) oxides and transformed into stable forms known collectively as occluded P. The P remaining in old, highly weathered soils is partitioned between a tightly cycling organic pool and a geochemically occluded pool that is largely unavailable to plants and soil organisms. There is a high diversity in the pools of both mineral and organic P that occur in soils, with marked differences in the mineral fractions as a result of soil mineralogy. For example, soils formed from volcanic ash parent materials (Andisols) contain large quantities of amorphous (allophanic) clay minerals which stabilise organic materials and have very high P fixation capacities (Velásquez et al. 2016). In agricultural systems, P inputs also occur through fertiliser, manure and other waste materials such as sewage sludges, which can modify the natural equilibria. Hence, site-specific processes and interactions occur and effective management of P availability for crop growth requires an understanding of the possible capacity for P fixation and the relative importance of the mineral and organic pools in plant P supply.

The cycling of P and consequently the amount of plant-available P in soil are governed both by soil P chemistry and by cycling through the microbial biomass and more stable organic forms of P (Fig. 4). Studies of P cycling in soils are beginning to untangle the complex interactions between pools of P often deploying fractionation approaches (e.g. Hedley et al. 1982; Cross and Schlesinger 1995) and/or spectroscopic and spectrometric speciation approaches (Kruse et al. 2015) to distinguish between P pools and using isotope techniques (^{32}P, ^{33}P) and modelling approaches to trace P flows and estimate the fluxes between pools (Di et al. 1997; Kruse et al. 2015). The chemical equilibria controlling the availability of P in the soil solution depend on soil mineralogy and pH. In alkaline soils, the P concentration in soil solution gradually decreases through precipitation of less soluble Ca-phosphates and Ca-oxides bound P. Soil solution P is strongly sorbed on to the edges of silicate clay minerals and to secondary Al- and Fe-oxides. In acidic soils rich in Al- and Fe-oxides, sorption may become gradually stronger by a slow diffusion of phosphate into micropores forming 'occluded P' with Al- and Fe- oxides or even transitions to precipitated Al- and Fe-phosphates. Organic P in soil is commonly found as phytates and other sugar phosphates, which may themselves be stabilised on soil surfaces. Soil microbial biomass P commonly makes up 2–5% of the organic P in arable soils and up to 24% in grassland soils (Turner et al. 2001). The microbial biomass powers the release of P from organic forms by mineralisation as a result of the decomposition of organic materials, as discussed in Section 3. The net release of P through mineralisation is determined by the C:P ratio of the resource pool, the C:P ratio of the decomposer/predator organisms and the assimilation efficiency of the decomposer/predator organisms, which controls the extent of immediate immobilisation. Some bacteria, fungi and plant roots excrete organic acids which modify the chemical equilibria controlling P dissolution from calcium phosphates by reducing pH and/or chelating calcium, thus increasing the solubilisation of P.

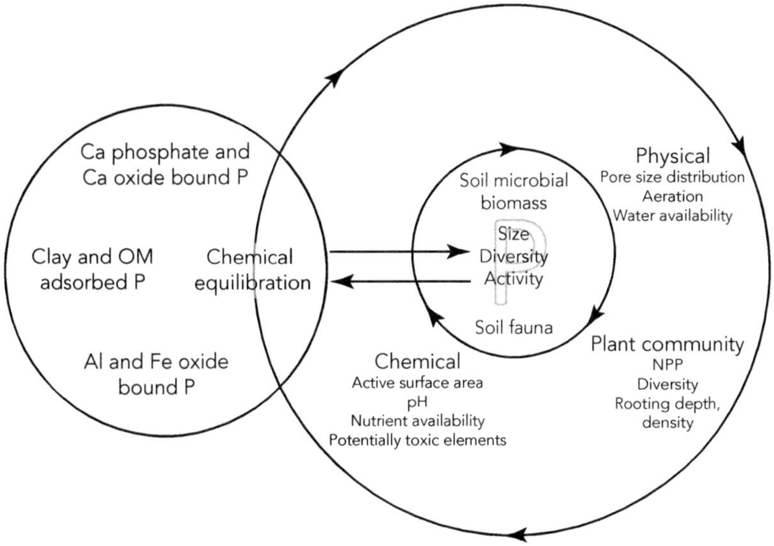

Figure 4 The soil P cycle showing the internal P cycle driven by decomposition and the interaction with geochemical equilibria, which stabilise P. The main proximal controls (soil community, physical, chemical, plant community) affecting the P cycle within the soil are also indicated.

The overall P dynamics in soil are therefore determined by the rates of, and the interactions between, these biological enzyme-mediated transformations and the complex chemical equilibria. A distinct but similar set of interactions between biological processes and chemical equilibria controls the availability of sulphur in soil (Eriksen et al. 1998); though for S inputs from the atmosphere and losses of S by leaching, especially on sandy soils, are important in controlling S availability.

Work carried out in Kenya by Ayaga et al. (2006) shows that management of P cycling through manipulation of the microbial biomass may be possible. Use of organic fertilisers (farmyard manure) in combination with triple superphosphate fertiliser increased the size of the microbial biomass P pool, which later acted as a slow release P source, protecting fertiliser P from fixation and possibly also from leaching (Ayaga et al. 2006). Soil organic matter also increases the number of low energy bonding sites available for P, which are critical to regulate short-term P availability (Johnston et al. 2001). Further understanding of nutrient transfer processes (from soil pools to plant uptake) and rates of these processes is particularly important, and it is most lacking where biological and chemical processes interact in the regulation of P availability.

The rate of supply of nutrients to any root depends on the rates of mass flow and diffusion – and where soil solution concentrations of nutrients are low, as for P, mass flow makes a relatively small contribution to plant uptake. Therefore, for P uptake the distribution of roots within the soil and the extent to which root hairs have formed are important and these factors control the total surface area over which uptake can occur. The ability of the plant root to find space in which to extend through the soil is often an important factor limiting plant growth, and the preparation of seedbeds with a tilth, and consequent network of pores, holding the right balance of air and water appropriate for the seed to be sown, is an art underpinned by the science described elsewhere in this volume. Root branching may also be regulated by external nutrient concentrations with local stimulation of lateral root growth observed in zones of high N, phosphate and K availability (Drew 1975; Forde and Lorenzo 2001), though not all plants respond in the same way (Hodge et al. 2008). Physical contact between a root tip and a zone of low available phosphate concentration causes inhibition in root growth as a result of reduced cell elongation and meristem activity (Reymond et al. 2006; Svistoonoff et al. 2007). Shane and Lambers (2005) highlighted the role that low soil P concentrations play in stimulating the development of cluster (proteoid) roots, for example, by *Lupinus albus*. These are major sites of H^+ and citrate excretion and consequently are able to mobilise sparingly soluble soil P (Cramer et al. 2005) for plant uptake. Plant uptake of P is also facilitated where arbuscular mycorrhizal fungal (AMF) associations with roots are present. The hyphae of AMF external to the host-plant root form a network that is in intimate contact with soil outside the rhizosphere. This network allows AMF-associated roots to acquire nutrients or explore nutrient-rich patches which non-AMF roots cannot (Finlay 2004; Jakobsen et al. 1992). The symbiosis between plants and AMF may alleviate plant P deficiency and enhance mild drought resistance (Suriyagoda et al. 2014). This results from the extension of the apparent rooting surface rather than any chemical solubilisation processes specific to the fungi; however, there is some evidence that AMF increase phosphatase concentrations and hence may increase the mineralisation of organic P sources (Suriyagoda et al. 2014). As a result of the importance of plant–soil interactions in controlling P availability, there is increasing interest in the use of crop selection and sequencing within rotations to manipulate P efficiency for the rotation as a whole (Stockdale et al. 2006).

6 Other nutrient cycles – potassium (K) and micronutrients

Most of the simple cationic forms of nutrients present in soil at any time are in exchangeable forms associated with clay minerals and the organic fraction of the soil; these can rapidly exchange with the cations in soil solution; for example, of the 1–2% of the total soil K that is readily available to plants approximately 90% is in exchangeable forms on soil surfaces (Sparks 2000). Low soil solution concentrations of any cation will lead to a simple exchange with release to soil solution from the exchangeable pool; the reverse will of course also occur where fertiliser has recently been added (Barber 1995). Most K sources (even farmyard manures, compost and crop residues) contain K in a simple cationic form; K is not converted into organic forms by plants. Johnston and Goulding (1990) suggested that inorganic fertilisers and organic manures are interchangeable as sources of K. Hence the cycling processes within the soil for these nutrients can be drawn simply (Fig. 5). Hence it could be argued that plant availability of these nutrients are not directly affected by biological processes, such that maintaining soil health will have relatively little impact on the cycles of these nutrients. However, the indirect impacts of a focus on soil health, for example, on soil structure, water movement, rooting and so forth, will even affect plant availability of exchangeable nutrient pools. In some soils, there may also be a significant pool of non-exchangeable K; while feldspars only weather very slowly, the micas and their weathering products are generally considered to be more important contributors to the supply of plant-available K. The annual release from temperate loams and clays is thought

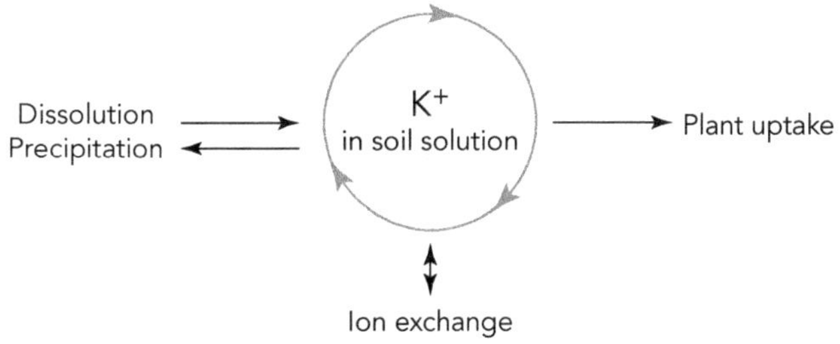

Figure 5 Cycling of K within the soil is largely controlled by the geochemical equilibration of K^+ with exchange surfaces and slower equilibration within mineral-bound K in some soils.

to be equivalent to about 50 kg K ha^{-1} during the growing season (Goulding and Loveland 1987). Inclusion of non-exchangeable K (from both residual fertiliser and soil minerals) to the exchangeable pool when deriving estimates of plant-available K may significantly improve the correlation between soil supply and crop uptake in some soils (Syers 1998). The soil cation exchange capacity (CEC) is dependent on the amount of clay present and its mineralogy and also the size of the soil organic matter pool. In temperate soils, the organic fraction of mineral soils normally constitutes 30–65% of the total cation exchange capacity. Where soil organic matter is increased (as discussed above) the capacity of the soil to retain cations, such as K, will be increased (e.g. Arden-Clarke and Hodges 1988).

Plant essential micronutrients include iron, zinc, manganese, calcium, boron, molybdenum and nickel; animals also require cobalt, chromium and tin. Most can also be toxic, if present in high concentrations. These trace elements are dominantly geologically derived, but in many soils, the most recent, and often largest, sources are from industrial processes. Metal limitations for microbial growth in soil are considered a rarity as soil microbes possess highly efficient mechanisms for metal uptake and many microorganisms excrete chelating compounds which help to maintain soil solution concentrations of metals (Tate 1995). Each of these micronutrients has a distinct cycle with specific solubilisation/precipitation reactions driven by acidity and redox conditions in the soil, which control the balance between soil solution concentrations, precipitation/stabilisation with soil minerals and uptake by microorganisms and plants (for specific detail see Hooda 2010). In agricultural systems, given that N is usually the main limiting nutrient, N availability will largely control crop demand for other nutrients from soil solution, which will then drive re-balancing of the chemical equilibria.

7 Case study – approaches in integrated nutrient management

In many parts of the world, fertiliser use is limited because of low availability, high transport costs and consequently the lack of purchasing capacity by smallholder farmers. But even in the developed world, inputs, such as N fertilisers and pesticides, are expected to become increasingly expensive and scarce; these increasing constraints together with environmental concerns about the impacts of fertiliser-derived nutrients in aquatic ecosystems are likely to necessitate an increased focus on maintaining soil health and the adoption of changed practices within intensive input-focused farming systems (Kirschenmann 2007). Consequently, uptake of practices such as attention to crop rotation, maintenance of soil organic matter and use of local organic wastes is likely to increase in all farming systems (Hanson et al. 2007) and be integrated with new approaches to nutrient management.

In the developing world, a range of renewable soil fertility replacement strategies have been developed (Ajayi et al. 2007, Vanlauwe et al. 2015). Significant attention has been given to the development of legume agro-forestry (fertiliser tree) systems, integration of green manure and dual-purpose legumes, careful use of animal manures often integrated into conservation farming approaches which integrate reduced tillage, retention of crop residues, targeted planting and farm input use and the use of legumes in the rotation (Sanchez and Salinas 1981). The impact of integrated nutrient management are broadly applicable (Janssen 1993) and the concept of combining organic with inorganic N sources can either be integrated into low-input systems to increase soil nutrient supply or high-input systems to potentially reduce N fertiliser requirements and improve soil health (Fig. 6).

Chivenge et al. (2010) compiled a comprehensive quantitative meta-analysis of the impact on maize yield, soil organic matter and N use efficiency. They showed markedly increased yield with the combined use of N fertiliser and organic resources compared with no amendment and with sole use of either input. However, they linked this to increased short-term N supply rather than to any observed increase in N efficiency or other benefits such as the supply of other nutrients. However, they did note improved efficiency where rates of application were low (<60 kg N ha^{-1} combined) and that other effects may become more evident in the longer term, especially as a result of the indirect impacts of increases in soil organic matter. Heijboer et al. (2016) investigated the impacts of organic amendments in combination with fertiliser and confirmed the importance of resource quality in driving increases in microbial activity (high quality = low C:N materials) or immobilisation of fertiliser (low quality = high C:N materials). Palm et al. (2001) developed a decision support system for use of organic resources which divided materials into four groups by both N content (>/<2.5%) and low/high polyphenol/lignin contents; for example, the lowest quality resources are recommended only for surface application as mulches to reduce raindrop impact and erosive losses. In the field, high polyphenol/lignin contents seem to have relatively little impact where %N is high (Chivenge et al. 2010); hence, the current recommendation is to consider the use of organic resources with reference to three classes (Fig. 6). Integration of fertiliser application with organic amendments has been shown to reduce N_2O emissions, especially for amendment of high quality, compared with the use of organic amendments alone (Graham et al. 2017).

Manipulation of N release in soil through the application of combinations of residues and organic fertilisers have been more often studied in tropical than in temperate climates (e.g. Tian et al. 1995). However, using typical temperate crop residues, Gunnarsson and Marstorp (2002) showed that it was possible to change both the total amount of N mineralised and the time-course of mineralisation by combining different quality residues during decomposition. In temperate systems, the management practices combined in integrated nutrient management approaches may have a range of both short-term (this season) and longer-term effects that are more difficult to quantify. For example, Magdoff et al. (1997) reported data showing that although crop recovery of fertiliser N (40%) was higher than legume residue N (17%), much more legume N (47%) was retained in the soil compared to fertiliser N (17%) and that the size of the microbial biomass in the soil, but not necessarily its activity, was also increased. While a range of timing and incorporation

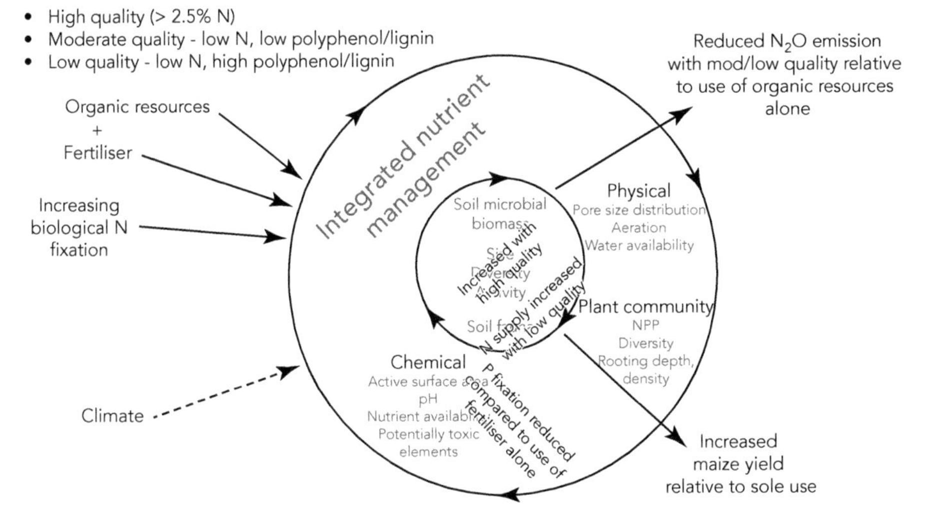

Figure 6 The effectiveness of integrated nutrient management approaches is determined by the quality of the residues, climate and soil factors (soil community, physical, chemical, plant community). These interactions may lead to different outcomes for the same management approach on different soil types or in different seasons. However, on average, the benefits of such integrated approaches are a result of both direct effects on nutrient cycling processes and also indirect effects through impacts on the soil factors, for example changes in soil structure.

options have been considered for organic materials and leguminous residues in arable systems (Korsaeth et al. 2002), there is clearly less flexibility to alter the synchrony of N supply and demand in legume/grass pastures themselves. However, the management and timing of cutting and grazing, particularly with respect to the significant spatial impact of excretal returns on N cycling, can be an important tool in manipulating the balance between N fixation, soil N supply and plant N uptake (Ledgard 2001). Recent advances in plant nutrient decision support models, improved access to high-resolution satellite images and the improved interpretation of crop and soil spectral signatures make site-specific recommendations possible (Scharf et al. 2002). Within sustainable agricultural systems it is also likely that a broader integrated approach to the assessment of soil fertility, considering the integration of soil physical, biological and chemical aspects together with inputs of water and nutrients from a range of sources, is likely to be needed both to target new management practices appropriately and thereafter to ensure sustainable management within farming systems.

Although these general principles have been researched and recommended for farmer uptake in sub-Saharan Africa for a number of years, adoption rates remain low. Kamau et al. (2014) carried out a survey of 1001 households in western and central Kenya and found that, unsurprisingly, farmers were price-responsive and improved access to market services enhanced demand for both inorganic fertilisers and soil amendments; policy interventions to educate women-headed households and new entrants and to open up land for rental were also considered to be important in promoting adoption of more sustainable nutrient management. It is essential to also engage farmers in the diagnosis and design of farm nutrient management and facilitate farmer learning. There is good evidence that as farmers become familiar with the principles underpinning soil health in sustainable agricultural systems, they begin to innovate with different strategies for integrated nutrient management in their own environments (Stockdale and Watson 2012).

8 Conclusions

Studies of microbial processes in soils have been a key part of soil science research since the beginning of the twentieth century and methods have been developed that allow the observation and quantification of soil organisms and biologically mediated processes at a range of scales. Over this period, the understanding of the roles that microorganisms play in C and nutrient cycling processes such as N fixation, decomposition of organic substrates and transformation of N, P and S has increased beyond recognition (Stockdale and Brookes 2006). However, in contrast to the world of medicine, where understanding of microbiology developed in the same period has stimulated a rapid development in new treatments and practice, there is yet to be a similar revolution in practice for agricultural systems. As described above, there is increasing recognition of the role of biological processes in supporting nutrient cycling, soil fertility and soil health but we still have limited ability to manipulate agricultural systems in a predictable and controllable way with regard to soil biological fertility and function. However, as summarised below, scientific research to date has provided a strong foundation of principles and theoretical framework against which farmers and their advisors can test their observations of site-specific practice and also identify opportunities for the development of innovative and locally adapted practices:

- the cycling of C through soils as a result of decomposition is the keystone process for all soil biological function and overall soil health;
- the pool of C in soil and its dynamics are an important but still relatively poorly understood component of the global C cycle and hence critical in our attempts to mitigate climate change;
- the key to maintaining soil health seems to be the regular input of fresh/active plant residues via roots, litter and managed inputs (crop residues, fertilising materials) rather than the total size of the organic matter pool *per se*;
- chemical equilibria and biochemical processes, mediated by a huge delivery of soil organisms; both have important roles within the complex network of interactions that result in the nutrient cycling processes in soils;
- biological nutrient cycling in soil has a key role in the supply of nutrient for plant uptake, especially, but not only, N, P and S;
- decomposition and release of nutrients by mineralisation from soil organic matter and added organic materials depend on a range of plant, soil and management factors. Crop selection and mixing of organic residues of different types can give some control to the rate and timing of nutrient release;
- active biological turnover processes in soil help promote and improve soil structure and its stability and may also help reduce the risks of soil-borne disease;
- the dynamics of the nitrate pool in soil and the balance between mineralisation and nitrification processes are important in determining the risk of nitrate losses from soil by leaching or as gaseous nitrogen oxides, both of which have important

environmental impacts. There is some evidence that crop/variety choice, rotation design and soil management practices can be used to modify this balance and increase synchrony of N release and crop demand;

- P cycling in soil is complex and P availability for crops is determined to a large degree by chemical equilibria strongly related to the age and intensity of weathering in soil and pH. Maintaining tightly coupled cycling of P in organic forms may be key to optimising P availability in strongly P fixing soils;
- even for nutrients where the crop supply is not determined mainly by biological processes, such as K, Mg, maintaining soil health will affect the availability of these nutrients to the crop because of the indirect impacts on soil structure, water movement and rooting;
- soil health improving practices such as attention to crop rotation, maintenance of soil organic matter and use of local organic wastes can be adopted in all farming systems and in industrialised farming systems these should be integrated with new precision approaches to nutrient management.

9 Future trends

Most biochemical-based techniques for studying nutrient cycling and microorganism–plant nutrient transfers are applied at scales several orders of magnitude greater (i.e. cm and mm, g of soil) than those at the cellular level at which the processes actually occur (Herrmann et al. 2007). For example, the available nutrient concentration in the extraction of a few grams of soil may bear little relation to the concentrations of the nutrient that microorganisms may experience at the microscale, which could range from effectively zero, in some microsites, to very high in the proximity of organic matter or fertiliser granules. Simultaneous analysis of identity (who is there?) and function (what are they doing?) of soil organisms with consideration for maintaining the soil three-dimensional structure (where are they located?) requires study of the links between soil physiochemical heterogeneity and biochemical processes. Yet, this is the cellular scale at which the microbial biomass and plants interact, decompose organic matter, cycle nutrients and produce greenhouse gases. Understanding C and N cycling at the soil–plant interface still remains one of the greatest challenges limiting our ability to predict the extent to which climate and anthropogenic disturbance alter nutrient flows between soil, plant and atmosphere. However, there are an emerging range of techniques that allow greater precision and may provide opportunities to link molecular identification (microbial phylogeny) and function (metabolic activity) concurrently (Li et al. 2008; Eichorst et al. 2015; Mayali et al. 2012). These and other such technological breakthroughs in the complementary fields of proteomics (analysis of protein structure and function) and metabolomics (analysis of intermediates and products of metabolism) open the microbial 'black box' and offer endless opportunities to explore soil organisms within agro-ecosystems. Bioinformatics, the science of analysing and interpreting large-scale complex biological data such as genetic codes, goes hand in hand with these techniques. Here, the challenge is to identify the pathways of significance amongst these complex networks of ecological interactions (Almeida 2002).

While new analytical methods and science findings will continue to increase our understanding of the soil nutrient cycles and the biological and geochemical interactions of which they are comprised, the challenge is, and may always be, to translate this knowledge into information that can be applied by farmers to increase the sustainability of farming system practices. Both top-down (e.g. government legislation) and bottom-up (e.g. land-holder participatory action research) approaches have been used to implement change, both have strengths and weaknesses but both are predicated on effective knowledge exchange with the research community. Given the increasing interest and active engagement of the farming industry with the concepts of soil health/quality, it is appropriate and timely to take the existing underpinning research knowledge and translate it into practical tools to support decision-making on-farm. At farm scale, management factors (tillage, crop choice, fertilisation, etc.) also play a key role in determining nutrient flows and farmers themselves are often leading innovation in the development of integrated approaches to management of crops and soils to increase nutrient use effectiveness and efficiency. In particular, precision approaches to the diagnosis and management of nutrient management are becoming increasingly available, but there is a need to provide approaches that turn the wealth of data available into useful information to support decision-making. In many cases, the key knowledge exchange step that is needed is not from research to farm but from farmer to researcher, so that research is embedded within a robust understanding of the practical demands and constraints of nutrient management. Farmers and agronomists have a key role as partners in the adaptive research approaches that are needed to develop effective site-specific nutrient management to support each unique sustainable agricultural system so that agricultural systems can concurrently meet the need to feed an increasing population from finite resources and maintain the whole 'ecological factory' of goods and services on which humanity depends.

10 Where to look for further information

Soil fertility is a topic that has been covered extensively in the soil science and agricultural management literature over the years. Hence, any introductory soil or agricultural science text book will include sections on organic matter and nutrient cycling. However, in the mid-late twentieth century, the focus of agricultural science was on maintaining crop health through external inputs, fertilisers, pesticides, herbicides, and hence such sources should be supplemented with reading that complements this with an understanding of the biological processes in soil. Many key sources are given in the reference list that follows this section but readers might also find the following useful:

Anderson, J. M. and Ingram J. S. I. (Eds). (1993). *Tropical Soil Biology and Fertility: A Handbook*. CAB International, p. 221.

Duiker, S., Myers, J. C. and Blazure, L. S. (2016). Soil Health in Field and Forage Crop production. USDA Natural Resources Conservation Service, Penn State University Extension, Capital Resource Conservation & Development, and Clinton County Conservation District (2016), p. 27.

Lal, R. (2016). Soil health and carbon management. *Food and Energy Security*, 5, 212–22.

Patra, A. K., Lenka N. K. and Biswas A. K. (2015). Soil Health Assessment and Management: Issues and Strategies. *Indian Journal of Fertilisers*, 11(12), 16–25.

Stirling, G., Hayden, H., Pattison, T. and Stirling, M. (2016). *Soil Health, Soil Biology, Soilborne Diseases and Sustainable Agriculture. A Guide*. CSIRO Publishing, p. 280.

In addition there is now a wide range of on-line resources which support increased understanding of soil health and nutrient management. The best-developed independently provided examples that are currently available are:

- Soil Health Portal. USDA Natural Resources Conservation Service https://www.nrcs.usda.gov/wps/portal/nrcs/main/national/soils/health/ Last accessed 1 September 2017.
- Soils are Alive, The complete soil health reference for farmers, consultants and researchers. Professor Lyn Abbott. Online book. www.soilhealth.com Last accessed 1 September 2017.

11 References

Adl, S. M. (2003). *The Ecology of Soil Decomposition*. CABI publishing, Wallingford, p. 368.

Ågren, G. I. and Weih, M. (2012). Plant stoichiometry at different scales: Element concentration patterns reflect environment more than genotype. *New Phytologist*, 194, 944–52. doi:10.1111/j.1469-8137.2012.04114.x.

Ajayi, O. C., Akinnifesi, F. K., Sileshi, G. and Chakeredza, S. (2007). Adoption of renewable soil fertility replenishment technologies in the southern African region: Lessons learned and the way forward. *Natural Resources Forum*, 31, 306–17.

Almeida, J. S. (2002). Predictive non-linear modeling of complex data by artificial neural networks. *Current Opinion in Biotechnology*, 13, 72–6.

Andrén, O., Brussaard, L. and Clarholm, M. (1999). Soil organism influence on ecosystem-level processes – bypassing the ecological hierarchy? *Applied Soil Ecology*, 11, 177–88.

Andrén, O., Kätterer, T. and Hyvönen, R. (2001). Projecting soil fauna influence on long-term soil carbon balances from soil faunal exclusion experiments. *Applied Soil Ecology*, 18, 177–86.

Angus, J. F. (2001). Nitrogen supply and demand in Australian agriculture. *Australian Journal of Experimental Agriculture*, 41, 277–88.

Arden-Clarke, C. and Hodges, R. D. (1988). The environmental effects of conventional and organic/biological farming systems. II. Soil ecology, soil fertility and nutrient cycles. *Biological Agriculture & Horticulture*, 5, 223–87.

Ayaga, G., Todd, A. and Brookes, P. C. (2006). Enhanced biological cycling of phosphorus increases its availability to crops in low-input sub-Saharan farming systems. *Soil Biology and Biochemistry*, 38, 81–90.

Bais, H. P., Weir, T. L., Perry, L. G., Gilroy, S. and Vivanco, J. M. (2006). The role of root exudates in rhizosphere interactions with plants and other organisms. *Annual Reviews in Plant Biology*, 57, 234–66.

Barber, S. A. (1995). *Soil Nutrient Bioavailability: A Mechanistic Approach*. 2nd Edition. John Wiley & Sons, New York, NY, p. 384.

Bardgett, R. D. and Wardle, D. A. (2010). *Aboveground-Belowground Linkages. Biotic Interactions, Ecosystem Processes, and Global Change*. Oxford University Press, Oxford, pp. 1–301.

Berry, P. M., Sylvester-Bradley, R., Philipps, L., Hatch, D. J., Cuttle, S. P., Rayns, F. W. and Gosling, P. (2002). Is productivity of organic farms restricted by the supply of available nitrogen? *Soil Use Management*, 18, 248–55.

Bowles, T. M., Hollander, A. D., Steenwerth, K. and Jackson, L. E. (2015). Tightly-coupled plant-soil nitrogen cycling: Comparison of organic farms across an agricultural landscape. *PLoS ONE*, 10(6), e0131888. https://doi.org/10.1371/journal.pone.0131888.

Bradford, M. A., Davies, C. A., Frey, S. D., Maddox, T. R., Melillo, J. M., Mohan, J. E., Reynolds, J. F., Treseder, K. K. and Wallenstein, M. D. (2008). Thermal adaptation of soil microbial respiration to elevated temperature. *Ecology Letters*, 11, 1316–27.

Bradford, M. A., Wood, S. A., Bardgett, R. D., Black, H. I. J., Bonkowski, M., Eggers, T., Grayston, S. J., Kandeler, E, Manning P., Setälä, H. and Hefin Jones, T. (2014). Discontinuity in the responses of ecosystem processes and multifunctionality to altered soil community composition. *Proceedings of the National Academy of Sciences of the United States of America*, 111, 14478–83. doi:10.1073/pnas.1413707111.
Brussaard, L. (1998). Soil fauna, guilds, functional groups and ecosystem processes. *Applied Soil Ecology*, 9, 123–35.
Carson, J. K., Gonzalez-Quiñones, V., Murphy, D. V., Hinz, C., Shaw, J. A. and Gleeson, D. B. (2010). Low pore connectivity increases bacterial diversity in soil. *Applied and Environmental Microbiology*, 76, 3936–42.
Cassman, K. G., Dobermann, A. and Walters, D. T. (2002). Agro-ecosystems, nitrogen-use efficiency and nitrogen management. *AMBIO Journal of Human Environment*, 31, 132–40.
Chivenge, P., Vanlauwe, B and Six, J. (2010). Does the combined application of organic and mineral nutrient sources influence maize productivity? A meta-analysis. *Plant and Soil*, 342, 1–30. doi: 10.1007/s11104-010-0626-5.
Cooke, G. W. (1981). *Fertilizing for Maximum Yield*. 3rd Edition. Granada Publishing, London, p. 465.
Cramer, M. D., Shane, M. W. and Lambers, H. (2005). Physiological changes in white lupin associated with variation in root-zone CO_2 concentration and cluster-root P mobilization. *Plant Cell Environment*, 28, 1203–17.
Cross, A. and Schlesinger, W. (1995). A literature review and evaluation of the Hedley fractionation: Applications to the biogeochemical cycle of soil phosphorus in natural ecosystems. *Geoderma*, 64, 197–214.
Darwin, C. (1881). *The Formation of Vegetable Mould through the Action of Worms with Observations on their Habits*. John Murray, London.
Di, H. J., Condron, L. M. and Frossard, E. (1997). Isotope techniques to study phosphorus cycling in agricultural and forest soils: A review. *Biology and Fertility of Soils*, 24, 1–12. doi: 10.1007/BF01420213.
Drew, M. C. (1975). Comparison of the effects of a localized supply of phopshate, nitrate, ammonium and potassium on the growth of the seminal root system, and the shoot, in barley. *New Phytologist*, 75, 479–90.
Ebrahimi, N., Viaene, N., Vandecasteele, B., D'Hose, T., Debode, J., Cremelie, P. and De Tender, C. (2016). Traditional and new soil amendments reduce survival and reproduction of potato cyst nematodes, except for biochar. *Applied Soil Ecology*, 107, 191–204. doi: 10.1016/j.apsoil.2016.06.006.
Eichorst, S. A., Strasser, F., Woyke, T., Schintlmeister, A., Wagner, M. and Woebken, D. (2015). Advancements in the application of NanoSIMS and Raman microspectroscopy to investigate the activity of microbial cells in soils. *FEMS Microbiology Ecology*, 91, fiv106.
Ekschmitt, K., Kandeler, E., Poll, C., Brune, A., Buscot, F., Friedrich, M., Gleixner, G., Hartmann, A., Kästner, M., Marhan, S. and Miltner, A. (2008). Soil-carbon preservation through habitat constraints and biological limitations on decomposer activity. *Journal of Plant Nutrition and Soil Science*, 171, 27–35.
Elliot, L .F. and Lynch, J. M. (1984). The effect of available carbon and nitrogen in straw on soil and ash aggregation and acetic acid production. *Plant and Soil*, 78, 335–43.
Eriksen, J., Murphy, M. D. and Schnug, E. (1998). The soil sulphur cycle. In: *Sulphur in Agroecosystems. Nutrients in Ecosystems*, Volume 2 (Schnug, E. (Eds)). Springer, Dordrecht, The Netherlands, pp. 39–73. doi: 10.1007/978-94-011-5100-9_2.
Ettema, C. H. and Wardle, D. A. (2002). Spatial soil ecology. *Trends in Ecology & Evolution*, 17, pp. 177–83.
Fierer, N., Strickland, M. S., Liptzin, D., Bradford, M. A. and Cleveland, C. C. (2009). Global patterns in belowground communities. *Ecology Letters*, 12, 1238–49.
Fillery, I. R. P. (2001). The fate of biologically fixed nitrogen in legume-based dryland farming systems: A review. *Australian Journal of Experimental Agriculture*, 41, 361–81.
Finlay, R. D. (2004). Mycorrhizal fungi and their multifunctional roles. *Mycologist*, 18, 91–6.

Foley, J. A., Ramankutty, N., Brauman, K. A., Cassidy, E. S., Gerber, J. S., Johnston, M., Mueller, N. D., O'Connell, C., Ray, D. K., West, P. C., Balzer, C., Benett, E. M., Carpenter, S. R., Hill, J., Monfreda, C., Polasky, S., Rockström, J., Sheehan, J., Siebert, S., Tilman, D. and Zaks, D. P. M. (2011). Solutions for a cultivated planet. *Nature*, 478, 337–42. doi:10.1038/nature10452.

Forde, B. and Lorenzo, H. (2001). The nutritional control of root development. *Plant and Soil*, 232, 51–68.

Giller, K. E., Beare, M. H., Lavelle, P., Izac, A. M. N. and Swift, M. J. (1997). Agricultural intensification, soil biodiversity and ecosystem function. *Applied Soil Ecology*, 6, 3–16.

Gonzalez-Quiñones, V., Stockdale, E. A., Banning, N. C., Hoyle, F. C., Sawada, Y., Wherett, A. D., Jones, D. L. and Murphy, D. V. (2011). Soil microbial biomass–Interpretation and consideration for soil monitoring. *Australian Journal of Soil Research*, 49, 287–304.

Goulding, K. W. T. and Loveland, P. (1987). The classification and mapping of potassium reserves in soils of England and Wales. *Journal of Soil Science*, 37, 555–65.

Graham, R. F., Wortman, S. E. and Pittelkow, C. M. (2017). Comparison of Organic and Integrated Nutrient Management Strategies for Reducing Soil N_2O Emissions. *Sustainability*, 9, 510. doi:10.3390/su9040510.

Gregory, R., Ohlson, D. and Arvai, J. (2006). Deconstructing adaptive management: Criteria for applications to environmental management. *Ecological Applications*, 16, 2411–25.

Gunnarsson, S. and Marstorp, H. (2002). Carbohydrate composition of plant materials determines N mineralisation. *Nutrient Cycling in Agroecosystems*, 62, 175–83.

Hanson, J. D., Liebig, M. A., Merrill, S. D., Tanaka, D. L., Krupinsky, J. M. and Stott, D. E. (2007) Dynamic cropping systems: Increasing adaptability amid an uncertain future. *Agronomy Journal*, 99, 939–43.

Hartman, W. H. and Richardson, C. J. (2013). Differential nutrient limitation of soil microbial biomass and metabolic quotients (qCO_2): Is there biological stoichiometry of soil microbes? *PLoS ONE*, 8, 1–14.

Havlin, J. L., Tisdale, S. L., Nelson, W. L. and Beaton, J. D. (2014). *Soil Fertility and Fertilizers. An Introduction to Nutrient Management*. Eighth Edition. Pearson, Upper Saddle River, NJ.

Hedley, M., Steward, J. and Chauhuan, B. (1982). Changes in organic and inorganic soil phosphorus fractions induced by cultivation practices and by laboratory incubations. *Soil Science Society of America Journal*, 46, 970–6.

Heijboer, A., ten Berge, H. F. M, de Rutier, P. C., Jørgensen, H. B., Kowalchuk, G. A. and Bloem, J (2016). Plant biomass, soil microbial community structure and nitrogen cycling under different organic amendment regimes: A ^{15}N tracer-based approach. *Applied Soil Ecology*, 107, 251–60.

Herrmann, A., Ritz, K., Nunan, N., Clode, P., Pett-Ridge, J., Kilburn, M., Murphy, D. V., O'Donnell A. J and Stockdale, E. A. (2007). Nano-scale secondary ion mass spectrometry – a new analytical tool in biogeochemistry and soil ecology: A review article. *Soil Biology and Biochemistry*, 39, 1835–50.

Hinsinger, P. (2001). Bioavailability of soil inorganic P in the rhizosphere as affected by root-induced chemical changes: A review. *Plant and Soil*, 237, 173–95. doi:10.1023/A:1013351617532

Hodge, A., Stewart, J., Robinson, D., Griffiths, B. S. and Fitter, A. H. (2008). Root proliferation, soil fauna and plant nitrogen capture from nutrient rich patches in soil. *New Phytologist*, 139, 479–94.

Hooda, P. (Ed.) (2010). *Trace Elements in Soils*. Wiley-Blackwell, Oxford, p. 616.

Houghton, R. A. (2007). Balancing the global carbon budget. *Annual Review of Earth and Planetary Sciences*, 35, 313–47.

Ingram, J. S. I. and Fernandes, E. C. M. (2001). Managing carbon sequestration in soils: Concepts and terminology. *Agriculture Ecosystems & Environment*, 87, 111–17.

Jacinthe, P. A., Lal, R. and Kimble, J. M. (2002). Carbon budget and seasonal carbon dioxide emission from a central Ohio Luvisol as influenced by wheat residue amendment. *Soil & Tillage Research*, 67, 147–57.

Jakobsen, I., Abbott, L. K. and Robson, A. D. (1992). External hyphae of vesicular-arbuscular mycorrhizal fungi associated with *Trifolium subterraneum* L. *New Phytologist*, 120, 371–80. doi:10.1111/j.1469-8137.1992.tb01077.x.

Janssen, B. H. (1993). Integrated nutrient management: The use of organic and mineral fertilizers. In: *The Role of Plant Nutrients for Sustainable Food Crop Production in Sub-Saharan Africa* (van Reuler, H. and Prins, W. H. (Eds)). VKP, Leidschendam, The Netherlands, pp. 89–105.

Jarvis, S. C., Stockdale, E. A., Shepherd, M. A. and Powlson, D. S. (1996). Nitrogen mineralization in temperate agricultural soils: Processes and measurement. *Advances in Agronomy*, 57, 187–235.

Jenkinson, D. S. (1966). Studies on the decomposition of plant material in soil. *Journal of Soil Science*, 17, 280–302.

Jenkinson, D. S. (1971). Studies on the decomposition of C14 labelled organic matter in soil. *Soil Science*, 111, 64–70.

Jenkinson, D. S. (1977). The soil microbial biomass. *New Zealand Soil News*, 25, 213–18.

Jenkinson, D. S. (2001). Nitrogen in a global perspective, with focus on temperate areas – state of the art and global perspectives. *Plant and Soil*, 228, 3–15.

Jenkinson, D. S. and Parry, L. C. (1989). The nitrogen cycle in the Broadbalk wheat experiment: A model for the turnover of nitrogen through the soil microbial biomass. *Soil Biology and Biochemistry*, 21, 535–41.

Jenkinson, D. S. and Powlson, D. S. (1976a). The effects of biocidal treatments on metabolism in soil – I. Fumigation with chloroform. *Soil Biology and Biochemistry*, 8, 167–77.

Jenkinson, D. S. and Powlson, D. S. (1976b). The effects of biocidal treatments on metabolism in soil – V. A method for measuring soil biomass. *Soil Biology and Biochemistry*, 8, 209–13.

Jenkinson, D. S. and Rayner, J. H. (1977). The turnover of soil organic matter in some of the Rothamsted classical experiments. *Soil Science*, 123, 298–305.

Johnston, A. E. and Goulding, K. W. T. (1990). The use of plant and soil analyses to predict the potassium supplying capacity of soil. In: *Development of K-Fertilizer Recommendations*. International Potash Institute, Basel, Switzerland, pp. 177–204.

Johnston, A. E., Poulton, P. R. and Syers, J. K. (2001). *Phosphorus, Potassium and Sulphur Cycles in Agricultural Soils*. Proceedings No.465. The International Fertiliser Society, York, UK.

Jones, D. L., Hodge, A. and Kuzyakov, Y. (2004). Plant and mycorrhizal regulation of rhizodeposition. *New Phytologist*, 163, 459–80.

Kaiser, C., Franklin, O., Dieckman, U. and Richter, A. (2014). Microbial community dynamics alleviate stoichiometric constraints during litter decay. *Ecology Letters*, 17, 680–90.

Kaiser, C., Franklin, O., Richter, A. and Dieckmann, U. (2015). Social dynamics within decomposer communities lead to nitrogen retention and organic matter build-up in soils. *Nature Communications*, 6, Article number: 8960.

Kamau, M., Smale, M. and Mutua, M. (2014). Farmer demand for soil fertility management practices in Kenya's grain basket. *Food Security*, 6(6), 793–806. doi:10.1007/s12571-014-0398-5

Kidd, J., Manning, P., Simkin, J., Peacock, S and Stockdale, E. (2017). Impacts of 120 years of fertilizer addition on a temperate grassland ecosystem. *PLoS ONE*, 12(3), e0174632. https://doi.org/10.1371/journal.pone.0174632.

Kirschenmann, F. L. (2007). Potential for a new generation of biodiversity in agroecosystems of the future. *Agronomy Journal*, 99, 373–6.

Kofoed A. D. (1978). The potassium cycle in cropping systems. In: *Potassium Research - Review and Trends*. International Potash Institute, Basel, Switzerland, pp. 435–49.

Korsaeth, A., Henriksen, T. M. and Bakken, L. R. (2002). Temporal changes in mineralisation and immobilisation of N during degradation of plant material: Implications for the plant N supply and losses. *Soil Biology and Biochemistry*, 34, 789–99.

Kruse, J., Abraham, M., Amelung, W., Baum, C., Bol, R., Kühn, O., Lewandowski, H., Niederberger, J., Oelmann, Y., Rüger, C., Santner, J., Siebers, M., Siebers, N., Spohn, M., Vestergren, J., Vogts, A. and Leinweber, P. (2015). Innovative methods in soil phosphorus research: A review. *Journal of Plant Nutrition and Soil Science*, 178, 43–88. doi:10.1002/jpln.201400327.

Kumar, K. and Goh, K. M. (2000). Crop residues and management practices: Effects on soil quality, soil nitrogen dynamics, crop yield, and nitrogen recovery. *Advances in Agronomy*, 68, 197–319.

Langley, J. A., McKinley, D. C., Wolf, A. A., Hungate, B. A., Drake, B. G. and Megonigal, J. P. (2009). Priming depletes soil carbon and releases nitrogen in a scrub-oak ecosystem exposed to elevated CO_2. *Soil Biology and Biochemistry*, 41, 54–60.
Lavelle, P., Blanchart, E., Martin, A., Martin, S. and Spain, A. (1993). A hierarchical model for decomposition in terrestrial ecosystems: Application to soils of the humid tropics. *Biotropica*, 25(2), 130–50.
Lavelle, P., Spain, A., Blouin, M., Brown, G., Decaëns, T., Grimaldi, M., Jiménez, J. J., McKey, D., Mathieu, J., Velasquez, E. and Zangerlé, A. (2016). Ecosystem engineers in a self-organized soil: A review of concepts and future research questions. *Soil Science*, 181, 91–109.
Ledgard, S. F. (2001). Nitrogen cycling in low input legume-based agriculture, with emphasis on legume/grass pastures. *Plant and Soil*, 228, 43–59. doi.org/10.1023/A:1004810620983.
Lehmann, J. and Kleber, M. (2015). The contentious nature of soil organic matter. *Nature*, 528, 60–8. doi: 10.1038/nature16069.
Li, T., Wu, T. D., Mazéas, L., Toffin, L., Guerquin-Kern, J. L., Leblon, G. and Bouchez, T. (2008). Simultaneous analysis of microbial identity and function using NanoSIMS. *Environmental Microbiology*, 10, 580–8.
Loveland, P. and Webb, J. (2003). Is there a critical level of organic matter in the agricultural soils of temperate regions: A review. *Soil & Tillage Research*, 70, 1–18.
Ma, B. L., Dwyer, L. M. and Gregorich, E. G. (1999). Soil nitrogen amendment: Effects on seasonal nitrogen mineralisation and nitrogen cycling in maize production. *Agronomy Journal*, 91, 1003–9.
Magdoff, F., Lanyon L. and Liebhardt B. (1997). Nutrient cycling, transformations, and flows: Implications for a more sustainable agriculture. *Advances in Agronomy*, 60, 1–73.
Mayali, X., Weber, P. K., Brodie, E. L., Mabery, S., Hoeprich, P. D. and Pett-Ridge, J. (2012). High-throughput isotopic analysis of RNA microarrays to quantify microbial resource use. *The ISME Journal*, 6, 1210–21.
Mazzola, M. (2002). Mechanisms of natural soil suppressiveness to soil borne diseases. *Antonie van Leeuwenhoek International Journal of General and Molecular Microbiology*, 81, 557–64.
Meisinger, J. J., Bandel, V. A., Angle, J. S., O'Keefe, B. E. and Reynolds C. M. (1992). Pre-sidedress soil nitrate test evaluation in Maryland. *Soil Science Society of America Journal*, 56, 1527–32.
Moore, J. C. (1994). Impact of agricultural practices on soil food web structure: Theory and application. *Agricultural Ecosystems and Environment*, 51, 239–48.
Murphy, D. V., Fillery, I. R. P. and Sparling, G. P. (1998). Seasonal fluctuations in gross N mineralisation, ammonium consumption and microbial biomass in a Western Australian soil under different land uses. *Australian Journal of Agricultural Research*, 49, 523–35.
Murphy, D. V., Recous, S., Stockdale, E. A., Fillery, I. R. P., Jensen, L. S., Hatch D. J. and Goulding, K. W. T. (2003). Gross nitrogen fluxes in soil: Theory, measurement and application of ^{15}N pool dilution techniques. *Advances in Agronomy*, 79, 69–118.
Murphy, D. V., Stockdale, E. A., Poulton, P. R., Willison, T. W. and Goulding, K. W. T. (2007). Seasonal dynamics of carbon and nitrogen pools and fluxes under continuous arable and ley-arable rotations in a temperate environment. *European Journal of Soil Science*, 58, 1410–24. doi: 10.1111/j.1365-2389.2007.00946.x.
Nannipieri, P., Ascher, J., Ceccherini, M. T., Landi, L., Pietramellara, G. and Renella, G. (2003). Microbial diversity and soil functions. *European Journal of Soil Science*, 54, 655–70.
Nishio, T. and Fujimoto, T. (1989). Mineralization of soil organic nitrogen in upland fields as determined by a $^{15}NH_4^+$ dilution technique, and absorption of nitrogen by maize. *Soil Biology and Biochemistry*, 21, pp. 661–5.
Nortcliff, S. and Gregory, P. J. (2013). The historical development of studies on soil–plant interactions. In: *Soil Conditions and Plant Growth* (P. J. Gregory and S. Nortcliff (Eds)). Blackwell Publishing Ltd, Oxford, pp. 1–21. doi:10.1002/9781118337295.ch1.
Oades, J. M. (1993). The role of biology in the formation, stabilization and degradation of soil structure. In: Int. *Workshop on Methods of Research on Soil Structure/Soil Biota Interrelationships* (Brussaard, L. and Kooistra, M. J. (Eds)). *Geoderma*, 56, 377–400.

Ostle, N., Whiteley, A. S., Bailey, M. J., Sleep, D., Ineson, P. and Manefield, M. (2003). Active microbial RNA turnover in a grassland soil estimated using a (CO_2)-^{13}C spike. *Soil Biology and Biochemistry*, 35, 877–85.

Palm, C. A., Myers, R. J. K. and Nandwa, S. M. (1997). Combined use of organic and inorganic nutrient sources for soil fertility maintenance and replenishment. In: Buresh, J. R., Sanchez, P. A. and Calou, F. (Eds), *Replenishing Soil Fertility in Africa*. SSSA, Madison, pp. 193–217.

Palm, C. A., Gachengo, C. N., Delve R. J., Cadisch G. and Giller K. E. (2001). Organic inputs for soil fertility management in tropical agroecosystems: Application of an organic resource database. *Agriculture, Ecosystems & Environment*, 83, 27–42.

Rawlins, B. G., McGrath, S. P., Scheib, A. J., Breward, N., Cave, M., Lister, T. R., Ingham, M., Gowing, C. and Carter, S. (2012). *The Advanced Soil Geochemical Atlas of England and Wales*. British Geological Survey, Keyworth, Nottingham.

Recous, S., Aita, C. and Mary, B. (1999). In situ changes in gross N transformations in bare soil after addition of straw. *Soil Biology and Biochemistry*, 31, 119–33.

Renneson, M., Barbieux, S. and Colinet, G. (2016). Indicators of phosphorus status in soils: significance and relevance for crop soils in southern Belgium: A review. *Biotechnologie, Agronomie, Société et Environnement*, 20(S1), 257–72.

Reymond, M., Svistoonoff, S., Loudet, O., Nussaume, L. and Desnos, T. (2006). Identification of QTL controlling root growth response to phosphate starvation in *Arabidopsis thaliana*. *Plant Cell Environment*, 29, 115–25.

Reynolds, B., Chamberlain, P. M., Poskitt, J., Woods, C., Scott, W. A., Roce, E. C., Robinson, D. A., Frogbrook, Z. L., Keith, A. M., Henrys, P. A., Black, H. I. J. and Emmett, B. A. (2013). Countryside survey: National 'Soil Change' 1978–2007 for topsoils in Great Britain—Acidity, Carbon, and total nitrogen status. *Vadose Zone Journal*, 12(2), 15. https://doi.org/10.2136/vzj2012.0114.

Saby, N. P. A., Bellamy, P. H., Morvan, X., Arrouays, D., Jones, R. J. A., Verheijen F. G. A., Kibblewhite, M. G., Verdoodt, A., Üveges, J. B., Freudenschuß, A. and Simota, C. (2008). Will European soil-monitoring networks be able to detect changes in topsoil organic carbon content? *Global Change Biology*, 14, 2432–42. doi:10.1111/j.1365-2486.2008.01658.x.

Sanchez, P. A. and Salinas, J. G. (1981). Low-input technology for managing Oxisols and Ultisols in tropical America. *Advances in Agronomy*, 34, 279–406.

Scharf, P. C., Schmidt, J. P., Kitchen, N. R., Sudduth, K. A., Hong, S. Y., Lory, J. A. and David, J. G. (2002). Remote sensing for nitrogen management. *Journal of Soil and Water Conservation*, 57, 518–24.

Shaffer, M. J., Ma, L. and Hansen, S. (2001). *Modeling Carbon and Nitrogen Dynamics for Soil Management*. CRC Press, Boca Raton, FL, p. 672.

Shane, M. W. and Lambers, H. (2005). Cluster roots: A curiosity in context. *Plant and Soil*, 274, 101–25.

Shepherd, M. A., Stockdale, E. A., Powlson, D. S. and Jarvis, S. C. (1996). The influence of organic nitrogen mineralization on the management of agricultural systems in the UK. *Soil Use and Management*, 12, 76–85.

Sollins, P., Homann, P. and Caldwell, B. A. (1996). Stabilization and destabilization of soil organic matter: Mechanisms and controls. *Geoderma*, 74, 65–105.

Sparks, D. L. (2000). Bioavailability of potassium. In: *Handbook of Soil Science* (Sumner, M. E. (Ed.)). CRC Press, Boca Raton, FL, pp. D38–53.

Stockdale, E. A. and Cookson, W. R. (2003). Sustainable farming systems and their impact on soil biological fertility – some case studies. In: *Soil Biological Fertility: A Key to Sustainable Land Use in Agriculture* (Abbott, L. K. and Murphy, D. V. (Eds)). Kluwer Academic Publishers, Dordrecht, The Netherlands.

Stockdale, E. A. and Brookes, P. C. (2006). Detection and quantification of the soil microbial biomass – impacts on the management of agricultural soils. *The Journal of Agricultural Science*, 144, 285–302.

Stockdale, E. A. and Watson, C. A. (2012). Managing soil biota to deliver ecosystem services. Natural England Commissioned Reports, No. 100. Accessed at: http://publications.naturalengland.org.uk/publication/2748107 (12 June 2017).

Stockdale, E., Hatch, D., Murphy, D., Ledgard, S. and Watson, C. (2002). Verifying the nitrification to immobilisation ratio (N/I) as a key determinant of potential nitrate loss in grassland and arable soils. *Agronomie*, 22, 831–8.

Stockdale, E. A., Watson, C. A. and Edwards, A. C. (2006). Phosphate rock: Using biological processes to increase its effectiveness as a fertilizer. In: *International Fertilizer Society - Proceeding 592.* International Fertilizer Society, Colchester.

Stone, R. C. and Meinke, H. (2005). Operational seasonal forecasting of crop performance. *Philosophical Transactions of the Royal Society B* 360, 2109–24.

Suriyagoda, L. D. B., Ryan, M. H., Renton, M. and Lambers, H. (2014). Plant responses to limited moisture and phosphorus availability: A meta-analysis. *Advances in Agronomy*, 124, 143–200.

Svistoonoff, S., Creff, A., Reymond, M., Sigoillot-Claude, C., Ricaud, L., Blanchet, A., Nussuame, L. and Desnos, T. (2007). Root tip contact with low phosphate media reprograms plant root architecture. *Nature Genetics*, 39, 792–6.

Swift, M. J., Andrén, O., Brussaard, L., Briones, M., Couteaux, M-M., Ekschmitt, K., Kjoller, A., Loiseau, P. and Smith, P. (1998). Global change, soil biodiversity and nitrogen cycling in terrestrial ecosystems: Three case studies. *Global Change Biology*, 4, 729–43.

Syers, J. K. (1998). *Soil and Plant Potassium in Agriculture*. Proceedings No. 411. The International Fertiliser Society, York, UK.

Tate, R. L. (1995). *Soil Microbiology*. Wiley & sons, New York, NY, p. 398.

Tian, G., Brussaard, L. and Kang B. T. (1995). An index for assessing the quality of plant residues and evaluating their effects on soil and crop in the (sub-) humid tropics. *Applied Soil Ecology*, 2, 25–32.

Trumbore, S. (2006). Carbon respired by terrestrial ecosystems–recent progress and challenges. *Global Change Biology*, 12, pp. 141–53.

Turner, B. L., Bristow, A. W. and Haygarth, P. M. (2001). Rapid estimation of soil microbial biomass in grassland soils by ultra-violet absorbance. *Soil Biology and Biochemistry*, 33, 913–19.

van Wesemael, B., Paustian, K., Meersmans, J., Goidts, E., Barancikova, G. and Easter, M. (2010). Agricultural management explains historic changes in regional soil carbon stocks. *Proceedings of the National Academy of Sciences of the United States of America*, 107(33), 14926–30. doi:10.1073/pnas.1002592107.

Vanlauwe, B., Descheemaeker, K., Giller, K. E., Huising, J., Merckx, R., Nziguheba, G., Wendt, J. and Zingore, S. (2015). Integrated soil fertility management in sub-Saharan Africa: Unravelling local adaptation. *Soil*, 1, 491–508. doi:10.5194/soil-1-491-2015.

Velásquez, G., Calabi-Flood, M., Poblete-Grant, P., Rumpel, C., Demnet, R., Condron, L. and Mora, M. L. (2016). Fertilizer effects on phosphorus fractions and organic matter in Andisols. *Journal of Soil Science and Plant Nutrition*, 16, 294–304.

Vitousek, P. M., Aber, J. D., Howarth, R. W., Likens, G. E., Matson, P. A., Schindler, D. W., Schlesinger, W. H. and Tilman D. G. (1997). Human alteration of the global nitrogen cycle: Sources and consequences. *Ecological Applications*, 7, 737–50.

Waksman, S. A. and Tenney, F. G. (1927). The composition of natural organic materials and their decomposition in the soil: II. Influence of age of plant upon the rapidity and nature of its decomposition-Rye plants. *Soil Science*, 24, 317–34.

Walker, T. W. and Syers, J. K. (1976). The fate of phosphorus during pedogenesis. *Geoderma*, 15, 1–19.

Wall, D. H., Bardgett, R. D., Behan-Pelletier, V., Herrick, J. E., Jones, H., Ritz, K., Six, J., Strong, D. R. and van der Putten, W. H. (2012). *Soil Ecology and Ecosystem Services*. Oxford University Press, UK.

Wardle, D. A., Bonner, K. I. and Nicholson, K. S. (1997). Biodiversity and plant litter: Experimental evidence which does not support the view that enhanced species richness improves ecosystem function. *Oikos*, 79, 247–58.

Weiss, A., Hays, C. J. and Won, J. (2003). Assessing winter wheat responses to climate change scenarios: A simulation study in the U.S. Great Plains. *Climatic Change*, 58, 119–47.

Chapter 12

Plant–soil interactions: an overview

Richard W. Zobel, USDA-ARS, USA

1 Introduction

Soil health is critical to agricultural sustainability. This book discusses the nature of soil health and management techniques to maintain and enhance soil health. The chapters in Part 1 give an overview of the different aspects of soil health, whilst Parts 2 and 3 cover many aspects of soil-based activities which impact on plant/soil interactions. Chapters in Volume 2 in this collection then review ways of monitoring and managing to soils to maintain their health. This chapter will not attempt to reiterate the material covered in the other chapters, but will give an overview of some specific plant/root interactions within many of the soil health aspects discussed elsewhere in the book.

Several 'new' fields of research have recently appeared on the animal/plant/soil research scene (including human management in the animal component): 'microbiome' and 'phytobiome' amongst others. The term 'microbiome' was initially coined for the microbial ecology in the human gut, but has also been utilized to encompass other ecological niches such as the root/soil interface. Phytobiome is a much broader term, has been adopted by an international research group (www.phytobiomesalliance.org) and has spawned an open access journal (www.phytobiomes.org). In essence, the phytobiome is the ecology involving (but not restricted to) environmental (soil as well as aerial), animal (including soil fauna and human management) and microbial interactions with plants, representing this as an integrated continuum rather than multiple discrete areas of study. As this book demonstrates, soil health is, by its very nature, a multidisciplinary issue, and also a critical aspect of the integrations giving rise to the phytobiome. Plant/soil interactions are a major constituent of the phytobiome, and a critical effector of, and beneficiary of, soil health.

http://dx.doi.org/10.19103/AS.2017.0033.15
Published by Burleigh Dodds Science Publishing Limited, 2018.

These interactions involve a multitude of physical, chemical and biological interactions amongst different plant parts, and with the rest of the phytobiome. Plant roots are the interface through which the plant interacts with the soil and its components. We will therefore focus here on plant roots and their interactions with their environment.

Plant/soil interactions, per se, are not restricted to the rhizosphere, but much of the action does take place within that portion of the soil normally colonized by plant roots and their symbionts. The term 'rhizosphere' has frequently been used to refer to the root surface where direct contact with the microbiome takes place. Alternatively, it has referred to the zone of soil within which direct interaction between the root and soil takes place or to that region of the soil in which plant roots and their associated endo- or ecto-mycorrhizae routinely explore. The latter is the definition used here. Despite the opaque nature of the rhizosphere, it is significantly more complex and scientifically extensive than that of the plant shoot and its interaction with its environment, even when including microbe, insect, animal and the environment impact on the shoot. In this overview, a number of different aspects of plant root activity will be introduced, and some of the environmental, biological and physical factors involved in these activities will be discussed.

There are four major multidisciplinary aspects of the plant–soil interactions that integrate to effect rhizosphere effects on the soil and soil health: root/shoot interactions; root by microbiome interactions; root by soil interactions; root by environment interactions. Specifics about these can be found in numerous books and reviews, most notably Keister and Cregan (1989), Waisel et al. (1991, 1995, 2002), Box (1998), Zobel and Wright (2005), Gregory (2006), Eshel and Beeckman (2013). In addition, a major constraint to multidisciplinary, and therefore multifactorial, research is the lack of statistical procedures designed to characterize and verify the significance of multiple complex interactions. The future potential for development in this last area will also be briefly presented. First, however, plant root system architecture will be discussed so that the reader will be able to distinguish between differential effects that will be brought up in the body of the discussion.

2 Plant root system architecture

The International Society for Root Research (ISRR) has established a framework for plant root nomenclature (Zobel and Waisel, 2010) (Table 1, Fig. 1). Subsequently Zobel (2011, 2013) demonstrated, using developmental genetic data, that there are at least 9 distinct classes of roots in plants (both monocots and eudicots) (Table 1, Fig. 2). A mature plant root system will contain up to 7–9 of these classes of roots. Zobel et al. (1992), Bushamuka and Zobel (1998a,b), and Zobel (2011) demonstrated that three of the earliest classes of roots to develop can have, during their initial stages of development, distinctly different functionality patterns. Based on this, Zobel (2011) suggested that all classes of roots must be assumed to be (until proven otherwise) functionally distinct from each other. Unfortunately, other than two seedling classes of roots (primary and lateral), the remaining seven classes have, by many researchers, been lumped into an ambiguous 'adventitious root' class. Most past research addressing adventitious roots does not specify the exact location(s) of the roots studied, and is, therefore, imprecise and difficult to repeat in subsequent research. It is recommended that the ISRR/Zobel classification scheme be used for greater clarity. The architecture of a mature plant root system develops sequentially from seedling to mature plant, and therefore represents an historical perspective of the sequence of internal and external factor impacts on the pattern of development of the root system. Knowledge of the timing of external factor

Published by Burleigh Dodds Science Publishing Limited, 2018.

Table 1 Nine root classes organized within the International Society for Root Research root system framework (Zobel and Waisel, 2010) as expanded to sub-classes by Zobel (2011, 2013). Abbreviations were selected for descriptive simplicity and size to fit easily into the plant ontology

ISRR framework classes	Sub-class	Abbreviation
Tap root (first root to emerge)		TRT
Basal root (develops from the hypocotyl)	First stage (at the base of the taproot)	BRT (Basal root)
	Second stage (on the hypocotyl above the tap root)	HRT (Hypocotyl root)
Shoot-borne root (develops from shoot tissues)	Coleoptile node	CNRT (Coleoptile node root)
	Shoot nodes (most prevalent in monocots)	SNRT (Shoot-borne node root)
	Shoot internodes (most prevalent in eudicots)	SIRT (Shoot-borne internode root)
Lateral root (branches from another root)	On tap or basal root	Prefix main root abbreviation with an L: that is, LTRT or LHRT (a lateral branching from another lateral = LTRT2 or LTRT3 dependent on branching level)
	On shoot-borne root	Begins with LS (Lateral shoot-borne): LSNRT(similar to above)
	Dominant lateral (a large exploratory lateral root developing in a manner similar to shoot-borne and basal roots) (Zobel 2013)	DLSNRT, DLTRT, DLBRT, etc. (Dominant lateral ...)

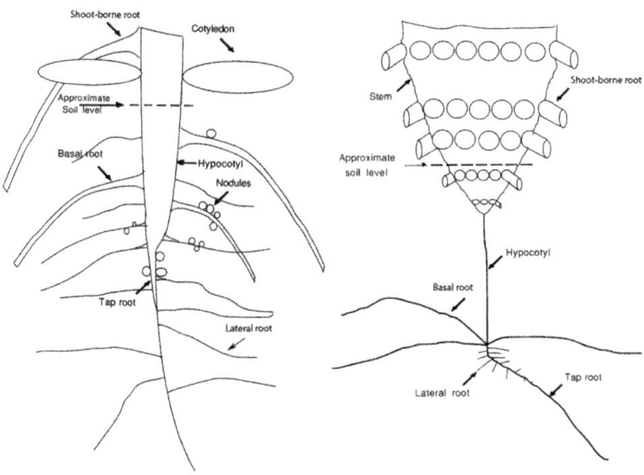

Figure 1 A composite drawing of a generic eudicot root system (left) and monocot root system (right), annotated with the ISRR (International Society for Root Research) nomenclature.

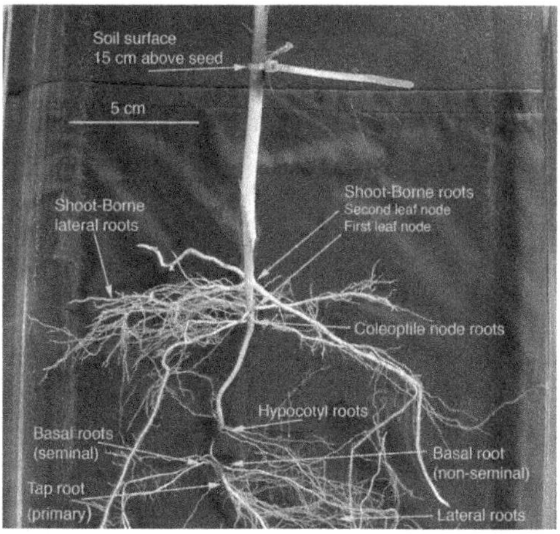

Figure 2 A maize (*Zea mays* L.) root system annotated with the ISRR (International Society for Root Research) recommended names and abbreviations from Table 1 and Zobel (2011).

impacts can lead to an understanding of the control of plant root system architecture, and improve the modelling of plant root system growth and development, and of management approaches to enhance sustainability. If a grower/producer monitors the root systems of his crops during the growing season, this knowledge of root system architecture will assist in the early identification of insipient soil health problems.

3 Root/shoot interactions

Root/shoot interactions are facilitated by the xylem (root effects on the shoot) and phloem (shoot effects on the roots). The most studied root effects on the shoot are nutrient uptake, water uptake and hormone effects. Marschner (1995) provided a thorough treatment of plant mineral nutrition. Calcium, taken up by the roots, is transported to the shoot where it is incorporated but not cycled back to the roots through the phloem as other minerals are. Thus root uptake of calcium is a one-way effect of the roots on the shoot. Root zone temperatures modify the ability of roots to take up water and nutrients, such that low soil temperatures can induce nutrient deficiencies and even wilting of the shoot due to poor nutrient and water uptake/transport (Mataruka et al., 1993; Miyasaka and Grunes, 1997). Marschner (1995, p. 38), however, suggested that the root zone temperature effect *in situ* is conditioned by modified root developmental patterns leading to low root/shoot ratios. This is a cause-and-effect (chicken and egg) relationship that needs to be untangled for better understanding of stress effects on root function and plant development. Concomitant with nutrient and water uptake, roots produce (amongst others) cytokinins and abscisic acid which are translocated to the shoot and modify flowering, growth and especially stomatal opening. The latter imposes control on the rate of photosynthesis, based on soil conditions, and thus feeds back on the supply of photosynthate for use in additional root growth and function. Root grazing by soil fauna and pathogenic action on roots cause a reduction in root surface area, and result in reductions in water, nutrient and hormone uptake/production and transport to the shoot (see Section 4), further regulating shoot growth and development. These regulations feedback, on root growth, by modifying the amount of photosynthate and hormones (auxins and gibberellins) transported from the shoot to the root. In some species, genetic differences in root system effectiveness at producing specific metabolites, or even disease resistance, for shoot development has led to the extensive development of grafting desired shoot genotypes onto selected root genotypes of the same or related species. Wine grapes (disease-tolerant root stocks) and dwarf fruit trees are prime, well known, examples of the use of this effect. Currently, for example, there is also a strong effort in grafting tomatoes, for commercial use, to provide for multiple disease resistance.

The reverse process (shoot effects on roots) effects (and affects) control of root interaction with the soil ecosystem. An obvious control is the rate and amount of photosynthesis and the movement of carbohydrates and other metabolites down to the root system through the phloem. This process provides for root growth and development, root metabolic products and root maintenance plus exudation to the rhizosphere. Specific compounds produced in the shoot, such as auxins and gibberellins, are noted for controlling developmental processes, for instance root initiation and outgrowth. A recent addition to the list of effective compounds is a non-hormonal compound, diageotropica (a cyclophilin A ortholog), which is produced in the shoot and moves to the roots via the phloem. This compound localizes the initiation sites for tomato lateral roots (Ivanchenko et al., 2015). In the absence of diageotropica, the taproot does not produce lateral roots, thus modifying (reducing) the structure of the root system.

Environmental effects on the shoot (ozone, carbon dioxide increases, temperature changes, pathogens, grazing by pests and animals) also impact root system development and maintenance. For example, grazing or clipping of the shoot of pasture species stimulates the loss of fine roots from the plant root system (Dawson, et al., 2000; Ritchey et al., 2004). This has been observed with both excavation and with minirhizotrons where the fine roots completely disappear within 24 hours of clipping or grazing (Zobel, unpublished). New roots begin to develop on these plants two to three days later. In some species the new roots appear closer to the soil surface and in others they appear deeper in the soil profile (Dawson, et al., 2000), effectively changing the overall root system structure. When orchardgrass (*Dactylis glomerata* L.) is grown in hydroponics and clipped, there is only a cessation of growth (no root death) which resumes approx. 48–72 hr later (Ritchey et al., 2004). The disappearance of roots after clipping, in the field, but not under controlled conditions demonstrates the extent, and rapidity, that clipping (grazing) affects the food availability for the soil microbiome. The intensity of clipping (amount of shoot biomass removed) modifies the magnitude of these responses (Dawson et al., 2000). In another case, it has been noted that aerial treatment with ozone conditions fewer and thicker fine roots in wheat, while treatment of the shoot with carbon dioxide (800 $\mu l\ l^{-1}$) conditions more, and thinner roots (Zobel et al., unpublished).

4 Root by microbiome interactions

For convenience the term 'microbiome', as used here, includes soil fauna and fungi as well as bacteria, protozoans and so on. The clipping examples discussed above demonstrate the presentation of raw organic material (FOOD) to the soil microbiome (including the microfauna). That the shed roots are utilized by the microbiome is represented by their extremely rapid visual disappearance from the soil profile (<24 hr – Zobel, unpublished;

Published by Burleigh Dodds Science Publishing Limited, 2018.

Bardgett et al., 1998; Dawson et al., 2000). (As a side note, this is the reason grazing patterns by cattle and sheep are often manipulated to reduce the extent of total biomass loss to a plant.) There are many examples of ongoing mutualistic interactions between the microbiome and plant roots, as well as pathogenic and symbiotic relationships. Many plants shed cells from the root tip as they grow into the soil, providing a food source for the microbiome. Plants that also develop a mucilaginous sheath around the root tip have been demonstrated to have live, apparently functioning, root cap-derived cells (termed border cells) within the sheath (Hawes, 1990; Hawes et al., 1998). Some species, that is, wheat, even develop mucilaginous sheaths near the base of the plant under the certain growth conditions. These mucilaginous sheaths and the entrained cells and cell metabolic products represent excellent sources of nutrients for the microbiome. Another source of nutrients is the active exudation of root metabolites into the rhizosphere where they can support microbiome populations (Bardgett et al., 1998). It has been demonstrated that support of specific elements of the microbiome can be differential amongst root classes (Ofek et al., 2007; Zobel et al., unpublished). Metabolic products of the roots also serve as attractants to pathogens, symbionts and saprophytes, for instance corn root worms, nematodes, mycorrhizae and legume rhizobia. It should be noted that the mutualistic interactions can be specific to plant cultivar, soil and strain of a microbiome species (e.g. VAM – Clark et al., 1999).

It has been suggested that in many species there is a continuous rapid turnover of the finest roots (e.g. a seven-day half-life in soybean – Zobel, 2005). As these fine roots are often rudimentary and without secondary tissues that would be more resistant to herbivory (McCully, 1987; Zobel, 2013), this would provide the microbiome with a constant supply of metabolic substrates. As roots age, they often shed their cortex (McCully, 1987), providing an additional source of substrates, or develop secondary thickening with much more complex molecular materials in the cell walls and other tissues. These older roots may slough off the root system and their more mature tissues would then provide a more resistant carbon source that can be utilized by different elements of the microbiome, while providing for increased organic matter for the soil.

As mentioned above, root exudates, including carbon dioxide, act as attractants for both symbionts and pathogens. The symbionts like the rhizobia directly utilize carbon sources from within the roots and fix nitrogen, releasing amino acids into the roots, thus increasing the nitrogen levels in the host plant. Other symbionts, for example, endo- and ecto-mycorrhizae, are able to utilize portions of the soil that are not available to plant roots, that is, very small micropores, or chemicals that plant enzymes are unable to metabolize, and provide the root with phosphorus and, often, water and other compounds. Part of the microbiome is active in converting materials into substrates such as nitrate or ammonia which the plant roots can take up and utilize, while others fix compounds into longer-lived compounds, thus enriching the soil.

5 Root by soil interactions

Soil physical structure and chemical makeup impact, and are impacted on, by plant species/cultivars. For example, acid (low pH, high aluminium) soils are often restrictive to root growth, but some cultivars within a species can be better adapted to deal with this problem. In both soybean and maize, some cultivars have full or partial tolerance to acid soils, and this tolerance can be restricted to specific root classes (Bushamuka and

Zobel, 1998a,b). The tolerance mechanism has been shown to be primarily the exudation of organic acids by the roots. Quantities of acids are low enough that there is little long-term effect on the soil, with most being either complexed or utilized by the microbiome in the immediate vicinity of the root. Soil structure has a strong effect on plant root growth. Compaction is one of the leading causes of soil-based reduced productivity by crops (Freitas et al., 1999). Plant root systems can deal with this issue when the level of compaction is within their ability to penetrate. Again the ability to penetrate compacted soil or soil layers is cultivar and root class dependent within a species (Bushamuka and Zobel, 1998a,b). Both the acid soil and soil compaction tolerance demonstrated in the literature suggest the availability of genetic variation that could be used in a directed breeding programme to further improve these tolerances.

Another soil impact is the possibility of high water tables or flooding. In many species, plant lines have been studied that have the ability to deal with this issue by developing aerenchyma in the root cortex. Species such as rice and eastern gamagrass always produce aerenchyma (Clark et al., 1998a), that is, the aerenchyma are constitutive, while in others such as maize, the aerenchyma are facultative, that is, maize roots develop aerenchyma after being subjected to a stress (Drew, et al., 1979; Lynch, 2015). A side benefit to the development of aerenchyma and resulting penetration of the saturated soil is the exudation/diffusion of oxygen into the surrounding soil, benefiting the microbiome as well as the plant (Kemper et al., 1998). In some cases, where the aerenchyma allow the penetration of saturated clay soil, the soil profile itself is modified such as to allow later crop roots to penetrate into the clay layers, significantly increasing yields (Clark, et al., 1998b) for a season or two. Many species can produce aerenchyma facultatively, although this may be restricted to only a few cultivars or lines, again suggesting the possibilities for breeding.

An additional impact of cultivar/species on the soil has been observed. In some soils of upstate New York, continuous maize cropping results in large soil aggregates, while continuous soybean cropping leads to a reduction in aggregation of the soil, and eventual soil loss through erosion (Ellsworth et al., 1991). Appropriate crop rotation resolves this problem apparently by balancing the two trends (Zobel, unpublished). Maize is noted for extensive exudation, and soybean is not. Also maize roots do not demonstrate secondary thickening, while soybean primary and secondary roots do demonstrate secondary thickening. With secondary thickening, if a root tip penetrates a soil pore in an aggregate then the root later expands, it can break down the aggregate, thus reducing the average size of the aggregates in the soil profile.

Within the soil, the microbiome and plant roots respire carbon dioxide which then diffuses to the surface and into the atmosphere. As the carbon dioxide concentration increases from ambient aerial levels (~400 $\mu l\ l^{-1}$) to 1%, microbiome and root growth and development are enhanced, and the gravitropic orientation of plant roots modified. Above 2% carbon dioxide, both the microbiome and plant roots are increasingly inhibited in their growth and functionality. Plant roots continue to grow more horizontally with increasing carbon dioxide levels. This increase in carbon dioxide is a normal effect of growth and development in the soil, but can be modified by temperature, and reduced by rainfall forcing the gas out of the soil as water fills the pore spaces. Observation of the angles of shoot-borne roots on different nodes of field-grown maize shoots often shows dramatic changes in root orientation that can be traced back to weather events (Zobel, unpublished).

A further soil/plant interaction is demonstrated with an analysis of the developmental patterns of grass root systems. Corn is often planted at ~5 cm depths, while smaller seeded grasses are planted at more shallow depths. It has been demonstrated that grass seeds planted in the soil first initiate their taproot (primary), followed by the basal (seminal) roots and the initiation of lateral roots. In most grass species, the hypocotyl (mesocotyl) then begins to elongate, pushing the coleoptile towards the soil surface. When the hypocotyl ceases elongation, roots form at the coleoptile node. When smaller seeded plants (e.g. the *Festucaceae*) are planted deep and the coleoptile has not reached the surface by the end of hypocotyl elongation, the first leaf internode elongates until the first leaf reaches the surface, then it ceases growth and roots form at the first leaf node (Avery, 1930). If grass seeds are planted at or very near the soil surface, the elongation does not take place due to the impact of red light on the germinating seed, thus causing the basal, coleoptile and first leaf node roots to develop at the same location (Avery et al., 1937). This effect reduces stand establishment and increases plant susceptibility to drought and surface activity such as animal or equipment disturbance of the soil. The exact pattern of this stem elongation and root initiation differs dramatically among the grass sub-families (Hoshikawa, 1969).

6 Future trends

By their very nature, plant/soil interactions take place in a dark and normally opaque environment. Extensive development of technologies and syntheses from both direct and indirect information has been required to achieve the current levels of knowledge in this area. Modelling, as one example, has been an effective approach towards developing insights leading to testable hypotheses about these interactions. Modelling has an unfortunate drawback when attempting to develop a systematic and complete understanding, the presence of 'empirical black holes' (Zobel, 2013), such as the existence of nine different classes of root with differential genetic controls of their development and function, cannot readily be detected. This is an empirical black hole, for all areas of root research in addition to modelling, in that few studies are structured to be able to demonstrate possible differences between root classes. Frequently this is due to studies being carried out on seedlings, and/or with single species and cultivars, followed by unjustified generalizations to higher levels. On other occasions, empirical black holes occur because adequate statistical procedures to analyse interactions above two-way interactions have not been adopted, resulting in the artificial lumping of different factors into combined groupings to allow for analysis. This problem may be rectified by the adoption of existing procedures that are used in other fields, and/or expansion of procedures currently in use.

The development of the AMMI (Additive Main effects and Multiplicative Interaction) procedure (Zobel et al., 1988; Gauch, 1988) introduced, to the ANOVA process, an effective two-way interaction analysis. Lee and Johnson (2006) expanded this to incorporate the use of regressions to better explain some specific interactions, as suggested by Tukey (1962) with his 'vacuum cleaner' process. Chemometrics research has developed and extended work by Tucker (1964, 1966), Kroonenberg (1994) and Eeuwijk and Kroonenberg (1998) in three-way analysis into a multi-way process (Bro, 2013). With the exception of the Lee and Johnson (2006) algorithm, current statistical practices (e.g. PCA, PARAFAC) require the use of 'orthogonal' algorithms, restricting assessment to multi-way analysis without consideration of inter-level interaction. The Lee and Johnson (2006), Tucker (1964) and Eeuwijk and Kroonenberg (1998) systems relaxed the orthogonal requirement into a more natural multi-way, multilevel interaction assessment. This line of statistics needs to be integrated into phytobiome (plant/soil) research because of its ability to handle large multi-way factorial experiments.

7 Conclusion

Plant/soil interaction and, as a consequence soil health, is a complex integration of many different factors and should be considered as a narrower aspect of the interaction between the factors that characterize the phytobiome. Changes in the nature of the aerial environment modify the physiology of the shoot, often resulting in changes in the physiology and development of the roots of the plant. The reverse effects are also obtained, resulting in soil and microbiome impacts on the plant shoot, and thus plant survival and productivity. Plant roots, through chemical and physical activity, modify soil structure, chemistry and the microbiome community, and are, in turn, modified by these aspects of the soil. The specific level or form of modification is dependent on the genetic makeup of both the microbiome and the plant, with differences predicated by different root classes within a root system and different strains within the microbiome. A thorough understanding of soil health and the resulting agricultural sustainability cannot occur without a multidisciplinary consideration of the whole phytobiome. Since this requires significant numbers of factor-by-factor interactions amongst and between many species, their efficient study will require the adoption of an enhanced statistical system that is capable of analysing multiple factor interactions in a multilevel environment.

8 Where to look for further information

There are many research articles that cover the issues of soil health that relate to roots. Gregory (2006) does an excellent job of pulling much of it together. For more detailed information, the 'Plant Roots: the Hidden Half' series (Waisel et al., 1991, 1996, 2002; and Eshel and Beekman, 2013) provides a wealth of information. Basically, research in this area is on the cusp of rapid expansion using the newer molecular techniques to access interactions taking place at the sub-micron level in physical space.

9 References

Avery, G. S. (1930), Comparative anatomy and morphology of embryos and seedlings of maize, oats and wheat, *Botanical Gazette*, 89, 1–39.

Avery, G. S., Burkholder, P. R. and Creighton, H. B. (1937), Polarized growth and cell studies in the first internode and coleoptile of *Avena* in relation to light and darkness, *Botanical Gazette*, 99, 125–43.

Bardgett, R. D., Wardle, D. A. and Yeates, G. W. (1998), Linking above-ground and below-ground interactions: how plant responses to foliar herbivory influence soil organisms. *Soil Biol. Biochem.*, 30, 1867–78.

Box, J. E. (1998), Root Demographics and their Efficiencies in Sustainable Agriculture, Grasslands and Forest Ecosystems. *Proc. Intl Soc. Root Research 5th Symposium*, Clemson SC. 14–18 July, 1996. Kluwer Academic Publishers: Dordrecht.

Bushamuka, V. N. and Zobel, R. W. (1998a), Maize and soybean tap, basal and lateral root responses to a stratified acid, al-toxic soil, *Crop Sci.*, 38, 416–21.

Bushamuka, V. N. and Zobel, R. W. (1998b), Differential genotypic and root type penetration of compacted soil layers, *Crop Sci.*, 38, 776–81.

Clark, R. B., Alberts, E. E., Zobel, R. W., Sinclair, T. R., Miller, M. S., Kemper, W. D. and Foy, C. D. (1998a), Eastern Gamagrass (*Tripsicum dactyloides*) root penetration and chemical properties in claypan soils, pp. 191–212. In J. Box (Ed.), *Root Demographics and their Efficiencies in Sustainable Agriculture, Grasslands and Forest Ecosystems*. Proc. Intl Soc. Root Research 5th Symposium, Clemson SC. 14–18 July 1996. Kluwer Academic Publishers: Dordrecht.

Clark, R. B., Alberts, E. E., Zobel, R. W., Sinclair, T. R., Miller, M. S., Kemper, W. D. and Foy, C. D. (1998b), Eastern Gamagrass (*Tripsicum dactyloides*) root penetration and chemical properties in claypan soils, *Plant Soil*, 200, 33–45.

Clark, R. B., Zeto, S. K. and Zobel, R. W. (1999), Arbuscular mycorrhizal fungal isolate effectiveness on growth and root colonization of *Panicum virgatum* in acidic soil. *Soil Biol. Biochem.*, 31, 1757–63.

Dawson, L. A., Grayston, S. J. and Paterson, E. (2000), Effects of grazing on the roots and rhizosphere of grasses, p. 61. In G. Lemaire, J. Hodgson, A. de Moraes, C. Nabinger and P. C. de F. Carvalho (Eds), *Grassland Ecophysiology and Grazing Ecology*. Kluwer: Dordrecht.

Eeuwijk, F. A. van and Kroonenberg, P. M. (1998), Multiplicative models of interaction in three-way ANOVA with applications to plant breeding. *Biometrics*, 54, 1315–33.

Ellsworth, T. R., Clapp, C. E. and Blake, G. R. (1991), Temporal variations in soil structural properties under corn and soybean cropping. *Soil Sci.*, 151, 405–16.

Eshel, A. and Beekman, T. (2013), *Plant Roots: The hidden half*. Fourth Edition. CRC Press: Boca Raton.

Freitas, P. L. de, Zobel, R. W. and Snyder, V. A. (1999), Corn root growth in soil columns with artificially constructed aggregates. *Crop Sci.*, 39, 725–30.

Gregory, P. J. (2006), *Plant Roots: Growth, activity and interaction with soils*. Blackwell: Oxford.

Hawes, M. C. (1990), Living plant cells released from the root cap: A regulator of microbial populations in the rhizosphere? *Plant Soil*, 129, 19–27.

Hawes, M. C., Brigham, L. A., Wen, F., Woo, H. H. and Zhu, Y. (1998), Function of root border cells in plant health: Pioneers in the rhizosphere, *Annu. Rev. Phytopathol.*, 36, 311–27.

Hoshikawa, K. (1969), Underground organs of the seedlings and the systematics of *Gramineae, Bot. Gaz.*, 130, 192–203.

Ivanchenko, M., Zhu, J., Wang, B., Medvecká, E. and *et al.* (2015), The cyclophilin A DIAGEOTROPICA gene affects auxin transport in both root and shoot to control lateral root formation, *Development*, 142, 712–21.

Keister, D. L. and Cregan, P. B. (1991), *The Rhizosphere and Plant Growth*. Beltsville Symposia in Agricultural Research, Kluwer Academic Publishers: Dordrecht, p. 386.

Kemper, W. D., Alberts, E. E., Clark, R. B. and Zobel, R. W. (1998), Aerenchyma, acid tolerance and associative nitrogen fixation enhance carbon sequestration in soil, pp. 221–34. In R. Lal, J. Kimble, R. Follett and R. A. Stewart (Eds), *Carbon Sequestration in Soil, International Symposium*. July 22–26 1996 Columbus (OH). CRC Press: Boca Raton, FL.

Kroonenberg, P. M. (1994), The TUCKALS line: A suite of programs for three-way data analysis, *Comp. Stat. Data Anal.*, 18, 73–96.

Lee, E. J. and Johnson, D. E. (2006), AMMI macros for multiplicative interaction models. *Proceedings, SAS Users Group International Conference*. 26–29 March 2006, San Francisco. Paper 049–31. Available at: http://www2.sas.com/proceedings/sugi31/toc.html Accessed 9 October 2016.

Lynch, J. P. (2015), Root phenes that reduce the metabolic costs of soil exploration: opportunities for 21st century agriculture, *Plant Cell Environ.*, 38:1775–84.

Marschner, H. (1995), *Mineral Nutrition of Higher Plants*. Academic Press: London.

Mataruka, D. F., Cox, W. J., Mt. Pleasant, J., Van-Es, H. M., Klausner, S. D. and Zobel, R. W. (1993), Tillage and nitrogen source effects on growth, yield, and quality of forage maize, *Crop Sci.*, 33, 1316–21.

McCully, M. (1987), Selected aspects of the structure and development of field-grown roots with special reference to maize, pp. 53–70. In P. J. Gregory, J. V. Lake and D. A. Rose (Eds), *Root Development and Function*. University Press: Cambridge.
Miyasaka, S. C. and Grunes, D. L. (1997), Root zone temperature and calcium effects on phosphorus, sulfur, and micronutrients in winter wheat forage, *Agron J.*, 89, 742–8.
Ofek, M., Ruppel, S. and Waisel, Y. (2007), Differences between bacterial associations with two root types of *Vicia faba* L., *Plant Biosyst.*, 1414, 352–62.
Ritchey, K. D., Zobel, R. W. and Snuffer, D. J. (2004), Nutrient Film Technique to Evaluate Clipping Height Effects on Orchardgrass Shoot Biomass Production and Root Dynamics, pp. 424–8. In: *Proceedings of the 2004 Conference of the American Forage and Grasslands Council*. Roanoke VA 12–16 June 2004.
Tucker, L. R. (1964) The extension of factor analysis to three dimensional matrices, pp. 110–82. In N. Frederiksen and H. Gulliksen (Eds), *Contributions to Mathematical Psychology*. Holt, Rinehart & Winston: New York.
Tucker, L. R. (1966), Some mathematical notes on three-mode factor analysis, *Psychometrika*, 31, 279–311.
Tukey, J. W. (1962) The future of data analysis, *Ann. Math. Stat.* 33:1–67
Waisel, Y., Eshel, A. and Kafkafi, U. (1991), *Plant Roots: The Hidden Half.* Marcel Dekker: New York, p. 948.
Waisel, Y., Eshel, A. and Kafkafi, U. (1996), *Plant Roots: The Hidden Half.* Second Edition. Marcel Dekker: New York, p. 1002.
Waisel, Y., Eshel, A. and Kafkafi, U. (2002), *Plant Roots: The Hidden Half.* Third Edition. Marcel Dekker: New York, p. 1120.
Zobel, R. W. (1991), Root Growth and development, pp. 61–71. In D. L. Keister and P. B. Cregan (Eds), *The Rhizosphere and Plant Growth.* Kluwer Academic Publishers: Dordrecht.
Zobel, R. W. (2005), Tertiary roots, pp. 35–56. In R. W. Zobel and S. A. Wright (Eds), *Roots and Soil Management: Interactions between Roots and the Soil.* Agronomy Society of America: Madison. Agron. Monogr. 48.
Zobel, R. W. (2011), A developmental genetic basis for defining root classes. *Crop Sci.*, 51, 1410–13.
Zobel, R. W. (2013), Modelling *Lolium perenne* L. roots in the presence of empirical black holes, pp. 155–72. In D. Timlin and L. R. Ahuja (Eds), *Advances in Modeling Agricultural Systems: Transdisciplinary Research, Synthesize, Modeling, and Applications. Vol 4. Enhancing Understanding and Quantification of Soil-Root Growth Interactions.* ASA-SSSA-CSSA: Madison, WI.
Zobel, R. W. and Wright, S. F. (2005), *Roots and Soil Management: Interactions between Roots and the Soil.* Agronomy Monograph No. 48. American Society of Agronomy. Madison, p. 272.
Zobel, R. W., Wright, M. W. and Gauch, H. G. (1988), Statistical Analysis of a yield trial. *Agron. J.*, 80, 388–93.
Zobel, R. W., Kochian, L. V. and Toulemonde, T. G. (1992), Plant Root Systems, pp. 30–40. In *Proceedings PPI Conference on Roots of Plant Nutrition*, Champaign IL.
Zobel, R. W. and Waisel, Y. (2010), A plant root system architectural taxonomy: a framework for root nomenclature. *Plant Biosyst.* 144, 507–12.

Chapter 13

Mechanisms of soil erosion/degradation

R. J. Rickson, Cranfield University, UK

1 Introduction

Healthy, functioning soils are able to 'sustain plant and animal productivity, maintain or enhance water and air quality, and support human health and habitation' (Karlen et al., 1997). They deliver a range of ecosystem goods and services which are associated with human health and well-being (MA, 2005; TEEB, 2010; NEA, 2011; Table 1). In turn, these goods and services have been directly linked to individuals' wealth and national economic status (Daily, 1997). Soil has been described as 'the thin layer of soil covering the earth's surface [which] represents the difference between survival and extinction for most terrestrial life' (Doran and Parkin, 1994). However, soil degradation is a major threat to soil health, reducing the soil's capacity to function (Karlen et al., 1997) and its ability to deliver vital ecosystem goods and services that are at the heart of sustainable development (European Commission, 2002; Table 1).

Soil degradation can be the result of human activities, such as intensive agricultural practices on susceptible lands, where land is used beyond its inherent capability (Lal, 2009). It is estimated to affect 12 million hectares of agricultural land every year, leading to potential losses of 20 million tonnes of grain (United Nations Convention to Combat Desertification, 2011; Bai et al., 2008). The European Environment Agency reports that 12% (about 120 million ha) of European land is affected by soil erosion by water, 5% by wind erosion and 45% has low organic matter content (European Environment Agency, 1995). Kendall and Pimentel (1994) estimate that soil degradation will lead to a 30% reduction in food production by 2040, undermining efforts to ensure food security, where 'all people, at all times, have physical and economic access to sufficient, safe and nutritious food to

http://dx.doi.org/10.19103/AS.2017.0033.16

Table 1 Ecosystem goods and services delivered by healthy, non-degraded soils

Ecosystem goods and services delivered by soil	Examples
Provisioning of material goods and services	Agricultural production (food, fibre, fodder and (bio)fuel) Water storage and supply Land for development (residential, industry, infrastructure)
Regulation of ecosystem processes	Flood control (water storage capacity) Carbon storage/sequestration (reduce CO_2 emissions and associated global warming, so mitigating climate change; retaining water, nutrient cycling, maintenance of physical structure and biodiversity)
Cultural, non-material services	Landscape aesthetic Recreation/amenity, protection of buried heritage (archaeology)
Supporting services	Habitats, biodiversity Soil formation processes

meet their dietary needs and food preferences for an active and healthy life' (FAO, 1996). According to Graves et al. (2015), '[soil] degradation, if continued and unabated, would eventually lead to complete loss in the stock value of the soil and the soil asset itself'.

According to the EC Thematic Strategy for Soil Protection (European Commission, 2002, 2006), the term 'soil degradation' refers to both the processes by which soil is degraded, as well as the state of the soil once these processes have taken place. Soil degradation processes (Table 2) adversely affect the physical, biological and chemical properties and functioning of soils. Soil degradation processes occur throughout the world to differing degrees (Bridges and Oldeman, 1999). Bringezu et al. (2010) estimated that 2–5 million ha of global arable land are lost every year to soil erosion alone. Estimates of other degradation processes, especially soil diffuse contamination, losses of soil biodiversity and soil sealing are sparse or very coarse in scale (Graves et al., 2015). Despite the gaps in knowledge/data, Graves et al. (2015) estimated the total costs of soil degradation in England and Wales alone to be between £0.9 and £1.2 billion per annum. Twenty per cent (=£200 million) of the estimated annual costs of soil degradation were associated with loss of provisioning services linked with agricultural productivity, both in terms of reduced output and increased costs of production. The remaining 80% were associated with loss of regulating services, such as greenhouse gas emissions (55%), flooding (20%) and detrimental effects on water quality (3%).

Many soil degradation processes occur naturally, but are dynamically balanced with natural rates of soil formation, regeneration or reclamation (Troeh and Thompson, 1993; Verheijen et al., 2009). However, soil degradation processes can be aggravated by human interference (both intentional and unintentional) which can accelerate the magnitude, frequency and extent of soil degradation processes, beyond natural soil formation rates, along with the associated environmental, social and economic impacts on food production. This was tragically demonstrated in the 1930s 'Dust Bowl' era in the United States (Schubert et al., 2004). On the positive side, soil conservation practices can control and even reverse degradation processes (Morgan, 2005).

Many soil degradation processes are interrelated, for example, loss of soil depth through soil erosion is associated with a decline in soil organic matter content and loss of soil biota/biodiversity. Soil compaction leads to greater run-off generation and higher rates of erosion. Similarly, desertification encapsulates many other soil degradation processes

Table 2 Types of soil degradation (from Huber et al., 2008)

Soil erosion	Removal of soil particles and/or aggregates by erosive agents, primarily rainfall, run-off, wind and ice, usually acting on the soil surface
Soil organic matter decline	Soil erosion Mineralisation
Soil contamination	Soils can be contaminated by inorganic contaminants, soil acidifying substances (such as sulphur or nitrogen) or heavy metals
Loss of soil biodiversity	Soil micro- and macrofauna include bacteria, fungi, nematodes, earthworms and invertebrates. Indicators of a loss of soil biodiversity include monitoring the diversity of earthworms, collembola and enchytraeid species present or rates of soil microbial respiration
Soil surface sealing	This is the covering of bio-productive, semi-natural or natural land by impermeable or semi-permeable materials such as asphalt, bitumen, concrete or tarmac. This process is associated with the built environment, especially human settlement and transport infrastructure such as roads and is associated with increasing rates of urbanisation
Soil compaction	Soil compaction is usually indicated by an increase in soil bulk density or packing density, with a concurrent reduction in air-filled pore volume in the soil
Salinisation	Soil salinisation refers to the vertical distribution of water-soluble salts in the soil profile and the accumulation of salts in the upper soil horizons. Applications of poor quality irrigation water and seawater ingress are associated with soil salinisation processes
Landslides	Mass movements such as landslides displace soil and regolith on relatively steep slopes, under the force of gravity. Over-steepening of engineered slopes (cuttings and embankments) can increase the risk of landslides
Desertification	Desertification occurs in arid, semi-arid and dry sub-humid areas and includes a number of soil degradation processes. Declining levels of rainfall and higher temperatures reduce vegetation cover, resulting in less organic matter added to the soil, causing a decrease in water retention capacity, and an increase in run-off and sediment yield (Boix-Fayos et al., 2005). Exposure of bare soil increases wind erosion risk. Kosmas et al., (1999) summarise the process thus: 'desertification of an area will proceed if certain land components degrade beyond specific thresholds, beyond which further change is irreversible. For example, soils may eventually become so stony that they can only degrade towards scree or bare bedrock'

including soil erosion, decline of organic matter, loss of soil biodiversity and salinisation, as described in Table 2. For some processes, rates will be affected by climate change, such as the increased risks of soil erosion and compaction associated with higher rainfall volumes and intensities, and greater oxidation of organic carbon associated with higher temperatures.

This chapter focuses on soil erosion, because it is arguably the most common form of soil degradation. The process involves the complete physical removal of the soil resource from its point of origin, so the effects of soil erosion are effectively irreversible, given the slow rates of soil formation (Verheijen et al., 2009). Emphasis is given to the 'on-site' consequences of erosion as these have direct impact on crop production and sustainable agriculture. However, it should be remembered that the 'off-site' consequences of erosion (e.g. sediment and associated contaminants reaching watercourses, causing poor water quality and disturbing aquatic habitats, sedimentation, increased flooding risk, etc.) may have greater impact on society as a whole and higher associated economic costs. Gorlach et al. (2004) estimated the total annual cost of erosion in Europe as €9.5 billion (2003 values), of which 9% was attributable to on-site costs and 91% to off-site costs. Graves et al. (2015) estimated the proportion of costs of soil erosion is 27% on-site and 73% off-site. However, for the purposes of this chapter, the off-site impacts are less directly connected with sustainable agriculture. The factors affecting the processes of soil erosion are discussed, followed by soil erosion impacts on crop production and the sustainability of agriculture.

2 Soil erosion processes: overview and erosion by water

2.1 Soil erosion: overview

Soil erosion is the removal of single soil particles or small aggregates by erosive agents, including rainfall, run-off (overland flow), wind, tillage/cultivations and co-extraction of soil on root crops or on-farm/field vehicles and machinery (Owens et al., 2006). Soil erosion is usually expressed in t ha^{-1} yr^{-1} or Mg ha^{-1} yr^{-1} or as a depth of soil removed (mm). Around 80% of the world's agricultural land suffers moderate to severe erosion (Pimentel and Burgess, 2013). It is estimated that 75 billion tonnes of soil are eroded worldwide each year, with about two-thirds originating from agricultural land (Myers, 1993). Arable and horticultural lands are particularly vulnerable to erosion due to row cropping and wheel tracks/tramlines (often oriented up/down the slope), extensive areas of bare soil exposed to erosive agents both before crop establishment and after harvest, and intensive soil disturbance from field operations such as tillage and cultivations.

The erosion process comprises three phases: detachment, transport and deposition. Once detached from the *in situ* soil mass, eroded material (sediment) is transported by the erosive agent until the transporting capacity of that agent is reduced either because of a decrease in flow velocity/energy (e.g. due to a reduction in slope gradient) or a decrease in flow discharge (e.g. due to infiltration). The distance over which soil particles are transported before deposition is a function of the erosive force and the size of the eroded particle/aggregates. This distance can range from a few millimetres (in the case of rainsplash) up to several kilometres (if the eroded material is transported by overland flow, for example). However, while transport distance may be limited, re-detachment of deposited sediment in the next erosive event (e.g. rainstorm or wind blow) means the erosion cycle can repeat over and over again, so that cumulative transport distance may be much greater. This remobilisation of sediment in the landscape has been referred to as cyclic sedimentary deposits or a sediment cascade (Collins and Walling, 2004; Fryirs, 2013; Harvey, 2002), likened to a jerky conveyor belt (Ferguson, 1981).

The detachment and transport phases of the erosion process are shown in Fig. 1. However, the complexity of the processes operating over space and time mean that 'No single, simple mathematical expression is likely to describe all the relationships between sediment, discharge, flow and soil particles' (Morgan, 2005).

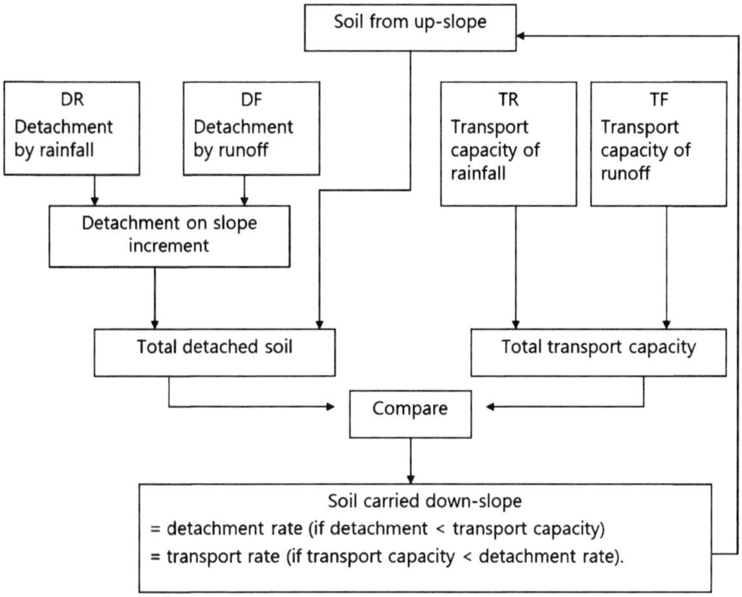

Figure 1 A simple model of soil erosion processes (after Meyer and Wischmeier, 1969).

2.2 Soil erosion by water

The three phases of soil erosion by water (detachment, transport and deposition) and the factors affecting them are shown in Fig. 2. A number of geomorphological processes, modified by hydrological, vegetation and soil factors, cause soil erosion by water, operating at different temporal frequencies and spatial magnitudes, resulting in a variety of erosional landforms.

Rainsplash erosion

Rainsplash erosion occurs when individual raindrops strike the soil surface. The ability of rainfall to cause erosion is termed 'erosivity' and is related to the rainfall intensity (mm/hr), fall velocity of drops, kinetic energy, duration of storm, drop size distribution and where relevant, wind velocity (Marzen et al., 2015). These factors are interrelated, for example, a 1 mm diameter raindrop has a terminal velocity of 4 m s^{-1} while a 5 mm diameter raindrop falls at 9 m s^{-1} (Morgan, 2005). If the shear stress/erosivity of the rainfall is greater than the shear strength of the soil, then detachment of particles or small aggregates occurs. The susceptibility of the soil to detachment (and subsequent transport) is known as 'soil erodibility' and is determined by soil properties such as texture, structure, permeability, moisture content, organic matter content and the stabilising effect of soil biota, especially soil microorganisms. Silt-sized particles are the most vulnerable to detachment by raindrops (Poesen, 1985). Smaller particles (clays) are more resistant to detachment because of inter-particulate cohesion, and larger particles (sands) are resistant because of their mass/weight. Soil detachment rate by rainfall is given in Equation 1 (Morgan, 2005):

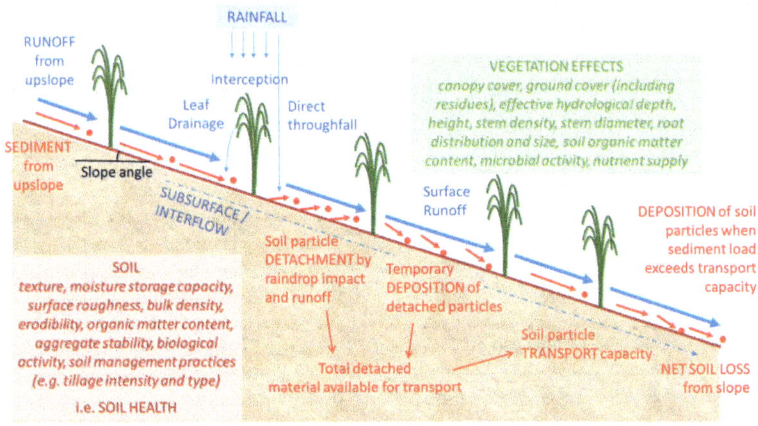

Figure 2 Schematic representation of the three phases of soil erosion (detachment, transport and deposition) and the factors affecting them (adapted from Morgan and Duzant, 2008).

$$DET = k * KE^{b} * e^{-a*h} \tag{1}$$

where
DET = detachment of material (g/m^2/sec)
k = index of soil detachability (g/J) (Brunori et al., 1989)
KE = kinetic energy of the rainfall (mass x velocity) (J/m^2/sec)
a, *b* = experiment coefficients related to soil texture
h = depth (mm) of surface water layer (if any)

If the depth of surface water is greater than the raindrop diameter, this can decrease the rate of detachment by buffering the impact of the raindrop on the soil. If the raindrop diameter is greater than the depth of water, then turbulence in the water layer can increase the rate of detachment (Gao et al., 2003).

Any detached material is then entrained in the jets of the striking raindrop and displaced from their origin. The rate of transport by rainsplash is given in Equation 2:

$$Tr \propto I^{j} * S^{f} \tag{2}$$

where
Tr = transport rate by splash
I = rainfall intensity (mm/hr)
S = slope (m/m or sine slope angle)
j = experimental exponent = 1.0
f = slope steepness exponent = 1.0; may be less on steeper slopes, for example, = 0.8 @ 20°; 0.75 @ 25°

On steep slopes, there is a net movement of splashed material downslope, as a function of slope gradient. However, rates of rainsplash erosion are limited by the small distances of the raindrop splash trajectories, although Marzen et al. (2015) recorded distances of 2000 mm for wind-assisted rainsplash.

Interrill erosion

Factors that determine the likelihood of overland flow generation include the infiltration capacity of the soil and the intensity and duration of rainfall. Interrill erosion (also known as sheet or wash erosion) can occur through unconcentrated surface overland flow, where the shear stress or energy of the flow is greater than the shear strength (or resistance) of the soil surface to detachment. The shear stress of the flow is related to its velocity, determined by the density of the fluid, the depth of flow (which for interrill overland flow is usually very shallow) and the slope gradient (Morgan, 2005; Equation 3):

$$\tau = \rho * g * r * s \tag{3}$$

where

τ = shear stress of overland flow
ρ = density of water
g = acceleration due to gravity (9.81 m/sec)
r = hydraulic radius/depth of flow in the rill
s = slope gradient

The amount of detached material that can be transported in the flow depends on the flow discharge, slope angle, resistance of the soil to detachment and the size of the detached material (Equation 4).

$$Qs = \frac{k * Qw^{a} * \sin\beta^{b}}{n^{c} * D_{50}} \tag{4}$$

where

Qs = net sediment transport by overland flow
k, a, b, c = empirically derived values
Qw = discharge of water
β = slope angle (degrees)
n = resistance to particle detachment
D_{50} = median particle size

On gentle slopes, the low velocity (and thus limited hydraulic energy) of the overland flow is reduced further because of high contact area (wetted perimeter) between water flow and soil surface, resulting in flow turbulence and reduced velocity [as expressed in coefficients such as Mannings 'n' (Morgan, 2005)]. Hence rates of interrill erosion tend to be limited: 'most of the sediment transported [by flow] is derived by raindrop impact, and that, except on steep slopes or on smooth, bare surfaces, grain shear velocity rarely attains the level necessary to detach soil particles' (Morgan, 2005).

Rill erosion

Rill erosion is initiated when overland flow becomes concentrated in microrills (Merritt, 1984), which then develop in size. This hydrological behaviour is due to local variations in surface microtopography, such as drill rows, tractor wheelings or surface alignment of crop residues. Flow velocities increase, so that shear stress can become greater than the resisting shear strength of the soil – sufficient for soil detachment to occur. Detached particles are then entrained in the overland flow and transported away from their origin. Rates of rill erosion are consequently higher than those associated with interrill erosion (Morgan, 2005; Whiting et al., 2001).

Gully erosion

Gully erosion occurs when rills enlarge through headwall retreat upslope and channel erosion. Gully erosion is 'the erosion process whereby run-off water accumulates and often recurs in narrow channels and, over short periods, removes the soil from this narrow area to considerable depths' (Soil Science Society of America, 2001). Gullies are large, permanent erosion features with intermittent flow. They tend to be deeper than they are wide and are associated with high rates of erosion and erratic hydrological behaviour. They cannot be traversed by agricultural equipment and pose a significant constraint to land management operations in agricultural fields. For example, they may dissect fields, preventing access and creating smaller, less economically viable land parcels. Gully erosion can also occur due to subsurface processes, as described below.

Subsurface erosion

Where permeable materials lie above relatively impermeable ones, rainfall can infiltrate to the impeded layer, where it concentrates. On sloping land, this may form concentrated subsurface flow paths that develop into small eroded pipes, which continue to enlarge during erosion events (Fox and Wilson, 2000). Subsurface erosion is often difficult to identify from the surface, although changes in vegetation pattern can reveal subsurface flow paths and pipes. As the pipes develop, their roofs become unstable and when they collapse, a deeply incised channel (gully) is formed (Soil Science Society of America, 2001).

3 Soil erosion processes: other mechanisms

3.1 Soil erosion by wind

The process of detachment and transport of material in wind erosion is determined by the combination of critical wind (shear) velocity (wind erosivity), the momentum imparted to *in situ* soil from soil particles already entrained in the airflow (saltation) and the soil susceptibility to movement (erodibility). Soil moisture determines the resistance of the soil to wind erosion, with wet soils being less erodible than dry soils, due to cohesive forces between particles. A protective cover of vegetation, crop residues, stones, erosion blankets and so on can reduce air velocity at the ground surface, and thus reduce wind speed and erosivity. Warren (2007) estimated that at the average rate of soil loss via wind erosion in the United States (6.4 tons/acre/annum), it would take twenty-five years to lose one inch of soil (Hughes-Popp et al., 2000).

3.2 Soil erosion by tillage

Downslope movement of soil particles by tillage operations is a redistribution process that results from gravity and the disturbance of soil by farm implements. Factors that control tillage erosion include slope gradient and variations in slope curvature, tillage operation (including plough depth, direction and speed) and tillage implement type (Owens et al., 2006a; Tsara et al., 2001; Table 3). The process of ploughing lifts and breaks up the soil mass. Downslope tillage displaces soil in a downslope direction by the drag of the plough and gravity. Upslope tillage drags the soil upslope but

Table 3 Effect of cultivation method, plough depth and tillage direction on soil displacement (after Tsara et al., 2001)

Cultivation method	Plough depth	Tillage direction	Soil displacement equation	Source
Mouldboard plough	40	Perpendicular to the contour lines (up/downslope)	= 0.38–1.18 slope gradient	Gerontidis et al., 2001
	30	Perpendicular to the contour lines (up/downslope)	= 0.30–0.99 slope gradient	
	30	Parallel to the contour lines (across slope)	= 0.37–0.78 slope gradient	
Disc harrow	Unspecified	Unspecified	= 0.115–0.33 slope gradient	Concalves-Ferreira et al., 1999
Cultivator	Unspecified	Unspecified	= 0.23–0.97 slope gradient	Govers et al., 1999

gravity draws some of it back downslope, so the net uphill movement is less than for downslope tillage. Tillage along the contour can cause both up- and downslope soil translocation, depending on the alignment of the tool in relation to the slope direction. Several studies have shown tillage erosion to be of equal or even greater importance than soil erosion by water in landscape evolution and soil degradation, particularly on fields with short slope lengths (e.g. Quine et al., 1999; Van Muysen et al., 1999; Turkelboom et al., 1999; Nyssen et al., 2000). However, unless the downslope field boundary is removed, it is unlikely that net export of soil from the field will occur due to tillage erosion alone.

3.3 Soil erosion through co-extraction with crops, vehicles and machinery

Soil can be removed from a field by co-extraction on crops during harvest (principally root vegetables such as potatoes, carrots, sugar beet and onions) and when attached to agricultural farm vehicles and implements. In the crop processing/agricultural industry, soil adhering to root crops is referred to as 'soil tare' or 'dirt tare'. Ruysschaert et al. (2004) provided an overview of factors that contribute to soil loss through co-extraction and on farm machinery. The most notable of these factors include soil wetness and high clay content which increase the adhesion of soil to crops and farm equipment. Therefore, soil type, timing and techniques of harvesting and weather conditions play a role in the amount of co-extracted soil. Poesen et al. (2001) reported that these amounts of soil loss can be responsible for an important soil profile truncation. For example, the >200 years of 'witloof' chicory root cultivation in Belgium may be responsible for a mean soil profile truncation of ca. 7 cm. In Owens et al.'s review (2006), no literature was found on the magnitude of soil co-extracted on farm machinery, but under certain conditions, it was estimated to be between 0.5 and 1.0 t ha^{-1} $harvest^{-1}$ (Owens et al., 2006).

Table 4 Comparison of the magnitude of soil loss for different erosion processes in England and Wales (from Owens et al., 2006)

	Wind	Tillage	Co-extraction with root crops and farm machinery	Water
Typical erosion rate range (Mg ha^{-1} year^{-1})	0.1–2.0	0.1–10.0	0.1–5.0	0.1–15.0
Land use affected	Arable, upland, some pasture	Arable	Arable	Arable, pasture, upland
Exported off field	Yes	No	Yes	Yes

3.4 Comparisons of the different forms of soil erosion

Owens et al. (2006) compared the typical annual erosion rates for different forms of soil erosion in England and Wales (Table 4). These authors point out that there will be significant temporal and spatial variations in the magnitude and extent of the different processes, with arable land being susceptible to all forms of erosion, and uncultivated land only at risk of water and, to some extent (i.e. exposed sandy and peaty soils), wind erosion. The rates in Table 4 exceed estimated rates of tolerable soil erosion, which is defined as 'any actual soil erosion rate at which a deterioration or loss of one or more soil functions does not occur' (Verheijen et al., 2009). The upper limit of tolerable soil erosion (equal to the rate of soil formation in Western Europe) was defined at around 1.4 Mg ha^{-1} yr^{-1} (Verheijen et al., 2009). This suggests that some of the erosion rates shown in Table 4 represent a net loss of the soil resource.

4 Consequences of soil erosion

All ecosystem goods and services delivered by soil (Table 1) are adversely affected by soil erosion, as this process results in an absolute reduction in the soil stock. A loss of soil depth via erosion processes is likely to cause deterioration in soil function, unless soil quality can be increased (Rickson et al., 2015). Sparovek and Schnug (2001) defined losses in soil depth in terms of the soil's 'life time'. This is the number of years before the critical soil depth needed for delivering given goods and services such as crop production is reached. In terms of sustainable agriculture, soil erosion affects the provisioning function of soil (i.e. crop production) by the reduction in soil depth (physical effects) and the losses of plant essential nutrients (including carbon, nitrogen, phosphorus and potassium).

4.1 Loss of soil depth

The effect of soil erosion on crop yields is frequently given in relation to a change in soil depth (Lal, 1998; Pimentel et al., 1995; Liu et al., 2013). The removal of soil by soil erosion processes reduces the depth and volume of soil (i.e. as a growing medium for crops) at the erosion site. The degree of soil depth loss will depend on a) the erosion rate (and thus the mass/weight of material eroded) and b) the bulk density of the soil, expressed in Equation 5. Soil erosion is usually measured as a 'loss of soil mass per unit area per unit time (Mg ha^{-1} yr^{-1})' (Huber et al., 2008). These values can then be converted into a depth of soil loss/change in soil depth due to erosion (mm). Pimentel et al. (1995) used a simplified method to estimate the impact of soil erosion on soil depth:

$$\text{Depth of soil loss per year (mm)} = \frac{\text{Mass of soil loss } \left(\text{Mg ha}^{-1}\ \text{yr}^{-1}\right)}{\text{Bulk density of soil } \left(\text{Mg/m}^{3}\right)} \tag{5}$$

The future impacts of different rates of erosion on soil depth are shown in Fig. 3.

Loss of soil depth through soil erosion affects plant growth principally in three ways. First, roots require anchorage for physical support. There is some evidence to suggest that a reduction in soil depth affects root anchorage and strength (Baker et al., 1998), which can lead to lodging and poor yields. Second, deeper soils have better water-holding capacity so that infiltrated water can either be stored in the soil profile (where it can be accessed by the crop) or drained away through soil infiltration processes during high rainfall events. A reduction in soil depth through erosion will reduce both functions, making crops prone to water stress during times of drought and waterlogging on poorly drained soils during wet periods. Finally, shallow, eroded soils usually provide a poor nutrient reserve for crop uptake. Crop yield potential is largely dependent on soil nutrient reserves in the topsoil (Spink et al., 2009).

The 'effective depth of rooting' that considers physical rooting depth and accessibility to water and nutrients varies with crop type (King et al., 2003). Most crops root to a depth of between 60 cm and 1.2 m, although some can exceed this range, for example sugar beet roots can reach up to 1.4 m (van Orshoven et al., 2013) and winter cereals can reach depths of over 2 m (Weaver, 1926; Gregory, 1976; Hoad et al., 2001; Fageria & Moreira, 2011).

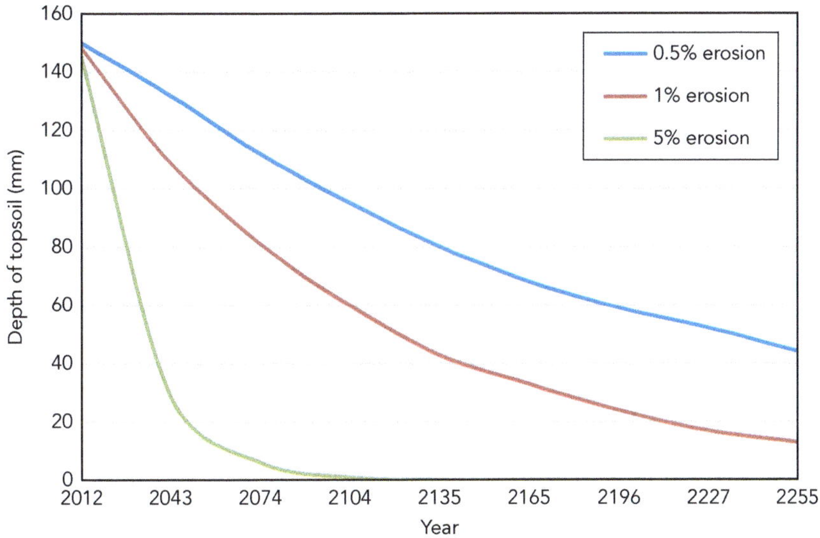

Figure 3 Predicted loss of fertile topsoil (redrawn from Young and Crawford, 2004).

4.2 Loss of soil nutrients and carbon

Soil erosion is a selective process that takes place in/on the top few millimetres of the soil profile, where there are relatively higher contents of organic matter, soil biota and nutrients, which are therefore preferentially lost with the eroded material (sediment). Soil erosion is also selective in detaching and transporting the particles in the silt to very fine sand fractions, which have neither the cohesion of clay nor the large particle size of sands. Unamunzaga et al. (2014) showed that soil erosion caused a variable distribution of particle size fractions in the topsoil. These most susceptible (erodible) particles are those with high specific surface areas, where absorption of carbon, nutrients and agrochemicals is greatest. As a result, eroded materials (sediment) are enriched with these properties, whilst the *in situ* soil is stripped of them (Schiettecatte et al., 2008; Wan and El Swaify, 1998; Wang et al., 2014; Quinton et al., 2006; Dungait et al., 2013). Eroded sediment can typically contain three times more nutrients than the soil left behind (Sharpley, 1980; Lal, 1998; Ali et al., 2006). Enrichment ratios describe the relative concentrations of nutrients in deposited sediment compared with the *in situ* (uneroded) soil. Graves et al. (2011) reviewed studies which reported enrichment ratios to derive typical values: N = 1.37; P = 2.15; K = 2.90; organic carbon = 1.56 (Table 5).

Losses of nutrients via soil erosion can be substituted (and therefore their effects masked) by additions of inorganic agrochemicals. However, this represents significant costs to production and may not be sustainable in the long term, given the reliance on fossil fuels for N production and the finite global supplies of P (Cordell and White, 2011). In the United States alone, $20 billion per annum is spent on replacement of nutrients lost through soil erosion (Troeh et al., 1991, 2004). In India, the figure is estimated at $245 million (Chaudary and Das, 1990). In any case, inorganic fertilisers cannot replace the actual depth of soil once it has eroded. Also, Hadas et al. (1989) have shown that unlike organic fertilisers, chemical fertilisers have no beneficial effects on soil structure or aggregation, making the soils susceptible to further erosion, with associated impacts on future yields. The poorer crop cover also increases erosion risk. Hence even higher application rates of nutrients are needed to mask these losses.

Table 5 Selected enrichment ratios (ER) showing the proportion of nutrients in eroded sediment relative to parent soils (from Graves et al., 2011)

Nutrient	Range of enrichment ratios	Source	Typical ER
P	1.32–3.04	Zheng et al., 2005	2.15
	1.47	Sharpley, 1985	
	2.79	Ali et al., 2006	
OC/OMC	1.08–1.4	Zheng et al., 2005	1.56
	0.9–2.6	Schiettecatte et al., 2008	
	2.001.23	Sharpley, 1985	
		Ali et al., 2006	
N	0.89–1.26	Zheng et al., 2005	1.37
	1.61	Sharpley, 1985	
	1.43	Ali et al., 2006	
K	2.90	Ali et al., 2006	2.90

4.3 Changes in soil quality/soil health

Following erosion, soil immediately below the topsoil is exposed. This material is usually relatively poorly structured, with low levels of carbon, nutrients and soil biology. Hence it tends to be less fertile and more erodible, and is associated with higher erosion rates (Romkens et al., 1977), due to poor development of vegetation cover and higher soil susceptibility to erosive agents. Removal of topsoil also exposes previously buried soil organic carbon sources, which mineralise when exposed to the atmosphere. Large-scale losses of CO_2 to the atmosphere is associated with higher atmospheric temperatures, global warming and climate change. It could be argued that the predicted increase in extreme weather events such as higher intensity rain storms will aggravate the erosion process further.

Baxter et al. (2013) reported on the impacts of erosion on soil biota and their role in soil functioning and delivery of ecosystem goods and services, including nutrient cycling for crop uptake. They discuss the role of erosion in transporting and redistributing soil biota (especially nematodes), as well as how erosion affects the physical, biological and chemical properties of their habitats (i.e. sites where erosion takes place and where eroded sediments are deposited). In other words, by changing the condition of soil over space and time, soil erosion has direct impact on soil health.

4.4 Depositional effects

Enrichment ratios show that deposited sediment is rich in nutrients and organic matter, and this has been exploited by agriculturalists to enhance crop production (e.g. Nakase et al., 2014). Schmitter et al. (2011) showed that, depending on the quality and quantity of the sediments, rice performance increased where finer nutrient-rich sediments were deposited. However, the linkage between sediment properties and crop productivity is complex, as other drivers are also important. On the other hand, deposited sediments within field can adversely affect rates of crop growth, especially if newly emerging seedlings are inundated (Awadhwal and Thierstein, 1985). Eroded particles, deposited on the soil surface can cause surface capping and sealing, leading to a thin layer of soil with higher penetration resistance than unaffected areas (Baumhardt et al., 2004). This cap reduces the ingress of air and water needed by developing roots, and inhibits the emergence of seedlings. The cap then has to be broken with additional tillage operations, and the crop redrilled, incurring additional production costs for the farmer. At later stages of growth, crop production can also be affected by plant stress as a result of inundation where eroded soil is deposited. Liu et al. (2013) reported that deposition of sediment eroded from gullies completely eliminated soybean yields in northeast China.

5 Relating soil erosion to crop productivity and sustainable agriculture

5.1 Evidence of the impacts of soil erosion on crop productivity and sustainable agriculture

As reported above, soil erosion is associated with reductions in crop yield, so threatening sustainable agriculture. Observed reductions in crop yields following soil erosion show wide

Table 6 Erosion-induced yield penalty (developed from Pimentel et al., 1995)

Yield penalty (%) associated with	20 year total	Average annual	Soil loss per mm
Water run-off	7	0.35	0.3
Losses of nitrogen, phosphorus and potassium	8	0.4	0.3
Soil depth	7	0.35	0.3
Loss of organic matter	4	0.2	0.1
Reduced water-holding capacity	2	0.1	0.1
Loss of soil biota	1	0.05	0.0
Total on-site penalty	20	1	0.74

variations, depending on crop rooting, nutrient storage and availability, soil water-holding capacity, soil profile characteristics, the crop grown, existing soil and crop management, and the microclimate (Lal, 1998; Pimentel et al., 1995). From a variety of measured data, Pimentel et al. (1995) found that crop yields declined by 20% over a period of 20 years. Using a mean erosion rate (17 t ha^{-1} a^{-1}) and a bulk density (1.25 t m^{-3}), the annual soil depth reduction was calculated to be 1.36 mm a^{-1}, giving a yield reduction of 0.74% mm^{-1} of soil loss (Table 6).

Other studies have found that yield reductions are typically between 0.03 and 0.05% per tonne soil lost (e.g. Biot and Lu, 1995; Arden-Clarke and Hodges, 1988; Owens et al., 2006). Larney et al. (1995) found a relationship between depth of topsoil removal (simulating soil erosion processes) with wheat grain yield, with R^2 values of 0.94 and 0.999 for dryland and irrigated lands, respectively. Young (1989) estimated that losses of the soils' water and nutrient-holding capacity due to erosion reduced crop productivity by 50–75%. Lucas et al. (1977) found that the yield potential of corn was reduced by 25% when erosion led to a decline in soil organic matter from 4.3 to 1.7. Liu et al. (2013) found that soil depth was the most important soil property that affected yields of soybean in the black soil region of northeast China. Ironically, eroded gullies were infilled using agricultural soil from adjacent soybean fields. In the fields affected, every 1 cm decrease in soil depth resulted in a 2% decrease in yield. Bakker et al. (2004) collated and quantified the results from a number of experimental studies investigating the effect of soil erosion on crop productivity. They found that soil erosion led to a reduction in crop productivity of 4.3–26.6% per 10 cm loss of soil depth. This was explained by reduced nutrient availability, water deficit and physical restrictions to root growth. Areas suffering from moderate erosion (~10 t ha^{-1} yr^{-1}) could lose an average of 0.4% yield per decade.

Since erosion-induced yield declines are of considerable importance, especially in low input agriculture, tools such as EPIC (the Erosion-Productivity Impact Calculator; http://epicapex.brc.tamus.edu/) have been developed to examine long-term effects of various components of soil erosion on crop production. EPIC is a mechanistic simulation model used to examine long-term effects of various components of soil erosion on crop production (Williams, 1985). EPIC is a public domain model that has been used to examine the effects of soil erosion on crop production in over 60 different countries in Asia, South America and Europe. The model has several components:

- soil erosion
- economic
- hydrologic

- weather
- nutrient
- plant growth dynamics and
- crop management.

The model provides output on crop yields and the economics of fertiliser use.

Alternatively, Graves et al. (2015) used an ecosystem services framework approach to estimate the impacts and total economic costs of soil degradation (including soil erosion) in England and Wales. This included the impacts of erosion on delivery of goods and services including agricultural production. First, typical rates of soil erosion were derived from field observations for given soil and land use combinations. Only soil erosion by water was considered, as Owens et al. (2006) demonstrated that this was more dominant than erosion by wind, tillage or co-extraction on farm machinery (Table 4). Ten different land covers (including arable, horticulture and grassland) and four soil types were used (sands, silts, clays and peats). Typical crop yields for each land cover/soil type combination were derived using parameters from the Silsoe Whole Farm Model (Annetts and Audsley, 2002). Where erosion occurred, the soil depth losses (mm a^{-1}) associated with the erosion rate for any given land use/soil type combination were used to assign a crop yield reduction (% a^{-1}), using the relationships between soil erosion rate, loss of soil depth as the growing medium and crop yield as described in Pimentel et al. (1995). The reduction in yield was converted into monetary values to derive the costs of soil erosion on crop yield. The costs of lost nutrients (i.e. chemical replacement costs) were added to this to give total costs of soil erosion on agricultural production (c. £40 million a^{-1}) (Table 7). This methodology

Table 7 Estimated erosion rates and associated on-site costs in England and Wales (after Graves et al., 2015)

Data category	Data description	Data value	Unit
Physical data			
Area at risk of soil loss	Total areas at risk	1022459	ha
	Total soil erosion	2920626	t a^{-1}
	Mean soil erosion	0.21	t ha^{-1} a^{-1}
Physical losses of soil nutrients and organic matter	Total N loss	18026	t a^{-1}
	Total P loss	4830	t a^{-1}
	Total K loss	38280	t a^{-1}
	Total C loss	225787	t a^{-1}
On-site costs			
Provisioning service	Crop productivity decrease	5358	£'000
Stock value of lost nutrients	N loss in soil stock	11176	£'000
	P loss in soil stock	3284	£'000
	K loss in soil stock	19906	£'000
	C loss in soil stock	151	£'000
Total on-site costs		39875	£'000

could be applied elsewhere, as long as accurate estimates of erosion rates and nutrient losses are available, and the relationships between soil depth and yield are validated for the area.

5.2 Critique of the methodologies used to relate soil erosion to crop production

The complex interactions between soil properties affected by erosion (e.g. depth, organic matter and nutrients) makes 'assessment of the direct impacts of soil erosion on productivity difficult' (Pimentel et al., 1995). Bakker et al. (2004) showed that the crop productivity response to soil erosion largely depends on the physical variables affected by soil erosion: soil erosion resulting in water deficits and physical root hindrance generally produce convex response curves, whereas nutrient deficits resulted in linear to concave response curves. Furthermore, the effects vary among crops, as demands for soil structure, rooting depth and water retention differ per crop. The degree of yield loss thus depends upon the soil profile characteristics, the crop grown, soil management and the microclimate (Lal, 1985; Posthumus and Stroosnijder, 2010). Also, the impact of erosion on soil productivity depends on the quality of the remaining soil and is thus location-specific.

There is also concern that the outputs of these studies might underestimate the true impact of soil erosion on crop production and sustainable agriculture. Some research studies simulate erosion by scraping off layers of topsoil '*en masse*' and observing subsequent effects on crop yields. This method does not reflect the selectivity of erosion processes, whereby nutrients and organic matter are preferentially eroded (Table 5). Also, in practice, yield declines due to soil erosion are easy to mask with additional (often inorganic) fertilisers, new crop varieties, use of pesticides and herbicides, and/or irrigation.

There can be large uncertainties about the effect of erosion on soil depth (Warren, 2007). Equation 5 assumes the process occurs over an area, but this approach is over-simplistic: in reality, bulk density of the soil will vary in space and time, so affecting the depth of soil lost for a given erosion rate. More importantly, the equation assumes the depth of soil loss is uniform over any given area: in reality erosion by water may be in the form of linear features such as rills and/or gullies that spatially occupy only a small proportion of the eroding field. It is less reasonable to extrapolate the loss of soil depth from a linear feature (covering, e.g. 10% of a field) to a larger areal extent.

It has been estimated that in most cases, <1% of a field is affected by erosion and deposition: more than 3% is rarely affected (Evans, 1996). Where these features occur there may be significant losses of soil depth, but when averaged over the whole field, this may amount to a few millimetres only. Where the erosion feature develops, there may be tangible impacts on crop growth, water regulation and carbon storage, whilst the rest of the field (specifically the soil's ability to function) is effectively unaffected. Other forms of erosion (wind, tillage and co-extraction on crops and agricultural machinery) tend to remove soil over an areal extent, whereas erosion by water tends to be concentrated as linear features such as rills or gullies (Morgan, 2005).

6 Mitigating soil erosion: soil conservation

Methods of soil erosion control can reduce erosion rates and mitigate the effects of soil loss on crop production. Mitigation measures act on one or more of the detachment, transport and deposition phases of soil erosion. The philosophy of 'control at source' is common in soil conservation design: this usually means control of the first phase of soil erosion – detachment of particles/aggregates. Methods of erosion control have been collated and reviewed extensively elsewhere (e.g. Newell Price et al., 2011; Maetens et al., 2012; Posthumus et al., 2011; Rickson et al., 2010; Graves et al., 2011; Rickson, 2014). Table 8 gives a list of the measures adapted from Cuttle et al. (2007) and Rickson et al. (2010). The effectiveness of measures to reduce soil erosion from agricultural soils is dependent upon a wide range of factors such as crop type, soil type, slope, drainage, vegetative cover, cultivation methods and timing of operations that affect carbon flow, biological activity and soil quality/soil health.

Table 8 Soil erosion mitigation measures (compiled from Cuttle et al. (2007) and Rickson et al. (2010))

Description	Rationale
Agronomic measures	
Land use change (e.g. arable to grassland)	Less soil disturbance and bare soil exposure over time
Cover/nurse/catch cropping	Maintain soil protection through crop cover, especially in vulnerable months
Mulching	Use of previous crop residues to protect soil surface
Timeliness of operations	Avoid disturbing the soil during rainfall/windy conditions
Buffer strips	Reduce slope length and intercept run-off and sediment
High-density planting	High seeding rates increase crop, stem and root density to limit erosion processes
Crop rotation	Compensate high erosion risk crops with low risk crops. Improves soil structure, soil resistance against detachment and surface cover
Shelterbelts	Planting rows of trees or hedgerows provides shelter and wind breaks around fields. Grow rows of trees or hedges to provide protection for soil and crops grown on the sheltered side and to trap airborne soil particles
In-field shelter (cover crops, straw planting)	Inter-row strips of residues or live cover crops reduce wind velocities. In-field shelterbelts exert drag on wind flow, reducing velocity and erosivity of wind. On peaty soils, mechanised straw planting in rows may provide shelter for early sown vegetable crops. Straw is planted between the crop rows just before or after drilling
Avoid overgrazing (e.g. outdoor pigs in lowland systems; sheep in upland areas)	Overgrazing leads to loss of vegetation, exposure of bare soil and compaction. Maintain vegetation cover
Avoid excessive burning of upland vegetation	Maintain protective vegetation cover
Strip cropping	Broad strips of alternating crops
Simulated vegetation (geotextiles)	Erosion control blankets simulate plant canopy, stem and root effects. Artificial windbreaks such as polyethylene netting or webbing can be appropriate for protecting small areas of high-value crops

(Continued)

Table 8 (*Continued*)

Description	Rationale
Field layouts	Avoid large fields/field consolidation to minimise fetch (uninterrupted distance over which wind blows) and slope length (water erosion control)
Field engineering measures	
Lined waterways and swales	Protect areas of concentrated flow
Earth banks/physical barriers	Interrupt slope length and intercept run-off
Drainage (surface and subsurface)	Intercept run-off, so erosion minimised. For wind erosion control, ensure soils are kept moist and cohesive
Terraces	Interrupt slope length and intercept run-off
Irrigation	Keep soil moist to reduce wind erosion
Soil management measures	
Reduced/minimum tillage	Maintain soil structure to resist erosive forces
Contour/across slope cultivation	Reduce slope length
Tramline management	Interrupt concentration of run-off in tramlines. Cover cropping can be used in tramlines and wheelings to reduce soil exposure to wind and water
Establish coarser seedbeds	Maintain soil structure to resist erosion
Increase soil organic matter	Correlation between SOM and aggregate stability/resistance to erosion (to 10% soil organic matter content)
Zero tillage	Maintain soil structure to resist erosive forces
Bed systems (in potatoes, etc.)	Avoid ridge systems which expose more soil surface to wind forces
Furrow press	Plough and furrow press the land when moist (to maintain microtopography created), using a 45° angle press to leave steep ridges (e.g. with a Cambridge roller). Drill as soon as possible, at an angle to the ridges to reduce wind velocities at the soil surface
Addition of clay-sized particles	Mixing of clay-rich soil with *in situ* soil
Synthetic stabilisers (polyvinyl alcohol)	Proprietary soil stabilisers including PVA (polyvinylacetate) emulsions or PAM (polyacrylamides) can provide protection from raindrop impact and wind shear when sprayed onto sands after drilling

One of the challenges associated with encouraging uptake of soil erosion mitigation measures is that to most farmers, rates of erosion are imperceptible. Only if gullies form will most farmers be inconvenienced and want to tackle erosion. Wen and Easter (1987) reported that adoption of effective soil conservation practices only becomes profitable when soil depth decreases to a critical level (where farmers perceive declines in crop yield). The on-site benefits of soil conservation are often insufficient to recover the related management costs. This highlights the importance of a more strategic, long-term view that will encourage the adoption of soil conservation practices before irreversible damage is done. The likely imposition of financial fines for discharging sediment into watercourses (and thus non-compliance of regulations such as the EU Water Framework Directive; European Commission, 2000) can also trigger action. Recent Farm Practice Surveys (e.g. Defra, 2007) reveal that the number of erosion incidents observed by farmers may be on the increase. At least one incidence of soil erosion happens on 12% of English holdings every year and on a quarter of holdings at least every three years (Defra, 2007). Soil erosion of some description is observed on over 50% of farms.

The effectiveness of these erosion control measures in one location should not be extrapolated to different soil type/land use combinations or spatial scales (e.g. field to farm to catchment to regional scale; Cherry et al., 2008). Indeed, reviews of reported effectiveness of these measures have brought cause for concern as to the reliability and performance of these measures (Maetens et al., 2012; Rickson, 2014). This is further compounded by the uncertainties of the effects of climate change on soil erosion processes and their mitigation. Even so, erosion rates have fallen in the United States as a result of better soil erosion control policies and practices (Pimentel and Burgess, 2013), but in places these rates can still exceed those of soil formation. This will give a net loss of the soil resource, with adverse consequences for sustainable agriculture.

7 Summary and conclusions

This chapter has focused on the effects of only one form of soil degradation (soil erosion by water) on agricultural production. There is robust evidence of direct relationships between soil erosion processes and crop production, via losses of soil depth and reduction in soil quality and soil health. Often these impacts of soil erosion on the delivery of ecosystem goods and services (e.g. loss of crop yield quantity and quality) can be masked with additional inputs to the farming system (such as increased fertiliser use, irrigation or new crop varieties). These interventions increase production costs to the farmer in the short term (so reducing already squeezed economic margins) and may undermine progress towards sustainable intensification in the longer term. The future economic viability of agricultural lands at risk of erosion is uncertain, putting many rural livelihoods at risk, as well as undermining the security of food, fodder, fibre and (bio)fuel supplies. In addition, there are also environmental impacts of soil erosion off-site.

Climate change will impact on these relationships: Changes in the intensity, amount, frequency and type of rainfall, with an increasing trend for 'heavy' (greater than 10 mm d^{-1}) and 'extreme' (>95th quartile) rainfall events will increase rainfall erosivity (Simmons and Rickson, 2008). For erosion by water, wetter soils are more likely to promote potentially erosive run-off and will be more erodible due to lower cohesion values. Changing temperature and rainfall regimes may also affect cropping patterns, in terms of varieties grown or even the introduction of new crops. These crops may have more or less ability to control the processes of erosion through interception of rainfall, run-off and airflow. It is possible that higher temperatures will promote more vigorous crop growth (if not limited by other factors such as water availability), which would offer soils greater protection from erosion. However, this is countered by the fact that higher temperatures are also associated with more intense and erosive rainfall events. With population growth, increasing competition for finite land resources between agriculture, industry, residential and infrastructure projects may force agricultural production to more marginal lands, including steeper lands where erosion risk is increased.

It is essential that soil erosion (and other forms of soil degradation) are managed to tolerable levels. Defining such levels is complicated and challenging (Verheijen et al., 2009). Even so, the very minimum requirement is to ensure current soil physical, chemical and biological conditions are maintained, or better still, improved to provide a healthy, living soil on which we depend.

8 Where to look for further information

Further details of the major types of soil degradation and how these can be measured and monitored over space and time can be found in Huber et al. (2008). Morgan (2005) is an excellent text book for non-specialists, covering many aspects of soil erosion processes and their control. If the reader requires information on soil erosion rates throughout the world, Verheijen et al. (2009) is a comprehensive review paper, which compares erosion induced soil loss with soil formation rates. From this analysis, these authors propose tolerable rates of soil erosion. Karlen et al. (1997) is a key text on the concept, definition and measurement of soil quality. Discussion of the impacts of soil erosion (especially on crop productivity) can be found in the classic papers by Lal (1998) and Pimental et al. (1995).

Readers wishing to develop links with other soil erosion scientists can make contact via a number of national and international soil science societies around the world. Many have members who specialise in soil erosion studies, including the International Union of Soil Sciences (IUSS; http://www.iuss.org/) and the International Erosion Control Association (IECA; http://www.ieca.org/). National societies of soil science also exist and will undoubtedly address a range of erosion topics. Key conferences that regularly include presentations on all aspects of soil erosion include the World Congress of Soil Science (WCSS; held every four years) and the annual European Geosciences Union General Assembly, held in Vienna, Austria every year.

Many scientific organisations carry out fundamental and applied soil erosion research throughout the world. For example, staff at the author's institution, Cranfield University, study soil degradation processes over a range of spatial and temporal scales, including the technical and socio-economic aspects of soil degradation, with special focus on soil erosion by water. Working internationally and across the rural and urban sectors, Cranfield staff engage with land managers, research collaborators and policy makers to gain better understanding of the causes and impacts of soil degradation, as well as their solution. The European Commission's Joint Research Centre and the European Soil Data Centre (ESDAC) investigate a number of soil degradation themes, including soil erosion. ESDAC aims to be the single reference point for all soil data and information at the European level. Another example of creating influential research outputs is the US Department of Agriculture Agricultural Research Service, which hosts the National Soil Erosion Research Laboratory. The mission of the laboratory is to develop the knowledge and technology needed by land users to conserve natural resources for future generations.

9 References

Ali, I., Khan, F. and Bhatti, A. U. 2006. Some physico-chemical properties of soil as influenced by surface erosion under different cropping systems on upland-sloping soil. *Soil & Environment*, 25(1), 28–34.

Annetts, J. E. and Audsley, E. 2002. Multiple objective linear programming for environmental farm planning. *Journal of Operational Research Society*, 53(9), 933–43.

Arden-Clarke, C. and Hodges, R. D. 1988. The environmental effects of conventional and organic/biological farming systems. II. Soil ecology, soil fertility and nutrient cycles. *Biological Agriculture & Horticulture*, 5(3), 223–87.

Awadhwal, N. K. and Thierstein, G. E. 1985. Soil crust and its impact on crop establishment: A review. *Soil and Tillage Research*, 5(3), 289–302. https://doi.org/10.1016/0167-1987(85)90021-2.

Bai, Z. G., Dent, D. L., Olsson, L. and Schaepman, M. E. 2008. *Global Assessment of Land Degradation and Improvement Identification by Remote Sensing*. International Soil Reference and Information Centre (ISRIC), Wageningen, NL.

Baker, C. J., Berry, P. M., Spink, J. H., Sylvester-Bradley, R., Griffin, J. M., Scott, R. K. and Clare, R. W. 1998. A method for the assessment of the risk of wheat lodging. *Journal of theoretical biology*, 194(4), 587–603.

Bakker, M. M., Govers, G. and Rounsevell, M. D. A. 2004. The crop productivity-erosion relationship: an analysis based on experimental work. *Catena*, 57, 55–76.

Baumhardt, R. L., Unger, P. W. and Dao, T. H. 2004. Seedbed surface geometry effects on soil crusting and seedling emergence. *Agronomy Journal*, 96(4), 1112–17.

Baxter, C., Rowan, J. S., McKenzie, B. M. and Neilson, R. 2013. Understanding soil erosion impacts in temperate agroecosystems: bridging the gap between geomorphology and soil ecology using nematodes as a model organism. *Biogeosciences*, 10(11), 7133–45.

Biot, Y. and Lu, X. X. 1995. Loss of yield caused by soil erosion on sandy soils in the UK. *Soil Use and Management*, 11(4), 157–62.

Boix-Fayos, C., Martinez-Mena, M., Calvo-Cases, A., Castillo, V. and Albaladejo, J. 2005. Concise review of interrill erosion studies in SE Spain (Alicante and Murcia): erosion rates and progress of knowledge from the 1980s. *Land Degradation & Development*, 16, 517–28.

Bridges, E. M. and Oldeman, L. R. 1999. Global assessment of human-induced soil degradation. *Arid Soil Research and Rehabilitation*, 13(4), 319–25.

Bringezu, S., O'Brien, M., Pengue, W., Swilling, M. and Kauppi, L. 2010. Assessing global land use and soil management for sustainable resource policies. Scoping paper for the International Panel for Sustainable Resource Management, UNEP.

Brunori, F., Penzo, M. C. and Torri, D. 1989. Soil shear strength: its measurement and soil detachability. *Catena*, 16, 59–71.

Chaudary, H. P. and Das, S. K. 1990. Nutrient status in relation to intensity of erosion in ravines of Yamuna. *Indian Social Soil Sciences*, 38, 126–9.

Cherry, K. A., Shepherd, M., Withers, P. J. A. and Mooney, S. J. 2008. Assessing the effectiveness of actions to mitigate nutrient loss from agriculture: A review of methods. *Science of the Total Environment*, 406, 1–23.

Collins, A. L. and Walling, D. E. 2004. Documenting catchment suspended sediment sources: problems, approaches and prospects. *Progress in Physical Geography*, 28, 159–96. doi:10.1191/0309133304pp409ra.

Concalves-Ferreira, A., Rafael Marques da Silva, J. and Manuel Carrilho Nunes Soares, J. 1999. Individual progress report, TERON project. 2nd annual report, contract number (FAIR3±CT96±1478), European Commission, p. 50.

Cordell, D. and White, S. 2011. Peak phosphorus: Clarifying the key issues of a vigorous debate about long-term phosphorus security. *Sustainability*, 3(10), 2027–49.

Cuttle, S. P., Macleod, C. J. A., Chadwick, D. R., Scholefield, D., Haygarth, P. M., Newell-Price, P., Harris, D., Shepherd, M. A., Chambers, B. J. and Humphrey, R. 2007. An Inventory of Methods to

Control Diffuse Water Pollution from Agriculture (DWPA). User Manual. January 2007. Prepared as part of Defra Project ES0203. IGER/ADAS.

Daily, G. C. 1997. *Nature's Services: Societal Dependence on Natural Ecosystems*. Island Press, Washington.

Defra. 2007. *An Introductory Guide to Valuing Ecosystem Services*. PB12852 Department for Environment, Food and Rural Affairs, Nobel House, 17 Smith Square, London, SW1P 3JR.

Doran, J. W. and Parkin, T. B. 1994. Defining and assessing soil quality. In *Proceedings of a Symposium on Defining Soil Quality for a Sustainable Environment (Minneapolis, 1992)*, edited by J. W. Doran, D. C. Coleman, D. F. Bezdicek and B. A. Stewart. Soil Science Society of America/ American Society of Agronomy, Wisconsin, pp. 3–21.

Dungait, J. A. J., Ghee, C., Rowan, J. S., McKenzie, B. M., Hawes, C., Dixon, E. R. and et al. 2013. Microbial responses to the erosional redistribution of soil organic carbon in arable fields. *Soil Biology and Biochemistry*, 60, 195–201.

European Commission. 2002. Communication of 16 April 2002 from the Commission to the Council, the European Parliament, the Economic and Social Committee and the Committee of the Regions: Towards a Thematic Strategy for Soil Protection. COM 2002, 179final. http://europa.eu.International/scadplus/printversion/en//lvb/l28122.htm.

European Commission. 2006. *Thematic Strategy for the Protection of Soil*. COM (2006) 231 final. Commission of the European Communities, Brussels.

European Commission. 2000. Directive 2000/60/EC of the European Parliament and of the Council of 23 October 2000 establishing a framework for Community action in the field of water policy (The Water Framework Directive 2000/60/EC). OJ L327, 2000, pp. 1–73.

European Environment Agency. 1995. Europe's Environment. The Dobris Assessment-Chapter 7 Soil.

Evans, R. 1996. *Soil Erosion and its Impacts in England and Wales*. Friends of the Earth, London.

Fageria, N. K. and Moreira, A. 2011. The role of mineral nutrition on root growth of crop plants. In *Advances in Agronomy Volume 110*, edited by D. L. Sparkes. Academic Press, Burlington, pp. 251–331.

FAO. 1996. *Rome Declaration on World Food Security and World Food Summit Plan of Action*. World Food Summit 13–17 November 1996. Rome.

Ferguson, R. I. 1981. Channel forms and channel changes. In *British Rivers*, edited by J. Lewin. G. Allen and U. London, pp. 90–120.

Fox, G. A. and Wilson, G. V. (2010). The role of subsurface flow in hillslope and stream bank erosion: A review. *Soil Science Society of America Journal*, 74(no. 3), 717–33.

Fryirs, K. A. 2013. (Dis)Connectivity in Catchment Sediment Cascades: A Fresh Look at the Sediment Delivery Problem. *Earth Surface Processes and Landforms*, 38, 30–46. doi:10.1002/esp.3242.

Gao, B., Walter, M. T., Steenhuis, T. S., Parlange, J.-Y., Nakano, K., Rose, C. W. and Hogarth, W. L. (2003). Investigating ponding depth and soil detachability for a mechanistic erosion model using a simple experiment. *Journal of Hydrology*, 277(1–2), 116–24. http://doi.org/10.1016/S0022-1694(03)00085-4.

Gerontidis, D. S., Kosmas, C., Detsis, B., Marathianou, M., Zafirious, T. and Tsara, M. (2001). The effect of moldboard plow on tillage erosion along a hillslope. *Journal of Soil and Water Conservation*, 56(2), 147–52.

Gorlach, B., Landgrebe-Trinkunaite, R., Interwies, E., Bouzit, M., Darmendrail, D. and Rinaudo, J. D. (2004). *Assessing the Economic Impacts of Soil Degradation Volume IV: Executive Summary'. Study commissioned by the European Commission*. Ecologic, Berlin, pp. 1–31.

Govers, G., Lobb, D. A. and Quine, T. A. (1999). Preface – Tillage erosion and translocation: emergence of a new paradigm in soil erosion research. *Soil and Tillage Research*, 51, 167–74.

Graves, A. R., Morris, J., Deeks, L. K., Rickson, R. J., Kibblewhite, M. G., Harris, J. A. … and Truckell, I. (2015). The total costs of soil degradation in England and Wales. *Ecological Economics*, 119, 399–413. https://doi.org/10.1016/j.ecolecon.2015.07.026.

Graves, A., Morris, J., Deeks, L. K., Rickson, R. J., Kibblewhite, M. G., Harris, J. A. and Farewell, T. S. 2011. *The Total Costs of Soils Degradation in England and Wales*. SP1606. Final Report to Defra, June 2011.

Gregory, P. J. 1976. The growth and activity of wheat root systems. PhD Thesis, University of Nottingham.
Hadas, A., Hadas, A. and Quinton, J. N. 1989. Long-term effects of high application rates of NPK fertilizer on tensile strength and water stability of the soil structure. *Geoderma*, 47, 381–92.
Harvey, A. M. 2002. Effective timescales of coupling within fluvial systems. *Geomorphology*, 44, 175–201. doi:10.1016/S0169-555X(01)00174-X.
Hoad, S. P., Russell, G., Lucas, M. E. and Bingham, I. J. 2001. The management of wheat, barley, and oat root systems. *Advances in Agronomy*, 74, 193–246.
Huber, S., Prokop, G., Arrouays, D., Banko, G., Bispo, A., Jones, R. J. A., Kibblewhite, M. G., Lexer, W., Möller, A., Rickson, R. J., Shishkov, T., Stephens, M., Toth, G., Van den Akker, J. J. H., Varallyay, G., Verheijen, F. G. A., Jones, A. R. (Eds). 2008. *Environmental Assessment of Soil for Monitoring: Volume I Indicators & Criteria*. EUR 23490 EN/1, Office for the Official Publications of the European Communities, Luxembourg, 339p.
Hughes-Popp, J. S., Huszar, P. C. and Hoag, D. L. 2000. Reducing wind erosion damages and the Conservation Reserve Program [M]. In: Napier, T. L., Napier, S. M. and Tvrdon, J. (Eds), *Soil and Water Conservation Policies and Programs*. CRC, Boca Raton, FL, pp. 143–54.
Karlen, D. L., Mausbach, M. J., Doran, J. W., Cline, R. G., Harris, R. F. and Schuman, G. E. 1997. Soil quality: A concept, definition and framework for evaluation. *Soil Science Society of America Journal*, 61, 4–10.
Kendall, H. W. and Pimentel, D. 1994. Constraints on the expansion of the global food supply. *Ambio*, 23, 198–205.
King, J., Gay, A., Sylvester Bradley, R., Bingham, I. A. N., Foulkes, J., Gregory, P. and Robinson, D. 2003. Modelling cereal root systems for water and nitrogen capture: towards an economic optimum. *Annals of Botany*, 91(3), 383–90.
Kosmas, C., Kirkby, M. J. and Geeson, N. (Eds). 1999. *The Medalus Project: Mediterranean Desertification and Land Use: Manual on Key Indicators of Desertification and Mapping Environmentally Sensitive Areas to Desertification*. Directorate-General Science, Research and Development.
Lal, R. 1998. Soil erosion impact on agronomic productivity and environment quality. *Critical Reviews in Plant Sciences*, 17(4), 319–464. doi: 10.1080/07352689891304249.
Lal, R. 2009. Challenges and opportunities in soil organic matter research. *European Journal of Soil Science*, 60(2), 158–69.
Larney, F. J., Izaurralde, R. C., Janzen, H. H., Olson, B. M., Solberg, E. D., Lindwall, C. W. and et al. 1995. Soil erosion crop productivity relationships for six Alberta soils. *Journal of Soil and Water Conservation*, 50, 87–91.
Liu, H., Zhang, T., Liu, B., Liu, G. and Wilson, G. V. 2013. Effects of gully erosion and gully filling on soil depth and crop production in the black soil region, northeast China. *Environmental Earth Sciences*, 68(6)2731–41. https://doi.org/10.1007/s12665-012-1863-0.
Lucas, R. E., Holtman, J. B. and Connor, L. J. 1977. Soil carbon dynamics and cropping practices. In *Agriculture and Energy*, edited by W. Lockeretz. Academic Press, New York, pp. 333–51.
MA. 2005. *Ecosystem Services and Human Wellbeing*. Millennium Ecosystems Assessment Synthesis report, 155p.
Maetens, W., Poesen, J. and Vanmaercke, M. 2012. How effective are soil conservation techniques in reducing plot runoff and soil loss in Europe and the Mediterranean? *Earth-Science Reviews*, 115(1–2), 21–36. doi:10.1016/j.earscirev.2012.08.003.
Merritt, E. 1984. The identification of four stages during micro-rill development. *Earth Surface Processes and Landforms*, 9, 493–6.
Marzen, M., Iserloh, T., Casper, M. C. and Ries, J. B. 2015. Quantification of particle detachment by rain splash and wind-driven rain splash. *Catena*, 127, 135–41. http://doi.org/10.1016/j.catena.2014.12.023.
Meyer, L. D. and Wischmeier, W. H. 1969. Mathematical simulation of the process of soil erosion by water. *Transactions of the American Society of Agricultural Engineers*, 12, 754–62.
Morgan, R. P. C. 2005. *Soil Erosion and Conservation*. 3rd edition. Blackwell Publishing, Oxford.

Morgan, R. P. C. and Duzant, J. H. 2008. Modified MMF (Morgan-Morgan-Finney) model for evaluating effects of crops and vegetation cover on soil erosion. *Earth Surface Processes and Landforms* 32, 90–106. Published online on 8 June 2007 in Wiley InterScience (www.interscience.wiley.com).

Myers, N. 1993. *Gaia: An Atlas of Planet Management*. Anchor/DoubleDay, Garden City.

Nakase, D. K., Hartshorn, A. S., Spielmann, K. A. and Hall, S. J. 2014. Eolian deposition and soil fertility in a prehistoric agricultural complex in Central Arizona, USA. *Geoarchaeology*, 29(2), 79–97. https://doi.org/10.1002/gea.21463.

NEA. 2011. *UK National Ecosystem Assessment: Synthesis of the Key Findings*. UNEP-WCMC, Cambridge.

Newell Price, J. P., Harris, D., Chadwick, D. R., Misselbrook, T. H., Taylor, M., Williams, J. R., Anthony, S. G., Duethmann, D., Gooday, R. D., Lord, E. I. and Chambers, B. J. 2011. 'Mitigation Methods – User Guide'. An Inventory of Mitigation Methods and Guide to their Effects on Diffuse Water Pollution, Greenhouse Gas Emissions and Ammonia Emissions from Agriculture. Prepared as part of Defra project WQ0106, 158p.

Nyssen, J., Poesen, J., Haile, M., Moeyersons, J. and Deckers, J. 2000. Tillage erosion on slopes with soil conservation structures in the Ethiopian highlands. *Soil and Tillage Research*, 57, 115–27.

Owens, P. N., Rickson, R. J., Clarke, M. A., Dresser, M., Deeks, L. K., Jones, R. J. A., Woods, G. A., Van Oost, K. and Quine, T. A. 2006. Review of the existing knowledge base on magnitude, extent, causes and implications of soil loss due to wind, tillage and co-extraction with root vegetables in England and Wales, and recommendations for research priorities. NSRI Report to DEFRA, Project SP08007, Cranfield University, UK.

Pimentel, D. and Burgess, M. 2013. Soil erosion threatens food production. *Agriculture*, 3(3), 443–63. doi: 10.3390/agriculture3030443.

Pimentel, D., Harvey, C., Resosudarmo, P., Sinclair, K., Kurz, D., McNair, M. and et al. 1995. Environmental and economic costs of soil erosion and conservation benefits. *Science*, New Series, 267(5201), 1117–23.

Poesen, J. 1985. An improved splash transport model. *Zeitschrift für Geomorfologie* 29, 193–211.

Poesen, J. W. A., Verstraeten, G., Soenens, R. and Seynaeve, L. 2001. Soil Losses due to harvesting of chicory roots and sugar beets: an underrated geomorphic process? *Catena*, 43, 35–47.

Posthumus, H. and Stroosnijder, L. 2010. To terrace or not: The short-term impact of bench terraces on soil properties and crop response in the Peruvian Andes. *Environment, Development and Sustainability*, 12(2), 263–76.

Posthumus, H., Deeks, L. K., Fenn, I. and Rickson, R. J. 2011. Soil conservation in two English catchments: Linking soil management with policies. *Land Degradation & Development*, 22(1), 97–110. doi: 10.1002/ldr.987.

Quine, T. A., Walling, D. E. and Zhang, X. 1999. Tillage erosion, water erosion and soil quality on cultivated terraces near Xifeng in the Loess Plateau, China. *Land Degradation and Development*, 10, 251–74.

Quinton, J. N., Catt, J. A. Wood, G. A. and Steer, J. 2006. Soil carbon losses by water erosion: Experimentation and modeling at field and national scales in the UK. Agriculture. *Ecosystems and Environment* 112(2006), 87–102.

Rickson, R. J., Deeks, L. K., Posthumus, H. and Quinton, J. N. 2010. Understanding the costs and benefits of mitigation measures used to reduce or prevent soil erosion. Sub Project C of Defra Project SP1601: Soil Functions, Quality and Degradation – Studies in Support of the Implementation of Soil Policy, Defra, London.

Rickson, R. J. 2014. Can control of soil erosion mitigate water pollution by sediments? *Science of the Total Environment*, http://dx.doi.org/10.1016/j.scitotenv.2013.05.057.

Rickson, R. J., Bhogal, A., Brewer, T. R., Burgess, P., Deeks, L. K., Graves, A., Hannam, J., Hess, T. M., Holman, I., Kirk, G., Newell-Price, J. P., Rollett, A., Tibbett, M., Truckell, I. J. and White, C. 2015. Examination of the current evidence base on 'how does a loss of soil depth impact on the ability of soils to deliver vital ecosystem services?' Work Package 1 Final Report. Defra Project SP1317.

Romkens, M., Roth, C. and Nelson, D. 1977. Erodibility of selected clay subsoils in relation to physical and chemical properties. *Soil Science Society of America Journal*, 41(5), 954–60.

Ruysschaert, G., Poesen, J., Verstraeten, G., Govers, G. 2004. Soil loss due to crop harvesting: significance and determining factors. *Progress in Physical Geography* 28(4), 467–501.

Schiettecatte, W., Gabriels, D., Cornelis, W. M. and Hofman, G. 2008. Enrichment of organic carbon in sediment transport by interrill and rill erosion processes. *Soil Science Society of America Journal*, 72, 50–5.

Schmitter, P., Dercon, G., Hilger, T., Hertel, M., Treffner, J., Lam, N., Vien, T. D. and Cadisch, G. 2011. Linking spatio-temporal variation of crop response with sediment deposition along paddy rice terraces. *Agriculture, Ecosystems & Environment*, 140(1), 34–45.

Schubert, S. D., Suarez, M. J., Pegion, P. J., Koster, R. D. and Bacmeister, J. T. 2004. On the cause of the 1930s Dust Bowl. *Science* 303, 1855–9.

Sharpley, A. N. 1985. The selective erosion of plant nutrients in runoff. *Soil Science Society of America Journal* 49, 1527–34.

Sharpley, A. N. 1980. The enrichment of soil phosphorus in runoff sediments. *Journal of Environmental Quality* 9, 521–6.

Simmons, R. W. and Rickson, R. J. 2008. Increasing rainfall and soil erosion. In: Förare, J. (Ed.), *Water for Food*. The Swedish Research Council for Environment, Agricultural Sciences and Spatial Planning (Formas), Stockholm, Sweden, pp. 89–96.

Soil Science Society of America, 2001. *Glossary of Soil Science Terms*. Soil Science Society of America, Inc., Madison, WI, p. 140.

Sparovek, G. and Schnug, E. 2001. Temporal erosion-induced soil degradation and yield loss. *Soil Science Society of America Journal*, 65(5), 1479–86.

Spink, J., Street, P., Sylvester-Bradley, R. and Berry, P. 2009. The potential to increase productivity of wheat and oilseed rape in the UK. See http://www.appg-agscience.org.uk/linkedfiles/ADAS%20Report%20-%20Increasing%20Wheat%20%26%20OSR%20Productivity.pdf.

TEEB. 2010. *The Economics of Ecosystems and Biodiversity Ecological and Economic Foundations*, edited by Kumar, Pushpam. Earthscan, London and Washington.

Troeh, F. R. and Thompson, L. M. 1993. *Soils and Soil Fertility*. 5th edition. Oxford Univ. Press, New York.

Troeh, F. R., Hobbs, A. H. and Donahue, R. L. 2004. *Soil and Water Conservation: For Productivity and Environmental Protection*. Prentice Hall, Upper Saddle River.

Troeh, F. R., Hobbs, J. A. and Donahue, R. L. 1991. *Soil and Water Conservation*. Prentice-Hall, Englewood Cliffs.

Tsara, M., Gerontidis, S., Marathianou, M. and Kosmas, C. 2001. The long-term effect of tillage on soil displacement of hilly areas used for growing wheat in Greece. *Soil Use and Management*, 17(2), 113–20.

Turkelboom, F., Poesen, J., Ohler, I. and Ongprasert, S. (1999). Reassessment of tillage erosion rates by manual tillage on steep slopes in northern Thailand. *Soil and Tillage Research*, 51, 245–59.

Unamunzaga, O., Besga, G., Castellon, A., Uson, M. A., Chery, P., Gallejones, P. and Aizpurua, A. (2014), Spatial and vertical analysis of soil properties in a Mediterranean vineyard soil, *Soil Use and Management*, 30(2), 285–96.

United Nations Convention to Combat Desertification (2011). *Desertification: a Visual Synthesis*. UNCCD Secretariat, Bonn. www.unccd.int/knowledge/docs/Desertification-EN.pdf.

Van Muysen, W., Govers, G., Bergkamp, G., Roxo, M. and Poesen, J. 1999. Measurement and modelling of the effects of initial soil conditions and slope gradient on soil translocation by tillage. *Soil and Tillage Research* 51(3–4), 303–16.

Van Orshoven, J., Terres, J.-M. and Tóth, T. (Eds). 2013. *Definition and Scientific Justification for the Common Biophysical Criteria*; Technical Factsheets. European Commission, Joint Research Centre, Publications Office of the European Union.

Verheijen, F. G. A., Jones, R. J. A., Rickson, R. J. and Smith, C. J. 2009. Tolerable versus actual soil erosion rates in Europe. *Earth-Science Reviews*, 94, 23–38.

Wan, Y. and El-Swaify, S. A. 1998. Sediment enrichment mechanisms of organic carbon and phosphorus in a well-aggregated Oxisol. *Journal of Environmental Quality*, 27(1), 132–8.

Wang, L., Shi, Z. H., Wang, J., Fang, N. F., Wu, G. L. and Zhang, H. Y. 2014. Rainfall kinetic energy controlling erosion processes and sediment sorting on steep hillslopes: A case study of clay loam soil from the Loess Plateau, China. *Journal of Hydrology*, 512, 168–76.

Warren, A. 2007. Sustainability: A view from the wind-eroded field. *Journal of Environmental Sciences-China*, 19(4), 470–4.

Weaver, J. E. 1926. *Root Development of Field Crops*. McGraw-Hill, New York.

Wen, F. H. and Easter, K. W. 1987. Soil erosion and the loss in productivity: An example of the Terril soil series in Minnesota. Station Bulleting 577-1987 (Item No. AD-SB-3200) Agricultural Experiment Station, University of Minnesota, 19p.

Whiting, P. J., Bonniwell, E. C. and Matisoff, G. 2001. Depth and areal extent of sheet and rill erosion based on radionuclides in soils and suspended sediment, *Geology*, 29(12), 1131–4.

Williams, J. R. 1985. The physical components of the EPIC model. In *Soil Erosion and Conservation*, edited by S. A. El-Swaify, W. C. Moldenhauer and A. Lo. Soil Conservation Society of America, Ankeny, IA, pp. 272–84.

Young, A. 1989. *Agroforestry for Soil Conservation*. CAB, Wallingford.

Young, I. M. and Crawford, J. W. 2004. Interactions and self-organization in the soil-microbe complex. *Science*, 304(5677), 1634–7.

Zheng, F., He, X., Gao, X., Zhang, C.-E. and Tang, K. 2005. Effects of erosion patterns on nutrient loss following deforestation on the Loess Plateau of China. *Agriculture Ecosystems and Environment*, 108, 85–97.

Index